GLENIS HEATH   HEATHER McKENZIE   LAUREL TULLY

FOOD *by* DESIGN

EDITION 3

FOR YEARS 7–10

NELSON
A Cengage Company

Australia • Brazil • Japan • Korea • Mexico • Singapore • Spain • United Kingdom • United States

Food by Design
3rd Edition
Glenis Heath
Heather McKenzie
Laurel Tully

Associate Publishing Editor: Tanya Wasylewski
Editor: Nadine Anderson-Conklin
Copyeditors: Duncan Campbell-Avenell and Angela Tannous
Proofreader: Valina Rainer
Indexers: Bruce Gillespie and Nadine Anderson-Conklin
Senior Image and Permissions Researcher: Helen Mammides
Art direction: Olga Lavecchia and Danielle Maccarone
Cover and text design: Leigh Ashforth
Cover images: shutterstock/secondcorner, © the food passionates/Corbis, shutterstock/Tim UR
Senior Production Controller: Erin Dowling
Typesetter: Cenveo Publisher Services
Reprint: Jess Lovell

Any URLs contained in this publication were checked for currency during the production process. Note, however, that the publisher cannot vouch for the ongoing currency of URLs.

© 2015 Cengage Learning Australia Pty Limited

Copyright Notice

This Work is copyright. No part of this Work may be reproduced, stored in a retrieval system, or transmitted in any form or by any means without prior written permission of the Publisher. Except as permitted under the *Copyright Act 1968*, for example any fair dealing for the purposes of private study, research, criticism or review, subject to certain limitations. These limitations include: Restricting the copying to a maximum of one chapter or 10% of this book, whichever is greater; providing an appropriate notice and warning with the copies of the Work disseminated; taking all reasonable steps to limit access to these copies to people authorised to receive these copies; ensuring you hold the appropriate Licences issued by the Copyright Agency Limited ("CAL"), supply a remuneration notice to CAL and pay any required fees. For details of CAL licences and remuneration notices please contact CAL at Level 15, 233 Castlereagh Street, Sydney NSW 2000, Tel: (02) 9394 7600, Fax: (02) 9394 7601
Email: info@copyright.com.au
Website: www.copyright.com.au

For product information and technology assistance,
    in Australia call **1300 790 853**;
    in New Zealand call **0800 449 725**

For permission to use material from this text or product, please email
aust.permissions@cengage.com

National Library of Australia Cataloguing-in-Publication Data
Heath, Glenis, 1945- author.
Food by design / Glenis Heath, Heather McKenzie, Laurel Tully.

    3rd edition.
    9780170358507 (paperback)
    Includes index.
    For secondary school age.

Food--Textbooks.
Food--Textbooks.

McKenzie, Heather, 1957- author.
Tully, Laurel, 1946- author.

641.3

Cengage Learning Australia
Level 7, 80 Dorcas Street
South Melbourne, Victoria Australia 3205

Cengage Learning New Zealand
Unit 4B Rosedale Office Park
331 Rosedale Road, Albany, North Shore 0632, NZ

For learning solutions, visit **cengage.com.au**

Printed in Singapore by 1010 Printing Group Limited
8 9 10 11 21 20 19

CONTENTS

ABOUT THIS BOOK vi
ABOUT THE AUTHORS viii
ACKNOWLEDGEMENTS viii

1 DESIGNING WITH FOOD 1

The role of food 2
Describing food 2
Tasting food 3
Analysing the properties of food 4
The design process 5
The design brief 5
Investigating 8
Generating 8
Collaborating and managing 10
Evaluating 12

2 PREPARING FOOD SAFELY 19

Safety in the kitchen 20
Fire safety 20
Personal hygiene 21
Kitchen hygiene 21
Food poisoning 24
Preventing food poisoning 25
Safe use of knives 26
Stoves, ovens and cooktops 26
Using small appliances 29

3 RECIPE BASICS 41

Tools of the trade 42
Making sense of a recipe 43
Abbreviations in recipes 44
Measurement in recipes 44
Commonly used food preparation terms 47
Designing new recipes 49
Muffins 49

4 EAT WELL, BE WELL 55

Food and me 56
What is food? 56
Nutrients in food 56
Water 57
Digestion 58
Selecting food wisely 59
The Healthy Eating Pyramid 59
Nutrition throughout life 64
Adolescent food needs 64
Five healthy eating tips for teenagers 66

5 GET UP AND GO! 79

The role of breakfast 80
Breakfast eating habits 80
Processed breakfast cereals 82
Milk 83
Breakfast drinks 85
Eggs 85

6 THE GREENGROCER 99

Fruit 100
Apples 101
Bananas 104
Citrus fruits 105
Oranges 105
Packaging and labelling fresh fruit 107
Processing fruit 109
Vegetables: the colours of the rainbow 110
Orange vegetables: sweet potatoes 113
Green vegetables: green beans 113

Processing of vegetables 114
Prepacked and ready-to-eat salad mixes 115
White vegetables: potatoes 116
Preparing fruit and vegetables safely 117

7 GRAINS ARE GREAT 131

Cereals 132
The structure of a wheat grain 133
Wheat: from paddock to plate 134
Cereals for good health 135
Gluten: the protein in wheat flour 136
Preparing yeast doughs 136
Bread 140
Couscous 142
Rice 142
Rice: from paddock to plate 143

8 MEATY IDEAS 163

What is meat? 164
Poultry 170
Fish 172
Top tips for working safely with meat, poultry and fish 172
Preparation for cooking meat, poultry and fish 173
Cooking meat, poultry and fish 175

9 EATING WELL FOR THE FUTURE 191

Influences on food choices 192
Food marketing 193
Eating for good health 195
The Australian Dietary Guidelines 195
The Australian Guide to Healthy Eating 195
Energy balance 196
Sources of energy 197
Fat 197
Carbohydrates 199
Using energy 200
Obesity 202
Cardiovascular disease 203
Diabetes 203
Maintaining a healthy weight 204
Osteoporosis 209
The importance of fibre in the diet 211
Individual dietary needs 213
Vegetarian diets 216

10 INDIGENOUS FOODS AND FLAVOURS 229

Indigenous food customs 230
The traditional diet of Aboriginal and Torres Strait Islander people 232

Wattleseed PAVLOVAS PAGE 242

European settlement 233
The Aboriginal and Torres Strait
   Islander Guide to Healthy Eating 233
Native foods in today's menus 235
Cooking with bush flavours 236

11 GLOBAL GOODIES 245

Influences on eating in Australia 246
Influences from Italy 248
Influences from Asia 251
Vietnamese cuisine 255
Thai cuisine 256
Mexican cuisine 257

12 BEST BAKING 273

The art of baking 274
The ingredients in baked products 274
How baking works 277
Pastry 280
Chocolate 282
Presenting and decorating with
   chocolate 285
Decorating with sugar 286

GLOSSARY 304
INDEX 306
RECIPE INDEX 315

Fragrant COUSCOUS SALAD PAGE **160**

ABOUT THIS BOOK

Working with *Food by Design*

Food by Design has been written to foster lifelong skills and enthusiasm for cooking, nutrition and general wellbeing.

Covering Years 7–10, this new edition includes stunning new food photography, extensively trialled recipes and is fully updated to match the cross-curriculum priorities of the new Victorian Curriculum and Australian Curriculum. Written by market-leading authors, *Food by Design* is designed to stimulate students' interest in Food Technology and to help them develop the relevant knowledge and skills required for VCE-level studies.

Key knowledge
Highlights the key topics and issues addressed in the chapter

Key terms
Highlights the key terms used in the chapter

Victorian Curriculum links
Points to elements being addressed

Visually engaging
Photographs, illustrations, maps and graphics provide a visual richness to text will captions incorporate questions to encourage engagement and reflection.

Activities
Help students build their food skills in a fun and engaging manner.

Diagrams
Visually summarise complex ideas or concepts

Testing knowledge
Questions help students to apply the knowledge they acquire.

Design briefs
Provide teachers with a tool to monitor student progress as they investigate, design, produce, analyse and evaluate.

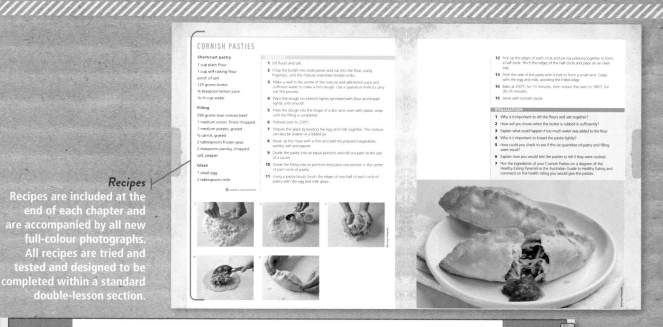

Recipes
Recipes are included at the end of each chapter and are accompanied by all new full-colour photographs. All recipes are tried and tested and designed to be completed within a standard double-lesson section.

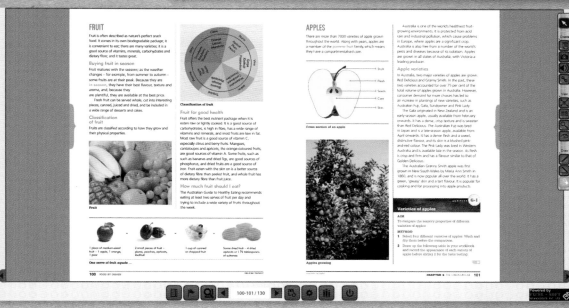

NelsonNet

Food By Design is a premium Cengage title and is fully supported by the NelsonNet platform.

NelsonNetBook

The NelsonNetBook is your digital textbook. Readable online and offline on desktops, laptops and tablets, it reproduces the student text in digital form. With annotations and reviewing tools, and the ability to add and customise your book, NelsonNetBook is accessible immediately via access codes. Please note that any notations made to the NelsonNetBook will expire two years after the access code is activated.

NelsonNet resources

Access to NelsonNet also provides students with additional web-based materials such as worksheets and skill sheets.

WEBLINKS

Students and teachers can link directly to external websites referred to in *Food by Design* via the free, unprotected weblinks site located at http://foodbydesign.nelsonnet.com.au.

DISCLAIMER

Please note that complimentary access to NelsonNet and the NelsonNetBook is only available to teachers who use the accompanying student textbook as a core educational resource in their classroom. Contact your sales representative for information about access codes and conditions.

ABOUT THE AUTHORS

Glenis Heath, **Heather McKenzie** and **Laurel Tully** are passionate, inspiring and highly experienced master teachers of Food and Technology. They combine their expert knowledge of curriculum and evidence-based learning strategies to present an integrated approach that engages and enriches student understanding of Food and Technology.

All of the authors have taught students who have achieved perfect scores, Premier's awards and many who have been selected for Top Designs.

Glenis Heath is an educational consultant who provides teaching and learning advice to teachers both in seminar and school-based settings. She is an experienced examination assessor and presents revision lectures for students in VCE Food and Technology. Glenis is a past president and life member of Home Economics Victoria.

Heather McKenzie is currently teaching at Strathmore Secondary College where she is the Leading Teacher of Instructional Practice. Previously she was the manager for the Technology KLA for 12 years. She has extensive experience as an examination assessor and presents examination revision lectures for students of VCE Food and Technology. Heather is a past president of Home Economics Victoria.

Laurel Tully is an experienced educational consultant who has worked on a range of projects with the VCAA. She has extensive experience in presenting professional learning activities to teachers and is actively involved in the assessment process for VCE Food and Technology. Laurel is a past president and life member of Home Economics Victoria.

ACKNOWLEDGEMENTS

The publisher would like to thank:
- Mark Fergus, photographer and stylist
- Melanie Vandegraaff of Red Courgette, chef and stylist
- Rhonda Fergus, photo shoot production
- Ellie Tomlin, model
- Robert Gordon Australia, provider of dinnerware and bakeware
- Belgrave Heights Christian School

The photographs appearing on the recipe pages of *Food by Design*, along with other selected photographs, were shot on-location at Belgrave Heights Christian School's Hospitality Trade Training Centre.

1 DESIGNING WITH FOOD

KEY KNOWLEDGE
- The role of food
- Describing food
- Tasting food
- Analysing the properties of food
  - Qualitative, or sensory measures
  - Quantitative measures
- The design process
- The design brief
  - Evaluation criteria
- Investigating
- Generating
- Collaborating and managing
  - Production plans
- Evaluating

KEY TERMS

considerations factors in the design brief, such as the season of the year or the skills of the chef, which are more flexible than constraints but may also influence the design and development of the product

constraints factors in the design brief with which the product must comply

design brief specific information about the type of product to be developed and the audience at which the new product is aimed

design process the process of investigating and designing, producing, and analysing and evaluating

preferred option the design option that best meets the requirements set out in the design brief

qualitative or sensory analysis the evaluation of the sensory properties of food, such as appearance, aroma, flavour and texture

quantitative measures ways to measure the physical, chemical or nutritional properties of food

sensory properties the appearance, aroma, flavour and texture of food

specifications the considerations and constraints within the design brief

AUSTRALIAN CURRICULUM LINKS

DESIGN AND TECHNOLOGIES
- Knowledge and Understanding
- Processing and Production Skills

GENERAL CAPABILITIES
- Critical and creative thinking

THE ROLE OF FOOD

Food plays a very important part in our lives. It is the essential fuel that keeps us alive. Food is frequently the focal point of our social lives; we share food with family and friends in our homes, in restaurants, at school, at sporting events and in a variety of other venues. Because food is so fundamental, it is important for us to understand how to prepare it in such a way that it will provide us with the essential nutrients needed to maintain good health, as well as be appealing to eat.

DESCRIBING FOOD

To work successfully with food, it is necessary to develop an understanding of its sensory properties: its appearance, aroma, flavour and texture. You need to be able to describe food and to explain it to others.

We all have certain foods that we particularly enjoy. What is it about these foods that makes us think of them as our 'favourites'?

The body's senses of taste, smell, touch, sight, and even hearing, are important in building up our knowledge of food. The tongue, eyes, ears, nose and fingers send messages to the brain to tell us which foods give us pleasure and which foods we find unpleasant. The sweet, smooth taste of chocolate, the smell of a roasting chicken or the sound of a crisp apple being bitten can give us a sense of excitement and anticipation.

ACTIVITY 1.1

Favourite foods

1. Sketch four of your favourite foods.
2. Annotate each sketch to describe the food's appearance, aroma, flavour and texture. Refer to the words in the sensory wheel to help you.

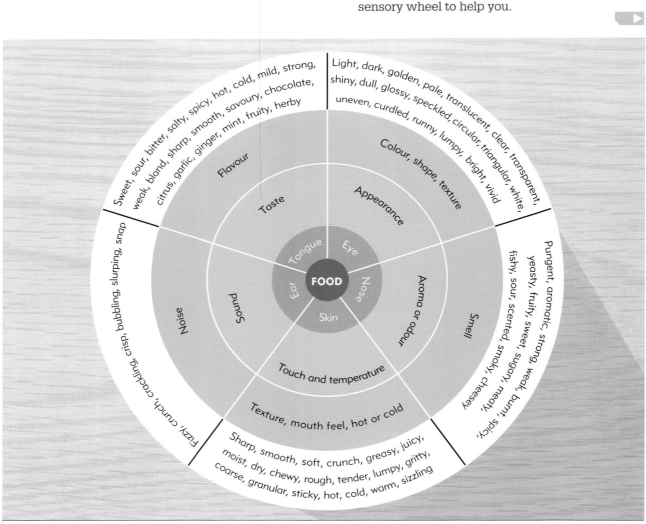

Sensory wheel

3 Look at the photographs of the Chicken and Noodle Stir-Fry and Peach and Apricot Crumble. Write down as many words as you can to describe the appearance of each dish, and how you imagine the aroma, flavour and texture of each dish would be.

4 Which of the properties of each of the two dishes appeals to you most? Why?

Chicken and Noodle Stir-Fry

Peach and Apricot Crumble

TASTING FOOD

The tongue's surface is covered with approximately 9000 tastebuds. These tastebuds detect the four main tastes: sweet, sour, bitter and salty. Other tastes are made up of a combination of these four. The tastebuds are located in four main zones on the surface of the tongue. Look at the diagram below to see where these taste zones are located. Our taste is also affected by our sense of smell. You will have noticed that when you have a cold and a blocked nose, your food seems tasteless.

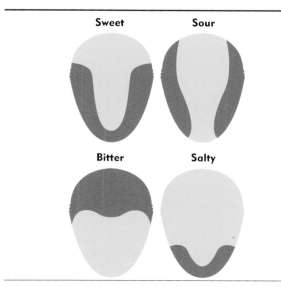

Taste zones on the tongue

Sweet foods that most people love include chocolate, strawberries, honey and ice-cream. Sour foods include vinegar and citrus fruits such as lemons and grapefruit. The pith of an orange, bitter melon, rocket and radicchio are examples of bitter foods. Bacon, potato crisps, soy sauce, fetta cheese and olives all have a salty flavour.

ACTIVITY 1.2

Taste test

For this taste test, you will need the following equipment:
- two trays, labelled Tray A and Tray B
- 12 different food samples, six on each tray
- samples cut into small pieces; the trays should be covered so that the foods cannot be seen
- enough blindfolds for half of the members of the class
- enough small plates, spoons and cups for every member of the class
- a copy of the following table for each class member to use to record the flavour, texture and identification of the food samples.

FOOD TASTING TABLE

Sample	Flavour	Texture	What I think the food is	What the food really is
Sample 1				
Sample 2				
Sample 3				
Sample 4				
Sample 5				
Sample 6				

Work with a partner to complete the taste test.
1 One partner should be blindfolded.
2 Select six food samples from Tray A. The blindfolded partner must not be told which foods have been selected.
3 Use a spoon to feed the blindfolded partner one of the six samples.
4 Ask them to describe the sample's flavour and texture, and to name what they think it is.
5 Record their answers on the table.
6 Continue to feed the blindfolded partner food samples and record their answers.
7 Remove the blindfold and share the results.
8 Change places, and repeat the process using Tray B.

ANALYSING THE PROPERTIES OF FOOD

After a food product is prepared and served, it is important to examine it closely to determine whether it was successful – that is, did people enjoy eating it and did the product meet the needs identified in the design brief? When evaluating the success of a food product, food manufacturers and cooks identify the ingredients used and how they worked together in a recipe, and determine whether or not improvements should be made to the processes and equipment used in production. Measuring the size, weight, volume and nutrient content of food enables comparisons to be made between individual products. Most importantly, what people remember about food is whether or not eating it was a pleasurable experience – and this hinges on the sensory analysis of appearance, aroma, flavour and texture.

Qualitative, or sensory measures

Qualitative, or sensory analysis is used to evaluate the sensory properties of food such as appearance, aroma, flavour and texture.

As consumers, we use all of our five senses – sight, smell, taste, touch and sound – to form opinions about food products. This form of sensory analysis can be very subjective, since some foods are very appealing to some people but not to others. In the food industry, however, specially trained testers carry out controlled sensory analysis tests on food to ensure that the end product will appeal to a wide range of consumers. Several systems are used to collect this data and collate it in the form of descriptors. Descriptors describe specific characteristics of the food – look at the sensory wheel on Page XX for some examples.

Facial hedonic descriptors enable the consumer to rate how much they like or dislike a particular food by using a scale based on a range of happy and sad faces.

Attitudinal descriptors indicate how people feel about a particular food. These descriptors provide more detailed statements about the food to allow the consumer to state their attitude to the product.

Hedonic scale

Attitudinal descriptors

Tick the statement that best describes your attitude to this product.	
I would eat this at every opportunity I had.	
I would eat this very often.	
I would eat this frequently.	
I like this and would eat it now and then.	
I would eat this if it was available, but would not go out of my way for it.	
I don't like this, but I would eat it occasionally.	
I would hardly ever eat this.	
I would eat this only if there were no other food choices.	
I would eat this only if I was forced to.	

Quantitative measures

Quantitative measures are ways of measuring the physical, chemical or nutritional properties of food. They enable consumers to compare similar food products. Features of food products that can be measured accurately using quantitative measures include the:

- ratio of ingredients
- weight of the finished product
- microbiological content (to ensure that food is safe to eat and to determine the shelf life of the product)
- colour
- nutrient content
- volume
- consistency
- texture.

Recording this type of data is an important part of the quality control process for food manufacturers. Products are analysed to determine whether they meet Australian Standards set out in the Food Standards

Code; to provide consumers with confidence about the food they purchase; and to provide a basis upon which consumers can compare competitors' products.

Testing knowledge

1. In one sentence, explain why food plays such an important part in our lives.
2. Name the four main flavours our tastebuds can detect, and give one example of each flavour.
3. Explain how people use their senses to evaluate food.
4. Suggest some factors that can affect a person's ability to identify specific flavours when eating.
5. Explain the difference between qualitative analysis and quantitative analysis of food products.
6. Describe how hedonic and attitudinal descriptors assist food manufacturers in product development.
7. List four properties of food that can be accurately measured using quantitative measures.
8. Identify two other flavours you could add to the sensory wheel.
9. Why do food manufacturers undertake an analysis of their food products?
10. Explain why hedonic descriptors are a useful tool when analysing food products.

THE DESIGN PROCESS

New food products and meal ideas seem to appear in the supermarket on an almost weekly basis. Food manufacturers respond to the consumers' needs by designing and producing new snack foods, frozen foods, convenience foods, instant meals and beverages.

Whenever a new food product is designed and produced, the manufacturer begins by developing a design brief. Once the design brief has been established, the manufacturer uses the design process, which involves the key stages – investigating, generating, collaborating and managing, producing and evaluating – to enable them to design and produce a product that best meets consumers' needs.

Just like food manufacturers, we also use the design process when we design and work with food. Look at the illustration above to see the way in which

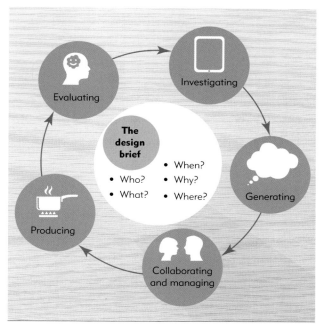

The design process

the design process can be used to develop recipes, either at school or at home.

THE DESIGN BRIEF

The design brief is the focal point of the design process. It includes all of the important information required to develop a new product, whether it is a single food item or a whole meal for a family. When designers are developing a design brief, they need to include information about the who, what, when, where and why of the new product. For example, the brief includes information about who the new product will be aimed at; for instance, children. It also provides information about what type of product is to be developed; for example, a new savoury snack food. These factors, or specifications, are often called constraints, because they are factors with which the product must comply.

Other factors, or specifications, may be considerations, such as the quality and sensory properties of the food, the season of the year, the skills of the chef or details about the consumer. These may also be included in the design brief and may influence the design and development of the product.

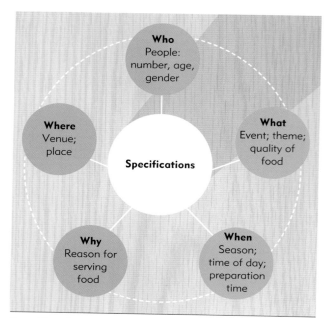

Specifications of the design brief

Evaluation criteria

Once the design brief has been established, a series of evaluation criteria or relevant questions are developed to ensure that any products that are designed will successfully meet the requirements of the brief.

The specifications – that is, the constraints and considerations within the brief – provide a starting point for developing criteria for evaluation, which in turn enable judgements to be made. Evaluation criteria questions that may be asked about a product include:

- Was the food suitable for the people consuming it?
- Did the food match the event, or theme?
- Were the quality and sensory properties of the food – appearance, aroma, flavour and texture – appropriate to the needs of the design brief?
- Were the food items appropriate to the season?
- Was the food prepared and served within the timeframe specified in the design brief?
- Did the food meet the needs of the design brief? What was the overall success of the product?

ACTIVITY 1·3
Specifications: constraints and considerations in the design brief

It is important to be able to identify the specifications – that is, the constraints and considerations – within a design brief so that you can then develop evaluation criteria.

Complete the following table by including other examples of constraints and considerations that may affect the design of a product.

Specifications – constraints and considerations in the design brief

Constraints	Considerations
• Nutrient requirements	• Food preparation knowledge and skill
• Age group of the consumer or client	• Fashion and trends in food

ACTIVITY 1·4
Developing evaluation criteria

DESIGN BRIEF

You and your friends are planning a movie marathon for Saturday night. Your friends have nominated you to organise the food, as you are the expert in the kitchen. They have requested finger food that is easy to eat during the movies, but is not the usual salty, high-fat fast food. They would like two savoury products – one to be served hot – and one sweet treat. You will need to prepare the food before the evening, but you will be able to reheat it at the venue. There will be four others at the event apart from you.

1. Highlight the constraints and considerations in the design brief, similar to those in the brief on Pages 7–8. These will be the focus for the criteria for evaluation.
2. Develop a question you can use to evaluate each specification in the brief.
3. Select your own client or occasion (not the movie marathon) and develop a design brief that includes a range of constraints and considerations. Write out the evaluation criteria to cover all aspects of the brief.

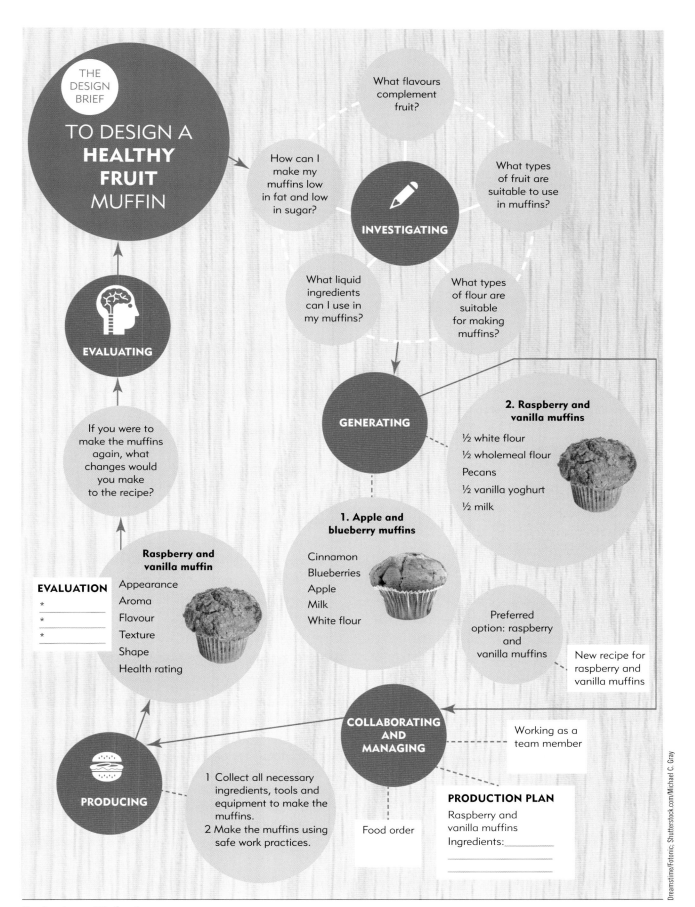

Designing with food

CHAPTER 1 DESIGNING WITH FOOD

Moomba food truck

Were there four savoury snack foods available for customers to purchase?

Was all of the food for sale in the food truck based on Mexican-style cuisine?

Moomba is a festival held annually in Melbourne that attracts large crowds, both to its street parade of floats and to the many activities that take place in the parklands around the Yarra River. It is a festival to which friends and families come to get together and have fun!

Melbourne City Council has granted permits for several food trucks to be positioned along the south walkway beside the Yarra River for the Moomba long weekend. One of the conditions of the permit is that each food truck should sell a different style of food. Maria and Joseph have been successful in their application to sell **Mexican-style food** from their food truck. Their menu will include **four savoury snack foods**. At least **one of the snack foods will include chicken, and another will be suitable for vegetarians**. Each of the snack foods will be **sold in individual serves** and **packaged in environmentally friendly material**. Maria and Joseph realise that they will have to compete for customers with all of the other food trucks, so their food must have **appealing sensory properties**.

Was there at least one chicken and one vegetarian food item on the menu?

Were all of the snack foods available for sale in individual serves?

Did Maria and Joseph package their snack foods in environmentally friendly materials?

Did the snack foods have appealing sensory properties?

Example of an annotated design brief and evaluation criteria

INVESTIGATING

The first stage of the design process involves food designers investigating information from a wide range of sources, such as technical journals, primary producers, competing manufacturers, magazines and surveys and focus groups. This provides them with up-to-date information about consumer demands, new ingredients that are available, new manufacturing techniques that have recently been developed, and even types of new packaging materials. They analyse this information to help them to develop a range of options that meet the specifications of the design brief, and to ensure that the design options are viable.

GENERATING

All the information that comes from the investigation is used to generate a range of possible options for new products. All of the options must be practical or viable solutions to the problem outlined in the design brief. At school, a simple method of developing design options is to use a recipe map, such as the one for muffins on Page 7 and the one for a Crustless Quiche on Page 10.

Once the design options have been developed, the designer may outline the advantages and disadvantages of each option, showing how it meets the needs of the client. Finally, each of the design options is evaluated, and the food manufacturer decides on the best solution to the problem outlined in the design brief. This solution is called the preferred option.

ACTIVITY 1·5
Investigating and generating

Food designers and home cooks can use a variety of sources to help them gather relevant information from which they can develop design options. Draw up a diagram like the one on the following page and include other examples of the types and sources of information that may be useful during the design stage.

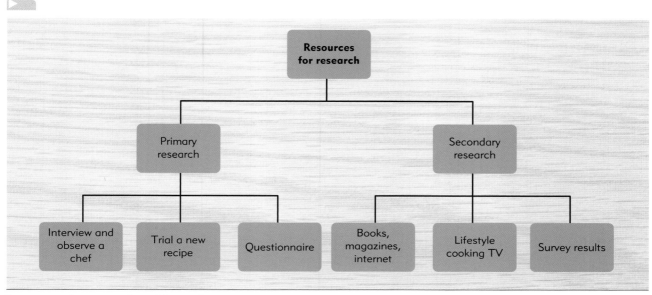

Resources for research and investigating

SWOT ANALYSIS

At school, a variety of strategies or thinking skills can be used to help you analyse and evaluate each option before deciding on the preferred option. For example, a strengths, weaknesses, opportunities, threats (SWOT) analysis is a useful tool for evaluating various options.

Focus of design brief:	
Design possibility 1	
Strengths:	Weaknesses:
Opportunities:	Threats:
Design possibility 2	
Strengths:	Weaknesses:
Opportunities:	Threats:
Preferred option:	
Justification:	

DECISION TABLE

Another useful strategy for helping you to evaluate each option is to use a decision table, which allows you to compare the advantages and disadvantages of various options.

Decision to be made		
Option 1:	Option 2:	Option 3:
Advantages:	Advantages:	Advantages:
Disadvantages:	Disadvantages:	Disadvantages:
Preferred option:		
Justification:		

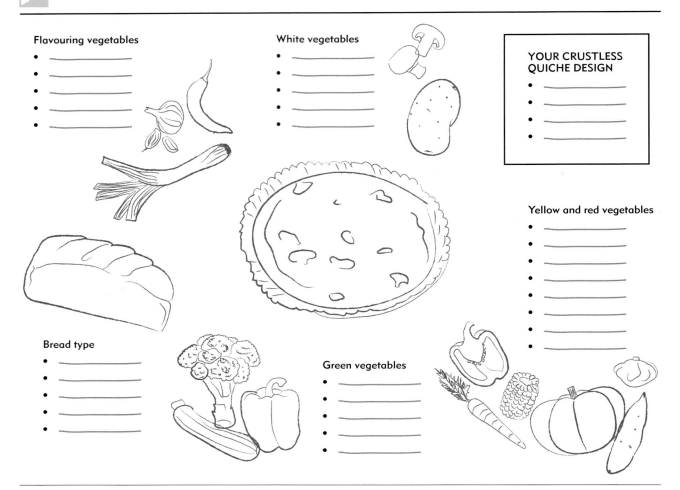

Recipe map for a crustless quiche

COLLABORATING AND MANAGING

Before producing the preferred option for the new product, the food manufacturer or home chef must undertake considerable planning. One important step in this planning involves completing a food order for all the ingredients required to make the product, so that they are all available at the time of production. It is also essential to list any specialist equipment that is required.

Food order for the home chef

FOOD ORDER		
Name:		
Production date:		
Recipe:		
Supermarket	Greengrocer (fresh fruit and vegetables)	Butcher
Items not available at other suppliers:		
Specialist equipment:		

Production plans

A production plan is another important tool that the food manufacturer or home chef can use to ensure that the product they plan to make is completed successfully and made within the time available. Production plans increase efficiency and eliminate time-wasting.

At school or at home, preparing a production plan enables the cook to read the recipe carefully, think about which job needs to be done first and consider ways to organise tasks so that there will be a smooth work flow in the kitchen. They also help the cook to consider the health and safety issues in the production process to avoid risks.

A production plan should contain:
- information about materials or ingredients needed
- a list of all the tools and equipment required
- the sequence of steps or processes involved
- an estimate of the time it will take to produce the product. At school or at home, it is important to note the time it will take to complete each of the steps, including the cooking of the product – you should break up these steps into 5–10-minute intervals – as well as the time needed to clean up
- information about all necessary health and safety regulations to be followed during the production phase.

During the production stage, the chefs and food manufacturer record any modifications or changes they have made to the production plan, so that they can consider these in future productions.

The following is an example of a simple production plan suitable for the home chef.

ACTIVITY 1·6

Preparing a production plan

1. Read the recipe for a Salad Roll-Up on Page 15.
2. Draw up a table similar to the one below and prepare a production plan for the recipe. Remember to list all the important safety considerations for each process.
3. Share your production plan with a classmate and ask them to give you ideas for processes, equipment or safety considerations you might have overlooked. Add these to your production plan.

A production plan

NAME OF RECIPE: Cheese Omelette (see recipe on Page 96)	Ingredients: 2 eggs, 1 tablespoon water, 30 grams cheddar cheese, 1 tablespoon parsley, 1 teaspoon butter		*Serves one*
Time (a.m.)	Important steps	Equipment required	Safety and hygiene considerations
9.00–9.10	Collect ingredients	Tray; bowls; spoons	Throughly wash hands.
9.10–9.20	Chop parsley; grate cheese	Cook's knife; grater	Keep fingers away from blade of knife while chopping.
9.20–9.25	Beat eggs and water	Bowl; whisk or fork	Make sure all equipment is clean and dry.
9.25–9.30	Prepare the omelette pan	Omelette pan	Use an oven mitt to handle the omelette pan.
9.30–9.40	Cook the omelette	Omelette pan; spatula	Do not leave the omelette pan unattended on the stove.
9.40–9.50	Serve and garnish	Clean serving plate	Place the hot pan in the sink so it cools before cleaning.
9.50–10.00	Clean up	Washing-up equipment	Do not put a sharp knife into a sink filled with soapy water.

CHAPTER 1 DESIGNING WITH FOOD

EVALUATING

Finally, the chef or manufacturer evaluates the final product using the previously established criteria. During this process, they will make judgements about whether the product addresses the specifications of the design brief and therefore solves the design problem. The evaluation of the product will also assess the sensory properties of the food, as well as the quality and finish of its presentation. This will help the chef or manufacturer to determine if any modifications are needed to make sure that the product will be something that the general public will buy.

Testing knowledge

11 Name the process that is used when new products are being designed and produced, and explain why the design brief is so important in this process.

12 Discuss the difference between a constraint and a consideration in a design brief.

13 Explain why it is important for the designer to develop evaluation criteria.

14 Outline the types of information that food manufacturers are likely to investigate before they begin to design new products.

15 Identify one reason why the investigation stage is such an important step in the design process.

16 Explain what happens during the design stage of the design process.

17 Explain what is meant by the term 'preferred option'.

18 List four types of information that a good production plan should contain.

19 Explain why it is important to include health and safety issues in a production plan.

20 Outline why it is important for food manufacturers to evaluate their product, and identify three factors that are analysed during the evaluation process.

THINKING SKILLS 1·1

COMPARE THE SENSORY PROPERTIES OF TWO POPULAR PASTA DISHES

Describe in detail the sensory properties of both spaghetti bolognese and lasagne using a table like the one below. In your descriptions, consider all elements of the food product; for example, spaghetti bolognese has pasta and a meat sauce, while lasagne has pasta, meat and cheese sauce.

Sensory properties	Spaghetti bolognese	Lasagne
Appearance		
Aroma		
Flavour		
Texture		

Discuss the similarities and differences of the sensory properties of spaghetti bolognese and lasagne. Identify the product that you like most, and justify your answer.

DESIGN ACTIVITY 1·1

A pancake stack dessert

DESIGN BRIEF

Design and produce a creative, exciting pancake stack dessert suitable for one person. The pancake base must incorporate at least one additional ingredient to give the dessert a new flavour. You must also include fruit in the dessert.

1 When writing your design brief, you should include:
  - a description of the event when the pancake stack will be served
  - an explanation of who will be eating the pancake stack
  - where and when the pancake stack will be served.

2 After writing your design brief, develop four evaluation criteria questions to judge the success of your pancake stack dessert.

INVESTIGATING

1 Research ingredients that would be suitable to use:
  a to add flavour and texture to the pancake batter
  b as a topping or sauce for the stack
  c as a garnish on top of the stack.

GENERATING

1 Use the information you have gained through your investigation to complete the recipe map for a pancake stack on the following page. Use the basic pancake recipe on Page 17.

2 Draw up two possible design options for a pancake stack. Include:
  a additional ingredients to add to the pancake base
  b topping and sauce ideas
  c a sketch of presentation ideas.

3 Select your preferred option, and explain why you chose it.

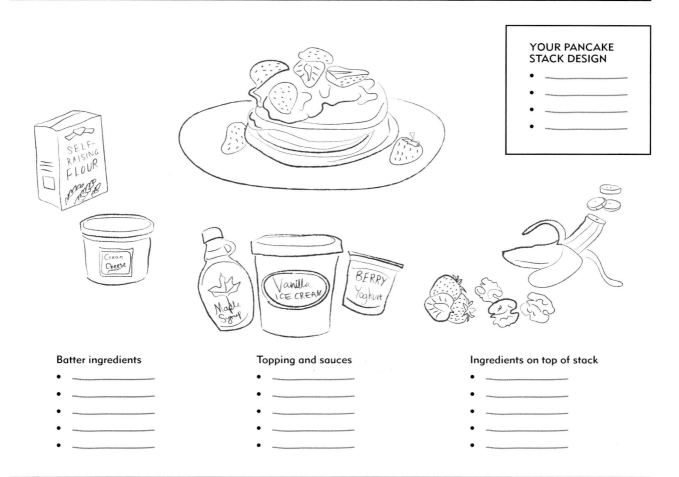

Batter ingredients

Topping and sauces

Ingredients on top of stack

YOUR PANCAKE STACK DESIGN

Recipe map for pancake stack

CHAPTER 1 DESIGNING WITH FOOD

Design options for a pancake stack dessert

	Design option 1	Design option 2
Additional ingredients for pancake base		
Topping and sauce ideas		
Sketch of design option		
Preferred option:		
Why you selected this option?		

PLANNING AND MANAGING

1. Write out your new recipe so that it is ready for production.
2. Sketch your presentation idea for the pancake stack.
3. Write up a production plan.

PRODUCING

1. Prepare the product.
2. Record any changes or modifications you made during production.
3. Photograph the final presentation of your pancake stack on a plate.

EVALUATING

1. Answer, in detail, the four evaluation criteria questions you developed earlier.
2. Annotate the photograph of your pancake stack – for example, by including the ingredients added to the basic batter and ingredients or products used as a garnish or decoration. Use arrows to link the annotations with the photograph.
3. Taste test your pancake stack and describe its sensory properties. Record your descriptions of the sensory properties in the table opposite.

	Sensory properties
Appearance	
Aroma	
Flavour	
Texture	

4. Rate the success of your pancake stack using the hedonic scale on Page 4.
5. Describe how including other ingredients in the batter affected the flavour and texture of the pancakes.
6. Were you able to produce pancakes of a similar size? If so, explain how you achieved this.
7. How did you test whether the pancakes were cooked?
8. Discuss two safety rules that should be followed when using a frying pan.
9. Plot the ingredients of your pancake stack recipe on a diagram of the Healthy Eating Pyramid (see Page 59) or the Australian Guide to Healthy Eating (see Page 195). Rate your product as:
   - very healthy (more than half of the ingredients come from the foundation layers of the pyramid)
   - healthy (half of the ingredients come from the foundation layers or middle layer of the pyramid)
   - not very healthy (a majority of the ingredients are low in nutrients or high in salt, sugar and unhealthy fats and/or are highly processed).
10. Would you to make this dish again? What changes would you make if you were to repeat it?

SALAD ROLL-UP

- 1 egg, or a slice of ham or turkey loaf
- 1 teaspoon mayonnaise
- 1 large lettuce leaf
- ½ tomato
- ¼ cucumber
- ¼ carrot
- 30 grams cheese
- 1 pitta bread or 2 slices mountain bread
- 2 tablespoons dip such as hummus or tzatziki

SERVES ONE

METHOD

1. Hard-boil the egg by placing it in a small saucepan with enough warm water to just cover it. Bring to the boil, then simmer for 8 minutes. Shell and mash the egg, with mayonnaise if desired.
2. Finely shred the lettuce and slice the tomato and cucumber very thinly.
3. Grate the carrot and cheese.
4. If using pitta bread, split it into halves, so that you are left with two thin rounds of bread.
5. Spread the bread with the dip.
6. Place half of the topping ingredients over each slice of bread.
7. Roll up the bread tightly in plastic wrap. Allow to rest in the refrigerator for 10–15 minutes so that the roll-up will hold its shape when cut.
8. Cut into serving portions.

EVALUATION

1. Explain why it is important to shred the lettuce and finely slice the tomato and cucumber for this recipe.
2. Describe how you would safely use the grater.
3. List three other ingredients you could use to spread over the bread instead of hummus or tzatziki.
4. Make a list of other ingredients you could use in the roll-up filling.
5. Write a paragraph to explain why the Salad Roll-Up is a healthy snack or lunch food.

POTATO SALAD WITH EGG AND TUNA

6 chat potatoes

2 eggs

1 carrot

50 grams snow peas

¼ green capsicum

¼ yellow or red capsicum

1 large ripe tomato

1 tablespoon parsley

1 small clove garlic, crushed (optional)

1 tablespoon red wine vinegar

2 tablespoons olive oil

100 grams lettuce or salad mix or rocket

170 grams canned tuna, drained

SERVES TWO

METHOD

1. Steam the chat potatoes for approximately 20 minutes or until tender.
2. Place the eggs into a saucepan filled with warm water. Bring the water to the boil; reduce the heat to simmer, and cook the eggs for exactly 8 minutes.
3. Remove the eggs from the heat and drop them into the sink to crack their shells. Place into a saucepan or bowl of cold water and allow to cool.
4. Shell the eggs, then slice into quarters.
5. Cut the potatoes into thick slices.
6. Cut the carrot into julienne sticks. Remove the tops from the snow peas.
7. Bring a small saucepan of water to the boil and blanch the carrots for two minutes. Add the snow peas and cook for a further 30 seconds.
8. Finely dice the capsicum.
9. Cut the tomato into wedges.
10. Finely chop the parsley.
11. Crush the garlic to a smooth paste.
12. Mix the garlic, vinegar, olive oil, chat potatoes, blanched carrots, snow peas, capsicum and tomatoes together in a small bowl.
13. Place the salad mix or rocket into two containers, or onto serving plates, and top with the tuna, capsicum, boiled eggs and finely chopped parsley.

EVALUATION

1. Explain why it is important to crack the shells of the eggs immediately after they are cooked.
2. Describe one safety rule to follow when blanching the snow peas.
3. Why does the recipe suggest cutting the chat potatoes into thick slices, rather than finely?
4. Ask a friend or member of your family to taste your salad and comment on the flavour and texture of the recipe.
5. Evaluate the Potato Salad with Egg and Tuna by plotting its ingredients on a diagram of the Healthy Eating Pyramid (see Page 59) or Australian Guide to Healthy Eating (see Page 195).
6. Describe the sensory properties – appearance, aroma, flavour and texture – of your Potato Salad with Egg and Tuna.
7. Rate the success of your salad using the hedonic scale on Page 4.

PANCAKES

¾ cup self-raising flour
2 teaspoons caster sugar
1 egg, lightly beaten
½ cup milk
extra milk (if required)
10 grams butter

SERVES ONE

METHOD

1. Sift flour and sugar into a large bowl.
2. Make a well in the middle; pour in egg and half a cup of milk.
3. Mix into a lump-free batter.
4. Allow the batter to rest for 10–15 minutes. Adjust consistency with extra milk if required.
5. Heat a frying pan. With a piece of paper towel, grease lightly with the butter.
6. Spoon the batter into the hot pan, or use a ¼ cup measure. Allow space to turn the pancakes.
7. When bubbles appear on a pancake's surface, turn it over.
8. Stack finished pancakes on a wire rack, inside a tea towel, to cool.

EVALUATION

1. Why is it important to add only half of the milk with the egg in Step 2?
2. Explain the importance of allowing the batter to rest for 10–15 minutes before making it into pancakes.
3. Briefly explain how you would use the frying pan safely.
4. Describe the sensory properties – appearance, aroma, flavour and texture – of your Pancakes.
5. Do you think that these Pancakes would be suitable to include as part of a healthy diet? Justify your answer.

2 PREPARING FOOD SAFELY

KEY KNOWLEDGE
- Safety in the kitchen
- Fire safety
- Personal hygiene
- Kitchen hygiene
  - Washing-up techniques
- Food poisoning
- Preventing food poisoning
  - When storing food
  - When preparing food
  - When cooking food
  - When shopping for food
- Safe use of knives
- Stoves, ovens and cooktops
  - Electric ovens
  - Gas ovens
  - Cooking in an oven
  - Microwave ovens
- Using small appliances
  - Using small appliances safely

KEY TERMS

cross-contamination the transfer of harmful bacteria from uncooked food to food that has been cooked or prepared

danger zone the temperature – between 5°C and 60°C – at which bacteria can multiply very quickly

electric oven uses radiant and convection heat produced by electricity to cook food

fire blanket an insulated blanket used to extinguish small fires in the kitchen

food poisoning an illness caused by eating food that has been contaminated with harmful bacteria

gas oven uses radiant and convection heat produced by gas to cook food

microwave oven an appliance used for cooking, thawing frozen foods and reheating precooked foods

small appliances pieces of equipment such as toasters, food processors, handheld beaters or blenders

AUSTRALIAN CURRICULUM LINKS

DESIGN AND TECHNOLOGIES
- Knowledge and Understanding
- Processes and Production Skills

HEALTH AND PHYSICAL EDUCATION
- Personal, social and community health

GENERAL CAPABILITIES
- Critical and creative thinking

SAFETY IN THE KITCHEN

Just like the home kitchen, the school kitchen is usually a very busy place. Whether you are preparing food at home or at school, you use a range of tools and equipment that, if not handled with care, have the potential to cause accidents. Workspaces can be dangerous places because of the equipment used in them.

Risk areas in the kitchen

1. With a partner, make a list of the main risk areas you can see in your school kitchen.
2. Draw up a risk and safety table for your school kitchen like the one below. Suggest a simple rule that should be followed to avoid accidents occurring in each risk area that you identify.

RISK AREA	SAFETY RULE
Floor	Wipe up all spills immediately.

3. Using the information from your risk and safety table, design and produce a safety poster for one of the risk areas – for example, floors, stoves, electrical appliances, or use of equipment or utensils. Use multimedia software to develop the poster.
   - Your poster should be bright and eye-catching.
   - Make sure that the appropriate safety rule for the risk area is highlighted on the poster.
   - Develop a symbol for use on your safety poster that will highlight the risk area.
4. As a group, look at all the safety posters produced by the class.
   a. Discuss the key features of each poster and the way each highlights a specific risk area.
   b. Does the safety symbol on each poster stand out?
   c. How could these symbols be used in other areas of the school?
5. Display the posters in your school kitchen near the relevant risk area for each poster. Laminate your poster so that it can be part of an ongoing health and safety campaign at school.
6. Write an item for your school newsletter or magazine, or the local paper, to highlight a safety issue in a food preparation area.

FIRE SAFETY

The possibility of fire is a major safety concern in the kitchen. The most common cause of kitchen fires is a piece of equipment, such as a frying pan or a wok, catching alight. If a fat fire occurs in the kitchen, it is important to smother the fire – this will stop oxygen from feeding the fire. To do this, you can use a **fire blanket**, which is extremely effective in extinguishing small kitchen fires.

It is important to follow these simple rules when installing and using a fire blanket:

- The fire blanket should be attached to the kitchen wall just above waist height, so that it is easy to access.
- Pull down firmly on the tabs of the fire blanket to remove it from its cover.
- If possible, gently place the fire blanket over the fire. Throwing it on the fire may spread the fire further.
- Turn off the gas or electricity heat source under the fire.
- Leave the fire blanket over the source of the fire until it has been extinguished and the frying pan or wok that contained the fire has cooled.
- A fire blanket can also be used to extinguish fire on a person's clothing. Place the blanket over the person and wrap it around them; the person should roll on the ground to extinguish the fire as soon as possible.
- It is important to replace a fire blanket after it has been used. You should never re-use a fire blanket.

Small fire extinguishers can be purchased for use in the home. A fat fire is the most likely type of fire to occur in a home kitchen. A dry chemical powder extinguisher can be used to extinguish this type of fire.

If you do not have a fire blanket or an extinguisher, the best method of putting out a fat fire is to cover it with a large saucepan lid or to pour flour or sand onto it. You should never pour water on a fat fire, as this will only cause the fire to spread.

ACTIVITY 2.2
Fire safety in the kitchen

1. Check to see if your school kitchen has any fire safety equipment, such as a fire blanket or fire extinguisher. If it does, where are they located?
2. Are the instructions on how to use the fire safety equipment easy to read?
3. Ask your teacher to demonstrate how you should use a fire blanket.
4. Look at the fire evacuation plan for your classroom. Is it clear and easy to follow?
5. Write a short article for your school newsletter or magazine about the effectiveness of your current school fire evacuation drill.

PERSONAL HYGIENE

Preparing food hygienically to avoid food contamination is just as important as taking care when working with equipment in the kitchen. Bacteria that can cause food poisoning seem to thrive on the warmth and moisture that the human body produces. They live on and in all parts of the body; the hands, fingernails, skin, hair, nose, mouth and even ears all provide a wonderful home for bacteria to thrive. The clothes you wear are also contaminated by the bacteria that naturally occur in the environment. When you sit down at home or when you are out, bacteria transfer from the environment onto your clothes.

KITCHEN HYGIENE

Working in a clean environment and using hygienic practices in the kitchen are equally important as following the rules for personal hygiene.

- Make sure that you use clean tea towels when drying the dishes. Tea towels should not be used to wipe up spills from the floor, flung over your shoulder while you are cooking, or used for drying hands.
- Dishcloths, like tea towels, must be kept clean. Replace your dishcloth regularly, or soak and sanitise it if it is re-usable.
- Keep the food preparation area clean and tidy while you work. Clean up the work area after each process of production by stacking and washing the dishes you have just used and wiping over the bench area.

Personal hygiene rules

1. Thoroughly wash hands with soap and water before preparing food. Disease-carrying bacteria live in all sorts of places, including the handles of school bags, on lockers, on the handrails of buses, trains and trams, and on desks, tables and chairs.
2. Make sure that you wash your hands after having been to the toilet. Harmful bacteria live in faeces, and they can be passed onto toilet paper and hands when you use the toilet.

5. Do not wear nail polish while you are preparing food – chips of nail polish can fall into food.
6. Take off rings and bracelets before you begin to prepare food – these can also be hiding places for bacteria.
7. Cover cuts with a clean, waterproof covering. If you cut yourself while preparing food, remember to sanitise the cut with disinfectant before you cover it.

9. Tie back your hair to make sure that any loose strands do not fall into the food.

3. Make sure you wash your hands after you have blown your nose; bacteria can pass onto your hands from the handkerchief or tissue.
4. Do not sneeze or cough over food; bacteria can easily be transferred through the air.

8. Wear a clean apron to cover your street clothes. Clothes pick up dirt and dust, and can therefore transfer bacteria onto food.

10. Use a clean teaspoon to taste food, and do not lick your fingers. Bacteria that live in your mouth can easily be passed onto the food from your fingers or a dirty spoon.

Washing-up techniques

Thoroughly washing and drying dishes is another important aspect of kitchen hygiene. Today, many families wash most of their dishes in a dishwasher and hand-wash only a few items, such as saucepans. The tables on the opposite page summarise some key points about using dishwashers and hand-washing.

ADVANTAGES OF WASHING DISHES IN A DISHWASHER	ADVANTAGES OF WASHING DISHES BY HAND

• Dishwashers can save the family time. • Dishwashers dry the dishes at a much higher temperature, and therefore more thoroughly disinfect the dishes. • Dishwashers are very convenient, and allow you to wash a large number of dishes at the same time. • Using a dishwasher once a day can be more water-efficient than washing dishes by hand after every meal.	• You can use environmentally-friendly detergent. • Hand-washing dishes is often quicker than using a dishwasher, which can have a very long washing and drying cycle. • You can wash the dishes as you use them, and therefore may not need as much crockery and glassware. • Washing up dishes with family members provides opportunities for family conversation.

USING A DISHWASHER	WASHING DISHES BY HAND
1 Scrape the food scraps from the dishes and rinse the dishes well.	1 Scrape the food scraps from the dishes. Rinse the dishes and carefully stack them into piles of similar kinds.
2 Stack the cutlery container so that the eating surface of the knife, fork or spoon is facing up.	2 Fill the sink with hot, soapy water. Grease will not come off in cold or lukewarm water.
3 Make sure glasses are carefully stacked and unlikely to tip over during the washing cycle.	3 Wash the glassware first, then rinse it in hot water. Allow to drain and air-dry, if possible.
4 Place heavy items on the lower shelf. Stack all plastic items so that they cannot flip over and fill with water.	4 Cutlery should be washed after the glassware. Carefully rinse. Remember not to put sharp knives into a sink filled with soapy water, as they may be come hidden from view.
5 Do not over-stack the dishwasher; if you do, some items may not be properly washed.	5 Next, wash all crockery, and then rinse in hot water. Allow to drain.
6 Regularly clean the filter. Secure the detergent's childproof lock and store detergent away from children.	6 Finally, wash the mixing bowls and saucepans. Rinse well.

Testing knowledge

1 Describe the correct procedure for using a fire blanket.
2 Describe the best procedure for extinguishing a fire in a frying pan or wok if you do not have a fire blanket.
3 Explain why it is important to wash your hands before preparing food.
4 Why should jewellery be removed before preparing food?
5 At school, before production begins, everyone puts on a clean apron. Why?
6 Discuss why 'double-dipping' is considered to be a health hazard.
7 Explain one environmental advantage of using a dishwasher to wash dishes and one environmental advantage of washing dishes by hand.
8 List what you think are the three most important rules when using a dishwasher.
9 Explain why a dishwasher may be considered a hazard for children.
10 Explain why you should wash glassware first when washing up by hand.

FOOD POISONING

Food poisoning is an illness caused by eating food that has been contaminated with harmful bacteria. It can occur when bacteria is transferred onto food, often unknowingly, because of poor personal hygiene or poor food handling. One of the most common causes of food poisoning is the **cross-contamination** of food. This occurs when harmful bacteria from uncooked food are transferred to food that has been cooked or prepared. For example, if raw chicken was cut on a chopping board, and then salad ingredients were cut on the same board without washing, harmful bacteria could be transferred. Washing the chopping board in hot soapy, water in between cutting the raw chicken and vegetables would prevent the transfer of bacteria.

It is very important to keep hot food hot (above 60°C) and cold food cold (below 5°C) so that it is out of the temperature range in which bacteria thrive. Between 5°C and 60°C is described as the **danger zone** because in this temperature range, one bacterium can multiply into approximately 17 million bacteria within eight hours. As well as the correct temperature, bacteria that cause food poisoning also need a moist environment, time to grow, a food supply, a low-acid environment; many bacteria also need oxygen.

It is important to note that poisoned food may look, smell and taste just like normal food. Symptoms of food poisoning can appear almost as soon as food is eaten (that is, within one hour) or take up to 36 hours to develop. Food poisoning symptoms can include diarrhoea, stomach cramps and vomiting. In severe cases, food poisoning can cause death.

A moist environment

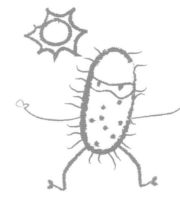

Warm temperatures – the danger zone is between 5°C and 60°C

Sufficient time to grow

A food supply – foods such as milk, cream, meat, poultry and rice

A low-acid environment

Oxygen (for many)

The conditions bacteria need to grow

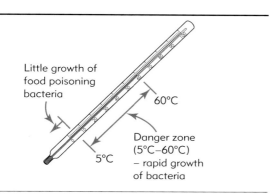

The danger zone

PREVENTING FOOD POISONING

The main thing to do when working with food is to prevent food from being contaminated in the first place and reduce the risk of bacteria in the food growing and multiplying. When storing, preparing, cooking and buying food, there are some strategies to follow that can minimise the risk of food poisoning.

When storing food

- Keep raw food separated from cooked food, and store raw food at the bottom of the fridge to avoid juices dripping onto and contaminating other food.
- Check that the refrigerator's temperature is below 5°C.
- Allow cooked foods to cool to room temperature (about 21°C) before storing in the refrigerator. (This should not take more than two hours – cooling will occur more quickly if you put the hot food into a number of smaller containers rather than leaving it in one large one.) This prevents the refrigerator temperature from rising and reduces the risk of bacterial growth in all food stored in the fridge.
- Cover all food with lids, tinfoil or plastic wrap.
- Do not store food in opened tin cans.

When preparing food

- Always wash your hands in warm, soapy water before touching the food.
- Do not cut salad ingredients on the same cutting board as raw meat without first washing the board in hot, soapy water. This rule also applies for chopping boards that are used to cut raw meat before being used for cooked meat; thoroughly cleaning the board in between uses reduces the chances of cross-contamination.

When cooking food

- Most food should be cooked to a temperature of at least 75°C. If you do not have a cooking thermometer, make sure to cook poultry until the meat is white, particularly near the bone. When cooking hamburgers, mince and sausages, the juices will run clear when they are ready; white fish will easily flake apart with a fork.

When shopping for food

- Keep potentially high-risk foods outside the temperature danger zone.
- Keep hot foods and cold foods separate, and buy them at the end of your shopping trip.
- Always check labels and do not buy food that is past its use-by or best before date.
- Avoid food in swollen, dented, leaking or damaged cans, containers or other packaging.
- Check that serving staff use separate tongs when handling separate food types, such as meats and vegetables.
- Take your shopping home quickly and immediately store it.

Testing knowledge

11 Define 'food poisoning'.
12 List the causes of food poisoning.
13 What physical symptoms might a person with food poisoning display?
14 Define 'cross-contamination' and explain how cross-contamination can be avoided.
15 What is the temperature range of the danger zone?
16 Outline the environmental conditions that bacteria need to grow.
17 Outline two strategies to follow to ensure food is stored safely.
18 What is the key safety strategy to follow when preparing food?
19 When cooking food, how can you tell if it is out of the danger zone?
20 List two safety strategies to follow when shopping for food.

SAFE USE OF KNIVES

Knives are among the most frequently used pieces of kitchen equipment, and are essential in cutting, slicing, dicing and peeling a wide range of foods. Although knives seem simple to use, it is important to use them safely to minimise the risk of injury.

- Select the most appropriate knife for the food you are preparing – for example, a cook's knife for large pieces of food or a vegetable knife for small pieces of fruit or vegetables.
- Keep your fingertips tucked under while using the 'spider' position.
- Make sure you keep the knife sharp. Sharper knives are safer because they cut through the food more easily and require less pressure to be used.
- Never run your finger along the cutting edge of the knife to test its sharpness.
- Make sure the handle of the knife is clean and dry, not greasy, so that the knife does not slip.
- Always cut food on a chopping board made from wood or polyethylene; this will help to protect the sharp edge of the blade. Do not use knives on glass boards, metal or plates, as these materials will blunt the knife.
- Make sure knives are kept away from the edges of benches and out of reach of small children.
- When passing a knife to someone else, remember to pass the handle of the knife, not the blade.
- If you need to move around the kitchen with a knife, hold it close to the side of your body with the blade pointing down.
- Do not put knives in a sink filled with hot, soapy water, as they may become hidden from view.
- Store knives in a knife block or a wall-mounted magnetic rack, not in a drawer with other kitchen utensils.

STOVES, OVENS AND COOKTOPS

In the past, the only option for a stove was an all-in-one cooker that consisted of an oven, a grill and cooktop. Today, there are many choices: for example, built-in or freestanding oven, internal or external grill, and wall oven or under-bench oven. Multifunction electric ovens are becoming increasingly popular because they allow the cook to use a combination of top, bottom, grill and, sometimes, a back element, as well as a fan, to achieve optimum results from their cooking. Combination ovens with convection heat and microwave options have also become popular in some homes. Steamer ovens are another new trend in home cooking.

Ovens are used to bake, roast, casserole and reheat food. Traditional ovens use radiant heat and convection heat to cook food, and include a thermostat that is used to measure the temperature in

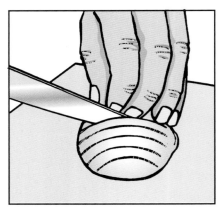

Using the 'spider' position for cutting and dicing

the oven. Convection ovens have a fan, which moves the heat around the oven using convection currents to ensure an even temperature throughout. If the fan does not come on automatically, you can turn it on after the temperature has been set.

Electric ovens

An electric oven has a coil that heats when the oven is turned on. The switch or dial is turned to the required temperature setting; the oven is at the correct temperature when the red light goes off. Most electric ovens take between five and 10 minutes to heat to the set temperature. Preheating the oven before cooking ensures that food does not dry out and that cakes, scones, muffins and bread rise to maximum size during cooking to achieve a light texture.

Gas ovens

When using a gas oven, light the oven according to the instructions; these are usually on the oven's doorplate. Set the oven to the correct temperature; when the fog has cleared from the glass, the oven is ready to use. Preheating a gas oven takes about 10 minutes.

Wok burner
- Has 2–3 rings of flame to produce intense heat
- Metal cradle holds the wok safely during cooking

Cooktop
- Saucepans should fit the hot plate or burner
- Saucepan handle should be turned out of the walkway
- Saucepans boil faster with lids on

Griller
- Pull the cooking tray out before preheating
- Place food on the cooking tray then return it to the grilling compartment to cook
- Adjust heat once griller is hot

Oven
- Arrange oven racks before preheating
- Preheat to recommended temperature before cooking food
- Stand to the side when opening oven door
- Always open oven door fully before removing food
- Always use oven mitts to carry hot food from the oven

Using an oven and cooktop safely

ACTIVITY 2.3

Making tiger toast

AIM
To help you to become familiar with using a griller by making tiger toast

INGREDIENTS
- one slice square bread
- one teaspoon Vegemite
- one slice melting cheese

Tiger toast

METHOD
1. Preheat griller on high.
2. Place bread under grill and cook until pale gold. Turn over and cook the other side to pale gold.
3. Spread the Vegemite over one side of the toast.
4. Cut the cheese into half-centimetre strips and arrange over the Vegemite, leaving half a centimetre between each strip.
5. Return to griller and cook until cheese has melted.
6. Admire the strips, and enjoy!

ANALYSIS
1. Why was the grill preheated before toasting the bread?
2. Why should you leave the griller door of an external griller open during the preheating and cooking of the tiger toast?
3. Explain how you safely removed the cooked toast from the griller.
4. Describe the process of safely cutting the cheese.
5. List some other foods that can be heated in the griller for a quick snack.

CONCLUSION
List two safety points to consider when using the griller to cook food.

Cooking in an oven

The oven is a very versatile appliance that is used to cook a wide variety of delicious recipes. The dry heat of the oven can produce beautiful cakes, biscuits, scones, crisp roast potatoes, golden roast chicken or delicious pastries.

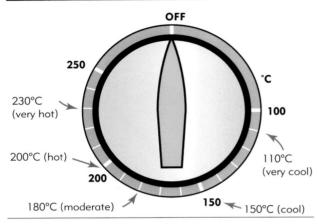

Oven temperatures

Microwave ovens

The microwave oven is an appliance that is found in many homes. Many families find a microwave very useful for cooking a wide variety of foods, thawing frozen foods and reheating precooked foods. Some microwaves have special feature panels that automatically calculate the power and cooking time required for foods such as root vegetables, rice, fish and casseroles. On some models, you can even calculate the kilojoule content of dishes. The types of features available depend on the brand and model of microwave.

While electric and gas ovens are based on radiant heat, microwave ovens use high-frequency radio waves that enter the food and cause its water molecules to vibrate. This creates heat, which in turn cooks the food. Food is not cooked by temperature, but by the number of watts the appliance has available; the power rating depends on the size of the microwave. The weight of the food, its water content and the power of the microwave all affect the cooking time required. Remember that the more food there is in an oven, the longer the cooking time will be.

It is important to allow standing time after cooking in a microwave, as the food continues to cook after the microwave has been switched off. A good rule of thumb is that the standing time should equal the cooking time. Note that food cooked in a microwave will not brown unless a browning plate is used.

Using a microwave oven safely

- Always carefully follow the manufacturer's instructions.
- Use dishes made of china, heat-resistant glass, ceramics, paper or appropriate plastics when cooking in a microwave.
- Do not use metal containers, as they may cause the microwave to arc and damage its magnetron. They also prevent the food from cooking because the microwaves are reflected away from the food.
- Do not turn a microwave on if it is empty; this can damage its magnetron.
- Use only microwave-safe plastic wrap or paper towel to cover food when cooking in a microwave. Never use aluminium foil, as this can cause the microwave to arc.
- Take care when removing plastic wrap from food after cooking – the steam can cause a severe burn.
- Clean the microwave regularly by placing a small amount of detergent on a damp cloth. Heat the cloth in the microwave for one minute, then wipe clean.

ACTIVITY 2·4

Microwaved pappadams

AIM

To make pappadams using a microwave oven

METHOD

1. Place one pappadam in a microwave.
2. Cook on high for 20 seconds. Observe the process, but stand well clear of the microwave.
3. If more cooking is required, increase the cooking time by a few seconds at a time.

ANALYSIS

1. Describe your observations of the cooking process in the microwave.
2. Where should you place food to achieve even cooking in the microwave?
3. Even though pappadams are not cooked in oil, they contain an ingredient that might cause some health concerns. Examine the information on the packaging and identify this ingredient.

CONCLUSION

List two points to consider when using a microwave oven to cook food.

Note: Prawn crackers can also be cooked this way.

USING SMALL APPLIANCES

Small appliances are used in the preparation of many food products. Most households have some small appliances, whether a toaster, a food processor, handheld beaters, a juicer, a blender, a sandwich maker, an electric kettle, an electric wok, a crepe maker or a coffee maker.

Using small appliances safely

Most small appliances are electrically powered, and many have sharp components. Therefore, considerable care is needed when using, cleaning and storing small appliances.

Food processors

Food processors are popular small appliances that come in many varieties, mainly because they can be used in a wide variety of ways; namely, to chop, slice or shred vegetables; to puree soups; to make pastry; to prepare breadcrumbs; to mix a 'quick-mix' cake; or to blend ingredients. However, different brands of food processors are often designed with slightly different features, and therefore each one will be operate in a slightly different way.

Food processor

It is important to follow the manufacturer's safety instructions when using a food processor. It is particularly important to make sure that you use the food plunger that is supplied with the machine for pushing food through the feeding tube (see page 30), rather than a knife or your fingers. Food processors are now designed so that they will not turn on unless the lid switch is in the safety position.

Handheld beaters

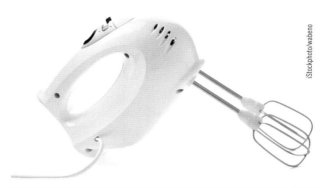

Handheld beaters

Handheld beaters are another very useful small appliance. They are generally very simple to use, since they are light to hold and have a series of speeds that can be easily adjusted, even while running. Handheld beaters can be used to cream together butter and sugar when making cakes or biscuits; to beat egg whites to a stiff foam for meringues; or to make batters for pancakes. When you use handheld beaters, it is important to observe the following safety precautions:

- Always make sure that the power is turned off when putting the beaters into the machine.
- Make sure the beaters are securely pushed into the machine.
- Do not operate the beaters near water.
- Remember to turn off the power and remove the beaters before washing.
- Remove the beaters and wash them in hot, soapy water. Throughly wipe the machine, wind up the cord and store it with the beaters in a clean, dry place.

Small appliances and safety

1. Carefully read through the instruction manual so that you understand how to correctly use the appliance.

Appliance being used near water (in this instance, a blender above a sink) is an electrocution hazard

Appliance is not resting on a flat, even surface, such as a kitchen bench

Cord hanging down to floor is a tripping hazard

2. Do not use the appliance near water or a stove.

3. Ensure that you have dry hands before plugging in, unplugging or operating an appliance.

Use the plunger or 'pusher' originally supplied with the machine.

4. Make sure you use the plunger supplied with a food processor for pushing food through the feeder tube – do not use your fingers or a knife or other utensil.
5. Carefully wash the blades of the food processor. Do not put them into a sink filled with water.

6. Never try to remove toast from a toaster with a knife or fork or other metal utensil.
7. Make sure the appliance is unplugged before you begin to clean it.
8. Store small appliances in a clean, dry place away from moisture and dust.

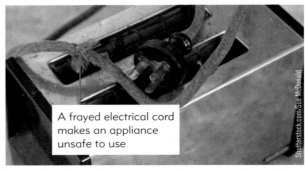

A frayed electrical cord makes an appliance unsafe to use

9. Check appliances frequently to make sure electrical cords do not become frayed.

Electric juicers

Electric fruit juicers are growing in popularity as people become more aware of the importance of fresh fruit and vegetable juices. Juicing one variety of fruit, such as apples or oranges, can make a beautifully refreshing drink for breakfast. Some people prefer to make an exotic 'fruit combo' by juicing a range of their favourite fruits, such as oranges, pineapple and mangoes. Equally delicious is a vegetable juice made by combining a variety of vegetables, such as carrot, celery and capsicum. You can even combine fruits and vegetables.

When using an electric juicer, remember to follow the same safety procedures as you do with a food processor.

Electric juicer

ACTIVITY 2.5
Taste testing apple juices

AIM
To compare the sensory properties of fresh apple juice made in an electric juicer with commercial apple juices

Worksheet

METHOD
1. Collect a sample of each of the following:
   - fresh apple juice made in a fruit juicer from Granny Smith apples
   - a clear commercial apple juice
   - a cloudy commercial apple juice.
2. Place a small amount of each juice into a glass.
3. Draw a table similar to the following. Taste test each of the apple juices. Record the ingredients and your sensory properties, and then compare the properties of each juice.

RESULTS

Property	Fresh apple juice made from Granny Smith apples	Clear commercial apple juice	Cloudy commercial apple juice
Appearance			
Aroma			
Flavour			
Texture			
Ingredients			
Overall rating			

ANALYSIS
1. Describe what is left in the fruit juicer after juicing the apples. Does this substance have any nutritive value?
2. Describe the difference in each of the fruit juices' levels of sweetness. Which was the sweetest and which was the least sweet?
3. Which fruit juice had the most intense apple aroma?
4. Which fruit juice had the most appealing appearance?
5. Did any of the apple juices contain additional ingredients such as preservatives? Why would these ingredients be included in this product?
6. Name one advantage and one disadvantage of using an electric fruit juicer for producing apple juice.
7. What is one environmental disadvantage of using commercial apple juice?

CONCLUSION
Which apple juice did you prefer as a refreshing drink? Why?

Testing knowledge

21. Describe the best method of safely dicing an onion.
22. Explain why it is recommended to store knives in a knife block or a wall-mounted magnetic rack, rather than in a drawer with other kitchen utensils.
23. Outline the best way to safely wash a cook's knife.
24. Identify two safe work practices you should follow when cooking food in a saucepan on a cooktop.

25. Describe the steps involved in safely removing food from an oven once it is cooked.
26. List two important safety rules to observe when using a separate griller.
27. Explain why standing time is important for food that is cooked in a microwave oven.
28. The electric toaster is one of the most commonly used small appliances. Briefly explain three important rules for safely using a toaster.
29. Identify one of the most important safety rules to follow when using a food processor.
30. List two rules for safely using an electric juicer.

DESIGN ACTIVITY 2·1

Swirly scones

SCONES

Scones are enjoyed in many parts of the world, and are usually cooked in an oven. They are thought to have originated near Scone, a village in central Scotland, in the early 16th century. The original version of scones were made in a triangular shape to represent the Stone of Scone (or the Stone of Destiny), a red sandstone block that was traditionally used when Scottish kings were crowned. The first scones were made from oats and cooked on a griddle.

Today, scones are usually made from flour, and can be either plain or flavoured with a wide range of ingredients to make a sweet or savoury snack. They are a popular treat for morning or afternoon tea because they are quick to prepare and can be made and baked in the oven within 20 minutes.

THINKING SKILLS 2·1

Complete the summary frames for the causes of food poisoning and for strategies to prevent food poisoning.

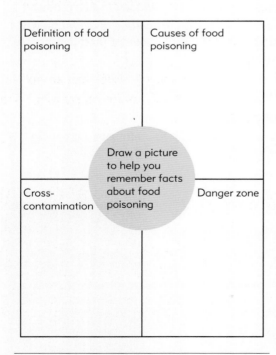

Causes of food poisoning

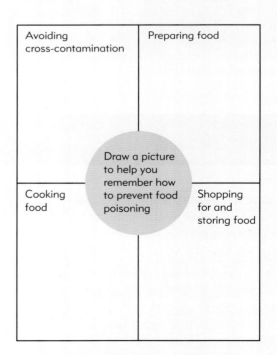

Preventing food poisoning

They are the major component of the well-known English Devonshire tea, which consists of warm scones, whipped cream and berry jam served with a cup of tea.

Devonshire tea

DESIGN BRIEF
Start with a basic scone, and then create an interesting new swirly scone that has a sweet or savoury filling. It could also have a healthy focus.

1. When writing your design brief, you should include:
   - a description of the occasion at which the scones will be served
   - an explanation of who will be eating the sconces
   - where the scones will be served
   - whether a sweet of savoury filling is required, and if so, whether it needs to be healthy.

2. After writing your design brief, develop four evaluation criteria questions from the specifications to help you judge the success of your product.

INVESTIGATING
1. Undertake a recipe search and make a list of ingredients that could be used to flavour a scone dough. Organise the information on a mind map, separating sweet and savoury filling ideas.
2. Sketch some ideas for shaping the flavoured scones, showing how the final product will be presented.

GENERATING
1. Use the recipe map table below to help you create three ideas for your fillings for the scones. Mix and match ingredients from the recipe map, or include some from your own research. You will need at least one ingredient in each functional role for the recipe to be successful.
2. Sketch how you intend to arrange or shape the scones for baking.
3. Explain how you made your decision for the final design.

PLANNING AND MANAGING
1. After selecting the filling option for your scones, write out the method you will use to combine the ingredients with the scone base to create your swirly scones.

PRODUCING
1. Preheat oven.
2. Prepare the filling ingredients for your scones.
3. Make up the Basic Scone recipe on Page 38, up to Step 8.

RECIPE MAP FOR FILLING OF SWIRLY SCONES

Functional role in the recipe	Quantity of ingredient	Filling ingredients option 1	Filling ingredients option 2
Ingredient that melts	40 grams butter		
Contrasting colour to dough	⅓–½ cup		
Flavouring ingredients	▪ 4 tablespoons sugar or sweetening **or** ▪ 1 teaspoon dried spice **or** ▪ 4 tablespoons fresh herbs		
Glaze or topping	1 tablespoon milk		
Selected option:			
Explanation for selection:			

4 Instead of patting the dough, roll out the dough to a rectangle (30 × 20 centimetres) and use the ingredients from your preferred option for the filling.

5 Soften the filling ingredients that will melt in a microwave oven for 10 seconds. Gently spread over the dough, leaving a two-centimetre strip on a long side uncovered.

6 Evenly sprinkle over the ingredients that will provide flavour and contrast in colour.

7 Brush the 2-centimetre strip with milk, then roll the dough from the other end.

8 Cut the roll into equal portions and arrange on a greased oven tray. Glaze before baking.

EVALUATING

1 Answer the four evaluation criteria questions you developed earlier.

2 Describe the sensory properties – appearance, aroma, flavour and texture – of your scones. Share your scones with two other people and record their comments.

3 Explain why it is important to have an ingredient in the filling of the scone that melts during baking.

4 List three health and safety issues you followed while preparing or baking the scones.

5 Discuss your organisation during production. Identify areas for improvement.

6 If you were to make your scones again, what changes would you make, to either the ingredients or the method?

7 Write a paragraph to explain why a swirly scone is a healthier option than a cupcake for afternoon tea.

Swirly scones

SUMMER OR WINTER FRUIT KEBABS WITH YOGHURT AND COCONUT DIP

Fruit kebabs

4 or 5 seasonal fruits

summer fruits: peach, nectarine, grapes, melon, pineapple, strawberries

or

winter fruits: orange, mandarin, apple, pear, kiwifruit, banana

1 teaspoon lemon juice

bamboo skewers

SERVES TWO

METHOD

1. Cut the fruit into bite-sized pieces.
2. Sprinkle with small amount of lemon juice to prevent browning.
3. Thread onto the bamboo skewers.
4. Serve with yoghurt and coconut dip.

Yoghurt and coconut dip

1 tablespoon shredded coconut

1 tub (200 grams) fruit yoghurt

METHOD

1. Place the coconut on an oven tray.
2. Toast the coconut in the oven at 200°C for approximately 2–3 minutes. Remove and check if it is pale golden in colour. Alternatively, toast the coconut under a griller or in a dry, Teflon-coated frying pan.
3. Allow the coconut to cool.
4. Stir the coconut through the yoghurt.

EVALUATION

1. Explain how you would use safely a cook's knife when cutting the fruit.
2. List all of the fruits you used for the kebab. Explain why these fruits were selected.
3. If you needed to substitute canned fruit for some of the fresh fruit in the kebab, list two canned fruits that would be most suitable to use.
4. Explain two safety rules you observed when browning the coconut in the oven.
5. If you were to make this product again, what changes would you make to the ingredients you selected or the processes you used to improve the finished product?

FROZEN BANANA WHIZ

1 very ripe banana

1 cup milk

1 scoop vanilla ice-cream

2 tablespoons vanilla yoghurt

pinch cinnamon or nutmeg (optional)

METHOD

1. Slice the bananas into 5-millimetre slices and seal in a freezer bag in a single layer.
2. Freeze overnight.
3. Place frozen banana, milk, ice-cream and yoghurt in a food processor and whiz until smooth.
4. Pour into tall glasses and sprinkle with a pinch of cinnamon or nutmeg.

MICROWAVED BANANA

1 ripe soft banana

2 teaspoons unsalted butter

1 tablespoon brown sugar

pinch cinnamon

METHOD

1. Peel and slice banana into 2-centimetre chunks.
2. Place in a microwave-safe bowl and cover with other ingredients. Do not mix.
3. Microwave on high for 1–1 ½ minutes.
4. Allow the dish to stand for 1 ½ minutes before serving.
5. Serve with custard, on pancakes or with ice-cream.

BANANA TOAST

1 slice fruit bread
½ ripe banana, sliced
pinch cinnamon
2 teaspoons honey

METHOD

1. Preheat grill and toast one side of the fruit bread.
2. Arrange the banana slices on the uncooked side of the fruit bread, sprinkle with cinnamon, then drizzle with honey.
3. Return to grill for 3 minutes or until the bananas are just warm and the honey has caramelised.
4. Remove from the grill and stand for 2 minutes to allow the honey to cool before eating.

EVALUATION

1. Why is it important that ripe bananas are used for these recipes?
2. In the Frozen Banana Whiz, why is it advisable to freeze the banana in a single layer?
3. Identify one safety consideration when using a food processor.
4. Why are both of the cooked banana dishes allowed to stand for a few minutes before they are eaten?
5. Identify a safety issue you should be aware of when using:

    a a microwave oven
    b a griller.

BASIC SCONES

2 cups (300 grams) self-raising flour
1 tablespoon (20 grams) butter
1–1 ¼ cup milk (approximately)
1 tablespoon milk, for glazing

MAKES 12 SCONES

METHOD

1. Arrange oven shelves and preheat oven to 230°C.
2. Collect ingredients.
3. Grease oven tray.
4. Sift flour into a large bowl.
5. Rub butter into the flour using fingertips until the mixture resembles fresh breadcrumbs.
6. Add 1 cup of milk all at once. Mix with a spatula until a soft dough is formed. Add a little extra milk if the dough is too dry.
7. Turn onto a lightly floured board and lightly knead for 30 seconds. Handle the dough as little as possible to prevent it becoming tough.
8. Gently pat out the dough to 2.5-centimetre thickness and cut scones out.
9. Place scones on the oven tray and glaze with milk.
10. Bake for 10–12 minutes or until golden brown.
11. Remove from oven and wrap in a clean tea towel to cool.

EVALUATION

1. How did you know when the butter was sufficiently rubbed into the flour?
2. Why are the scones patted out rather than rolled with a rolling pin?
3. Why were the scones glazed before going into the oven?
4. Identify two safety rules you followed when baking your scones in the oven.
5. Discuss which part of the production was the most successful and which you found the most challenging.

FOOD-PROCESSOR SWEET SHORTCRUST PASTRY

- 1 cup plain flour
- 1 cup self-raising flour
- 2 tablespoons caster sugar
- 125 grams butter, directly from the refrigerator
- 1 teaspoon lemon juice
- ⅓ cup cold water (approximately)

METHOD

1. Place the flours and sugar in the bowl of the food processor and pulse five times.
2. Chop the butter into small pieces and add to the dry ingredients in the food processor.
3. Process until the mixture resembles fine breadcrumbs.
4. Add the lemon juice and water and blend for a further minute or until the mixture just comes together.
5. Remove from the blender and place on a lightly floured board. Bring together into a ball.
6. Wrap in plastic wrap and refrigerate for 20 minutes.

EVALUATION

1. Why is a mixture of plain flour and self-raising flour used in making this pastry?
2. Explain why it is important to pulse the flour and sugar before adding the butter.
3. Why is lemon juice added to this recipe?
4. How would you change the method for making the pastry if you didn't have a food processor?
5. What is the purpose of wrapping the pastry in plastic wrap and resting it in the refrigerator for 20 minutes before rolling out?

APPLE AND CINNAMON TURNOVERS

½ quantity Food-processor Sweet Shortcrust Pastry

⅔ cup pie apples or 1 apple, peeled and sliced

2 teaspoons caster sugar

¼ teaspoon cinnamon

2 tablespoons sultanas

1 tablespoon milk

1 tablespoon icing sugar

MAKES TWO TURNOVERS

METHOD

1. Preheat oven to 200°C.
2. Roll out the pastry to a square of approximately 24 × 24 centimetres. Cut the pastry into half so that you have two rectangles 24 × 12 centimetres in size. Place on a baking tray.
3. Mix the pie apples, caster sugar, cinnamon and sultanas together in a small bowl. Divide the apple mixture into half.
4. Place each portion of the apple mixture on the lower half of each rectangle of pastry.
5. Brush the edges of the pastry with milk.
6. Turn the top half of the pastry over the apple mixture and press the edges together firmly. Trim the edges of the pastry if necessary.
7. Use a fork to decorate the edges of the pastry. Cut a steam vent in the top of the turnover.
8. Glaze the pastry with the milk.
9. Bake for 15 minutes at 200°C.
10. Remove from the oven and dust lightly with the icing sugar. Return to the oven and continue to bake for a further 5 minutes.

EVALUATION

1. Describe the sensory properties – appearance, aroma, flavour and texture – of your turnovers.
2. Explain why it is necessary to make an air vent in the top of the turnover.
3. Identify two safety rules you followed when baking your turnovers in the oven.
4. Explain why the pastry is glazed with milk before baking.
5. Plot the ingredients of your Apple and Cinnamon Turnovers on a diagram of the Healthy Eating Pyramid and decide if they are a healthy dessert. Justify your answer.

FOOD BY DESIGN

RECIPE BASICS 3

KEY KNOWLEDGE
- Tools of the trade
  - Processes and tools of the cooking trade
- Making sense of a recipe
- Abbreviations in recipes
- Measurement in recipes
  - Dry ingredients
  - Liquid ingredients
  - Measurement by weight
- Commonly used food preparation terms
- Designing new recipes
- Muffins
  - Muffin ingredients

KEY TERMS

functional ingredients recipe ingredients that flavour, create texture, colour, help other ingredients to combine or increase overall nutrient value

metric measuring tools spoons, cups, jugs and scales that have been calibrated to accurately measure ingredients by weight and volume using the metric system

recipe a list of ingredients and instructions for preparing food

AUSTRALIAN CURRICULUM LINKS

DESIGN AND TECHNOLOGIES
- Knowledge and Understanding
- Processes and Production Skills

GENERAL CAPABILITIES
- Critical and creative thinking

TOOLS OF THE TRADE

Every skilled tradesperson has their own special tools of trade that are specifically designed for working with particular materials, such as wood, fabric, metal or clay. Working with food requires specialist tools, too. Many of the tools used for preparing food are small pieces of equipment called utensils, and each utensil usually has a specific task. Utensils are often grouped with other pieces of equipment that perform similar functions, such as cutting, peeling or measuring. The table below will help you to identify some of the tools required for working with food.

Peeling a carrot

Processes and tools of the cooking trade

PROCESS	TOOLS OF THE TRADE	USE AND SAFETY
Cutting and peeling	Knives: – cook's – vegetable – serrated	• Always cut downwards onto a chopping board. • Wash knives separately – never put knives into a sink of soapy water. • Carry knives close to your side. • Pass the handle of the knife, never the blade.
Measuring	• Spoons • Cups • Jugs • Scales	• Measure accurately – level dry ingredients with a spatula. • Measure liquid ingredients at eye level. • Reset the scales between ingredients.
Grating and crushing	• Grater • Meat mallet • Potato masher • Garlic crusher • Ricer • Food processor	• Keep your fingertips away from the cutting edge on the grater. • Rinse the equipment immediately after use. • Remove grated food with a pastry brush. • Always use a 'pusher' when feeding food through the feed tube of a food processor.
Mixing and beating	• Wooden spoon • Whisk • Handheld electric beater • Stick mixer • Food processor	• Follow the manufacturer's safety instructions when using electrical equipment. • Switch off the power supply and unplug before cleaning. • Wash the equipment thoroughly in hot soapy water immediately after use.
Sieving and straining	• Sieve • Colander • Slotted spoon	• Use oven mitts to hold equipment when straining hot food. • Thoroughly dry equipment for sieving before storing.
Lifting	• Tongs • Egg lifter • Wire skimmer	• Do not rest the handles of tools onto hot saucepans or frying pans – heat may be transferred and cause burns.

ACTIVITY 3·1

Using tools safely

1. Read the recipe for Spaghetti Bolognese on Page 263. List all the equipment that would be needed to produce this recipe.
2. Draw up the Spaghetti Bolognese recipe as a flow chart, highlighting the main stages in the recipe. Annotate the flow chart with the safety issues involved in each stage.

MAKING SENSE OF A RECIPE

A **recipe** is a list of ingredients and instructions for preparing food. A recipe has several components:
- a name
- a list of ingredients, including quantities and, sometimes, details about preliminary preparation
- a method that explains how to prepare the ingredients and the order in which the processes should be completed

Minestrone Soup *[Recipe name]*

Quantities of ingredients

½ onion, diced

⅓ carrot, diced

½ potato, diced

½ stick celery, sliced

4 green beans, sliced

⅓ zucchini, sliced

½ cup cabbage, shredded

1 tablespoon oil

1 clove garlic, crushed

1 tablespoon tomato paste

100 grams diced canned tomatoes

1 cup beef stock

1½ cups water

1 tablespoon canned cannellini beans

1 tablespoon small pasta (for example, rigatoni)

1 tablespoon grated parmesan cheese, for serving

salt and pepper

[Some processes that need to be completed before starting the 'method']

[Order in which the ingredients should be put together]

Method

1. After cutting, place each vegetable in a separate pile, as they will be cooked at different times.
2. Heat the oil in a large saucepan. Add onions and cook over medium heat until transparent.
3. Add carrot and cook for 1–2 minutes. Stir occasionally and take care not to brown the vegetables.
4. Repeat this process, adding the celery, then the green beans, zucchini and potato.
5. Add the cabbage and cook until it wilts.
6. Add garlic, tomato paste, diced tomatoes, stock and water, and bring to boil.
7. Reduce heat to simmer and cover saucepan with lid. Cook for 15–20 minutes until vegetables are soft. *[Cooking time]*
8. Add cannellini beans and pasta, and cook for another 10 minutes.
9. Adjust the amount of liquid if necessary. Season with salt and pepper.
10. Serve and garnish with parmesan cheese.

Minestrone soup goes well with fresh, crusty bread. *[Suggestion for complementary food]*

Serves two. *[Number of serves]*

Chicken and Sweet Corn Soup

Finely dice half an onion and one quarter of a green capsicum. Then slice one stick of celery into thin pieces. Take a large saucepan and sauté the onion, capsicum and celery in one tablespoon of vegetable oil over medium heat. Do not brown. Open a 220-gram can of creamed corn and add to saucepan. Stir in two cups of water and half a packet of chicken noodle soup. Simmer for 15 minutes. Serve and garnish with a finely sliced spring onion.

Serves two

- an indication of cooking temperature and time
- a list of any special equipment required
- an indication of the number of serves the recipe makes.

Ideas for garnishes or decorations are sometimes included in the recipe, or other foods that will be complementary to the finished product are suggested. Photographs can give an idea of how the finished product will look.

ACTIVITY 3·2
Recipe formats

Worksheet

1. Compare the formats of the two soup recipes on the previous page (for Minestrone Soup and Chicken and Sweet Corn Soup).
2. If you were preparing a shopping list in a hurry, which recipe format would be easier to work from? Why?
3. What are the advantages of working from the recipe format used for Minestrone Soup?
4. What are the advantages of working from the recipe format used for Chicken and Sweet Corn Soup?
5. Reorganise and rewrite the Chicken and Sweet Corn Soup recipe so that it is in the same format as the Minestrone Soup recipe.
6. Collect two examples of recipes from product labels, magazines or promotional leaflets. Draw up a table like the one below and identify the features of each recipe.
7. After comparing the features of each recipe, identify which one you would be most likely to make, and explain why.

RECIPE COMPARISON

	Recipe 1	Recipe 2
Name		
Photograph		
Format of recipe		
Number of ingredients		
Number of steps to complete recipe		
Time it will take to prepare and cook recipe		
Equipment required		
Number of serves		
Garnishes or decorations		
Serving suggestions		

ABBREVIATIONS IN RECIPES

Measurements are usually given in recipes, to ensure that a successful product can be made, eaten and enjoyed. To make recipes easier to read and quicker to write, some aspects of the recipe are often abbreviated. Some of the most common abbreviations are shown in the table below.

g	gram
kg	kilogram
mL	millilitre
L	litre
°C	degrees Celsius
tsp.	teaspoon
tbs.	tablespoon
c	cup
SR flour	self-raising flour
cm	centimetre
min	minutes

MEASUREMENT IN RECIPES

Accurate measurement in food preparation is important, to ensure success in recipes and to allow the same product to be made again. Correctly calibrated tools are essential for accurate measurement. In Australia, we use the metric measurement system, so it is important to ensure that your tools are labelled according to this system.

Measuring spoons and cups

Dry ingredients

METHOD TO ACCURATELY MEASURE INGREDIENTS	MEASURING TOOLS		DRY INGREDIENTS MEASURED BY SPOONS AND CUPS
	SPOONS	CUPS	
Measuring spoons and cups. The holding capacity of the item should be written on its handle. Dip the spoon or cup into the ingredient slightly overfilling, then level off with a spatula.	• 1 tablespoon = 20 mL • 1 teaspoon = 5 mL • ½ teaspoon = 2.5 mL • ¼ teaspoon = 1.25 mL	• 1 cup = 250 mL • ½ cup = 125 mL • ⅓ cup = 80 mL • ¼ cup = 63 mL	• Flour • Sugar • Cocoa • Coconut • Spices

Liquid ingredients

METHOD TO ACCURATELY MEASURE INGREDIENTS	MEASURING TOOLS	LIQUID INGREDIENTS MEASURED BY VOLUME
Measuring jugs. Sit the jug on a level surface, pour in the liquid and read the quantity at eye level.	Measuring jugs with cup and millilitre measurements are most useful.	• Milk • Stock • Water • Cream

Measurement by weight

METHOD TO ACCURATELY MEASURE INGREDIENTS	MEASURING TOOLS	INGREDIENTS MEASURED BY WEIGHT
Check that the scales are set on zero before starting to measure ingredients.	Kitchen scales are graduated in either one-gram or five-gram measures.	• Butter • Cheese • Meat • Whole nuts • Fresh fruit and vegetables

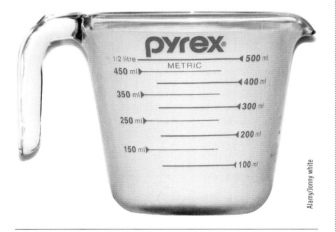

Measuring jug

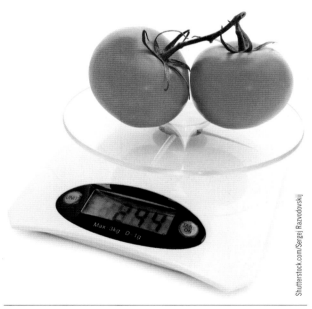

Scales

ACTIVITY 3·3

Measurement revision

1. Describe the method used to accurately measure dry ingredients using a measuring spoon or cup.
2. List three liquid ingredients other than those included in the table on Page 45.
3. List four dry ingredients other than those included in the table on Page 45.
4. Describe the method used to accurately measure liquid ingredients using a measuring jug.
5. What are the benefits of using measuring scales rather than measuring spoons and cups to measure meat or cheese?
6. When measuring dry ingredients, why is it more accurate to dip and level off with a spatula than to pack the ingredients into the measuring spoon or cup?
7. Describe one important rule to follow when using scales.
8. Explain how you would accurately measure the following ingredients:
   - ⅓ cup wholemeal flour
   - 150 millilitres milk
   - 1 ½ teaspoons curry powder
   - 100 grams mushrooms.
9. Draw up the following table in your workbook and fill in the equivalent measures.

EQUIVALENT MEASURES

1 tablespoon = ? millilitres	1 tablespoon = ? teaspoons
1 teaspoon = ? millilitres	1 cup = ? millilitres
teaspoon = ? millilitres	1 litre = ? cups
teaspoon = ? millilitres	1 kilogram = ? grams

Measuring liquid ingredients

Testing knowledge

1. Identify the piece of equipment you would use if you wanted to cut or chop:
   - stewing steak
   - celery
   - apples
   - whole pumpkin.
2. Identify the equipment you could use to cream together butter and sugar for a cake mixture.
3. List the information you would expect to find in a recipe.
4. List the abbreviations that are sometimes used in recipes for gram, litre, teaspoon, tablespoon, cup and degrees Celsius.
5. Describe the method you would use to measure one tablespoon of cocoa for a biscuit recipe.
6. List how many millilitres there are in 1 tablespoon, 1 teaspoon, 1 cup and 1 litre.
7. Describe the process of accurately measuring liquid ingredients.
8. Why is it more accurate to measure liquids in a metric measuring jug than in a cup?
9. Identify two ingredients that would be easier to measure with measuring scales than with cups.
10. Summarise the reasons why accurate measurement is important when following a recipe.

COMMONLY USED FOOD PREPARATION TERMS

Following is a list of terms that you will frequently find in recipes as you design and work with food, as well as their definitions and information on foods and equipment relevant to each process.

FOOD PREPARATION PROCESS	DESCRIPTION OF FOOD PREPARATION PROCESS	APPROPRIATE EQUIPMENT	FOOD EXAMPLES
Bake	Cook food using dry heat in an oven.	Baking tray; cake tin	Bread; biscuits; cakes
Beat	Vigorously mix ingredients to incorporate air or combine.	Wooden spoon; handheld and electric beater; whisk	Cream; egg whites
Bind	Stir ingredients to combine.	Bowl; wooden spoon; spatula	Hamburger mixture
Blanch	Plunge food into boiling water for 30 seconds. Drain and refresh in iced water.	Saucepan; sieve or colander	Almonds; snow peas
Blend	Mix a dry ingredient with a moist ingredient until it forms a smooth paste.	Bowl; wooden spoon	Cornflour and water
Boil	Heat a liquid to 100°C, or to boiling point.	Kettle or electric jug; saucepan	Water
Chop	Roughly cut food into small pieces.	Chopping board; cook's knife	Vegetables
Cream	Beat together sugar and butter until they resemble lightly whipped cream. The mixture will become lighter in colour.	Bowl; wooden spoon or electric beater	Butter cakes; biscuits
Dice	Cut food into small, even-sized cubes.	Chopping board; cook's knife	Onion
Fold	Gently combine a light, airy mixture into a heavier mixture; for example, beaten egg white into custard sauce. Use a metal spoon or spatula in short strokes to prevent loss of air or volume.	Bowl; metal spoon or spatula	Fluffy omelette; sponge
Fry	Cook food in hot fat or oil. Food may be deep-fried, shallow-fried or stir-fried.	Frying pan; lifter; wok	Potato chips; bacon and eggs
Garnish	Add edible decoration to a dish to enhance its appearance.	Vegetable knife	Fresh herbs such as parsley
Glaze	Brush a thin liquid such as milk or egg over food before baking to create a shiny, golden-brown surface.	Pastry brush; jug	Scones; pies; tarts
Grate	Reduce a piece of food into thin shreds by rubbing it against the serrated metal surface of a grater.	Grater; microplane grater	Cheese; vegetables
Grill	Cook small pieces of tender food by dry, radiant heat; for example, the griller on a stove or a barbecue.	Griller; barbecue	Small, tender cuts of meat or poultry; kebabs; satay sticks
Julienne	Cut food into thin, matchstick-sized pieces.	Chopping board; cook's knife	Carrot; celery; capsicum
Knead	Mix and shape a flour dough by hand. In bread-making, this process strengthens the gluten.	Floured board	Bread; scones; pastry
Marinate	Soak foods such as meat or poultry in a seasoned liquid to improve its the flavour and, sometimes, to tenderise.	Bowl	Meat strips for a stir-fry; tandoori chicken pieces
Mix	Combine ingredients so that they are evenly incorporated.	Bowl; spoon or spatula	Flour and sugar
Poach	Gently cook food in a simmering liquid.	Saucepan	Eggs; pieces of fresh fruit

FOOD PREPARATION PROCESS	DESCRIPTION OF FOOD PREPARATION PROCESS	APPROPRIATE EQUIPMENT	FOOD EXAMPLES
Purée	Make food into a smooth paste by passing through a sieve or by blending.	Sieve or food processor	Stewed apple; vegetable soups; tomato sauce
Roux	Mix melted butter or margarine and flour, blend, then cook. It is used to thicken a sauce.	Saucepan; wooden spoon	White sauce; gravy
Rub in	Mix butter or margarine through dry ingredients with fingertips until the mixture looks like breadcrumbs.	Hands; bowl	Scones
Sauté	Lightly toss food in fat or oil in a frying pan over direct heat. The process assists in flavour development, but does not brown.	Frying pan; saucepan; wooden spoon	Soups; casseroles
Sear	Brown food quickly over a high heat to seal in juices.	Frying pan; stovetop	Steak; lamb chops
Shred	Cut food into thin strands using a knife, a grater or a shredding disc in a food processor.	Chopping board; cook's knife; food processor	Lettuce; cabbage; carrot
Sift	Pass dry ingredients through a fine mesh sieve to mix, aerate and remove lumps.	Sieve	Sponges; cakes
Simmer	Bring liquid to just below boiling point so that small bubbles appear on the surface of the liquid.	Saucepan, sometimes covered with a lid	Stock
Slice	Cut food into thin pieces.	Cook's knife or serrated knife; chopping board	Processed meats; salad vegetables
Steam	Cook food over boiling water on a rack or in a special basket in a covered pan. This retains the food's shape and minimises nutrient loss.	Saucepan with tight-fitting lid; steaming basket made of metal or bamboo	Pork buns; dim sims; vegetables
Stew	Simmer food covered in liquid for a long time. Used in dishes containing tough cuts of meat with vegetables.	Saucepan with tight-fitting lid	Lamb; root vegetables; fruit
Stir	Use a wooden spoon to lightly mix ingredients.	Wooden spoon	Custard sauce; gravy
Toss	Mix ingredients by lightly lifting and folding several times.	Wok; salad servers	Vegetables (stir-fried in a wok); salad ingredients
Whisk	Incorporate air into ingredients such as cream, egg whites and sauces.	Bowl; whisk	Souffle sponge

ACTIVITY 3.4

Recipe terms

1. Read the recipe for Oriental Chicken Kebabs on Page 71. Make a list of the terms from the table on pages 47 and 48 that are used in this recipe.
2. List any other terms from the recipe that you think are important and write a definition for each of them.
3. Make a list of each of the ingredients in the recipe and write down the best method of storing each of them.
4. Select six other terms from the tables on pages 47 and 48 and find recipes in this textbook that use one of these terms. Use a table similar to the one following and write down a list of the terms and the matching recipe and page number.

RECIPE KEY TERMS

Term	Recipe name	Page number
Shred	Salad Roll-Up	15

DESIGNING NEW RECIPES

Sometimes, after you have prepared a recipe, you may decide that, while you quite liked the product, there were some aspects of it that you would change if you were to make it again. You may need to modify a recipe because some of the people you are cooking for have specific likes and dislikes or special dietary requirements. Alternatively, you may be wanting to showcase your creativity with food.

Recipes can be modified in a number of ways:
- The texture can be changed; for example, by cooking vegetables for less time so that they will be crunchier.
- The flavour can be changed; for example, by adding more herbs and spices for a stronger flavour.
- One ingredient can be substituted for another in the recipe that has a similar function; for example, by using wholemeal flour instead of white flour to increase the fibre content of a cake.

Once you have created a new product by altering ingredients and/or processes, it is important to give your new recipe a name and record what you did, so that it can be repeated at a later date. The name should reflect a feature of the product; for example, the main ingredient of the recipe could be included.

MUFFINS

Muffins are one of America's contributions to the culinary world. They are a small, cake-like bread that can be made with a variety of flours and often contain fruit and nuts. They use baking powder as their raising agent, and so are light and quick to make. They can be served hot or cold, and are great to serve for breakfast or a snack. Savoury muffins make a nice change from bread as an accompaniment to a dish such as soup.

Muffin ingredients

There are seven groups of ingredients in muffin recipes, all of which have specific roles or functions. To make changes to the recipe and produce a successful product, you should develop an understanding of the role of each *functional ingredient*.

Flour
- Gives volume and structure to the muffin
- Absorbs moisture
- High in starch
- Browns during the baking process, and so contributes to the golden finish of the muffin
- Available in a number of varieties – for example, plain, self-raising, wholemeal

Sweetenings
- Provide a sweet flavour
- Help to create a tender crumb
- Caramelise or brown during the baking process, and so contribute to the golden finish of the muffin
- Available in a number of varieties – in the case of sugar, for example, A1, caster, brown, demerara, raw

Eggs
- Help to bind the ingredients during mixing
- Expand and capture air during the beating process, increasing the muffins' volume during baking
- Coagulate and set during baking to form a firm structure
- Contribute to the muffins' rich flavour

Flavourings
- Give a specific flavour – for example, small amounts of herbs and spices
- Add texture and colour – for example, blueberries and flaked almonds
- Contribute to the muffins' volume
- May be sweet (for example, apple, banana, apricot, nuts or chocolate) or savoury (for example, cheese, bacon, dried tomato or olives)

Liquids
- Help to combine the ingredients
- Add flavour
- Create a moist end product
- A number of types can be used – for example, milk, buttermilk, sour cream and yoghurt

Shortening
- Helps to make a moist, rich texture
- Creates a tender crumb
- Helps keep the muffins fresh
- A number of products can be used – for example, vegetable oil, butter, margarine

Raising agents
- Make the muffins' texture light by creating bubbles of carbon dioxide, increasing the volume

- Available in a number of forms – for example, baking powder, self-raising flour (flour with baking powder added) and beaten egg or egg whites

Note: Chemical raising agents begin to produce carbon dioxide as soon as they are combined with moisture, so they should be mixed with liquid ingredients just before baking to maximise volume.

Muffins

Testing knowledge

11 Discuss three ways that you could modify or create a new variation of a recipe.

12 Describe how you would cream ingredients when making a cake mixture.

13 Explain the difference between chopping and dicing carrots.

14 Why is it important to be gentle when folding the ingredients in when making a sponge?

15 Define the term 'sauté', and explain why sautéing is considered to be an important cooking process.

16 Explain why steaming is considered to be a healthy method of cooking vegetables.

17 Explain why flour is an essential ingredient in muffins.

18 Outline why eggs are an important ingredient in cake or muffin recipes.

19 Explain what the effect would be if you did not include a shortening ingredient when preparing muffins.

20 Identify the ingredients that contribute to the light texture of muffins.

THINKING SKILLS 3·1

COMPARE TOOLS THAT CAN BE USED TO CARRY OUT SPECIFIC FOOD PREPARATION PROCESSES

Use the chart of commonly food preparation terms on Pages 47–8 to select a food preparation process and compare suitable tools that could be used as part of that process. Alternatively, draw up a table like the one below and complete the example shown in it.

1 Describe the characteristics of each tool.

Food preparation process: bind		
Description of process		
Suitable tool/s to carry out process	Wooden spoon	Plastic spatula
Sketch of tool		
Properties of materials used in tool		
Ease of use when carrying out the process		
How well the tool completes the process		
Ease of cleaning the tool		

2 Explain how the wooden spoon and plastic spatula are similar and different with respect to the characteristics you have observed.

3 Summarise your findings, including examples of food products for which you would use each tool.

DESIGN ACTIVITY 3.1

Designer muffins

DESIGN BRIEF

The manager of your school cafeteria will be promoting each curriculum learning area on the last Friday of each month with one specially designed menu item. There will be a competition for the best snack food that reflects the area. For Technology, the focus is a new flavoured muffin that includes fruit, vegetables or nuts.

Worksheet

1. Write a design brief that addresses the following points:
   - Who will be the target market for the new muffins?
   - Why and how will the new menu item reflect the Technology learning area?
   - When will the new muffin recipe be served at the cafeteria?
2. After writing your design brief, develop four questions to evaluate the success of your end product.

INVESTIGATING

1. Develop a mind map with as many flavouring ingredients as possible that could be used in a muffin recipe.
2. Classify the flavours into 'Sweet' and 'Savoury'.
3. Fill in the following table with examples of ingredients for each functional group in a muffin recipe.

FUNCTIONAL INGREDIENTS IN MUFFINS

Functional ingredient	Functional role of the ingredients in muffin recipes	Examples of types of ingredient
Flour		
Sweeteners		
Eggs		Eggs
Flavourings		
Liquids		
Shortening		
Raising agents		

GENERATING

1. Use the recipe map to develop two distinctly different recipe options that meet the needs outlined in your design brief. Remember to use your knowledge of functional ingredients in muffins when designing the recipe options. Remember that your muffin recipe must contain some fruit, vegetables or nuts.

RECIPE MAP FOR MUFFINS

Functional ingredient	Quantity required	Option 1 muffin recipe	Option 2 muffin recipe
Flour	2 cups		
Sweeteners	¾ cup		
Eggs	1 required		
Flavourings	1 cup		
Liquids	¾ cup		
Shortening	125 grams or millilitres		
Raising agents			

Preferred option:

Explanation for choice of preferred option:

PLANNING AND MANAGING

1. Write out a new recipe, incorporating your ingredients into the method. Remember to give your recipe an appealing name.
2. Complete a food order for your recipe.

PRODUCING

1. Prepare your preferred option, recording any changes you make to the ingredients or method during production.

EVALUATING

1. Answer, in detail, the four evaluation questions you developed earlier.
2. Describe the sensory properties – appearance, aroma, flavour and texture – of your muffins.
3. Share your muffins with two other people and record their comments about them.
4. Taking into account your tasters' comments and your own experience, suggest improvements you could make to the ingredients and/or method if you were to make the muffins again.
5. Discuss your level of organisation during production. Identify areas for improvement.
6. Plot the ingredients of your designer muffins on a diagram of the Healthy Eating Pyramid or the Australian Guide to Healthy Eating, and decide whether they are a healthy muffins. Justify your answer.

TOASTED MUESLI

1 cup rolled oats
45 grams shredded coconut
1 teaspoon sunflower seeds
2 teaspoons sesame seeds
¼ cup skim milk powder
¼ cup unprocessed oat bran
⅓ cup All-Bran
1 tablespoon peanuts, chopped
1 tablespoon vegetable oil
1 tablespoon golden syrup
2 teaspoons honey
2 dried apricots, chopped
½ cup sultanas

SERVES ONE

Muesli is great served with milk or yoghurt or nibbled on as a snack, since it is a very nutritious food. This recipe uses all of the types of measuring equipment used to prepare food.

METHOD

1. Preheat oven to 180°C.
2. In a large bowl, collect the rolled oats, shredded coconut, sunflower seeds, sesame seeds, skim milk powder, oat bran and All Bran.
3. Add peanuts to other ingredients in the bowl. Mix well.
4. Place the vegetable oil, golden syrup and honey in a small saucepan. (Hint: measure the oil first, then dip the tablespoon you used for the oil into the golden syrup; it will then off the spoon easily.)
5. Over a medium heat, bring the liquid ingredients to the boil. Remove from heat immediately and pour over dry ingredients. Mix well.
6. Spread muesli in a thin layer on a baking tray. Bake for 5 minutes.
7. Remove from oven, stir carefully, then return to oven for another 5 minutes or until golden brown.
8. Remove from oven and cool.
9. Mix in chopped apricots and sultanas.
10. When completely cool, package in an airtight container.

EVALUATION

1. Explain how you accurately measured the dry ingredients.
2. Which tool did you use to stir the muesli? Explain why it was the most suitable.
3. Why was it important to remove the liquid ingredients from the cooktop as soon as they boiled?
4. Describe the safest way to remove the hot tray of muesli from the oven.
5. Discuss how you could modify the muesli recipe to make it even healthier.

BASIC MUFFINS

1 cup flavouring ingredient/s
2 cups self-raising flour
¾ cup sugar
1 egg, lightly beaten
¾ cup liquid (usually milk)
½ cup shortening (usually oil)

MAKES 12 MUFFINS

Yummy dried fruit muffins

To make dried fruit muffins, just add one cup of dried fruit as the flavouring ingredient in the Basic Muffin recipe. Choose your favourite or use a combination of dried fruits such as sultanas, currants, dried apricots and cranberries; you could also use some raisins, dates, dried apples or dried pears as the flavouring ingredient. When the muffins are cooked, dust their tops with a little icing sugar to finish a really delicious treat!

METHOD

1. Preheat oven to 200°C. Grease or line a muffin tray (12 × ⅓ cup capacity).
2. Prepare flavouring ingredient/s.
3. Sift dry ingredients into a large bowl.
4. Combine the wet ingredients.
5. Mix the dry, wet and flavouring ingredients together. Do not over-mix, otherwise, the muffins will become tough.
6. Spoon evenly into muffin tray.
7. Bake for 15–20 minutes.
8. Test the muffins to see if they are ready; leave longer if necessary. (The muffins are ready if they spring back when lightly touched with your finger, or if a fine skewer comes out clean and dry from the centre of a muffin.)
9. Cool in tray for 5 minutes. Remove from pan and place on cake rack.

Easter magic muffins

To make a special treat for Easter, freeze 12 baby caramel Easter eggs, then make up the Basic Muffins recipe, using soft brown sugar. Once the eggs are frozen, remove their foil, then place the frozen eggs in the middle of each muffin. Use muffin papers for this recipe to prevent the mixture from sticking. Remember to allow the muffins to cool before eating, since the caramel will be boiling hot when just from the oven.

EVALUATION

1. Why are the wet and dry ingredients kept separate and mixed together just prior to baking?
2. Why is it important to have equal quantities of mixture in each muffin tin before baking?
3. Explain why muffins are cooked in a hot (200°C) oven rather than a cool oven.
4. Identify two safety rules you followed when baking your muffins in the oven.
5. Describe the sensory properties – appearance, aroma, flavour and texture – of your muffins.

MINI QUICHES

8 slices wholemeal bread
30 grams butter, melted
1 rasher bacon, finely diced
2 spring onions, finely sliced
30 grams tasty cheese, grated
2 eggs
⅓ cup milk

MAKES 8 MINI QUICHES

Pastry is traditionally used to encase an egg mixture in a quiche. In this recipe, bread replaces the pastry, because it has a much lower fat content and is quick and easy to prepare. During baking, the heat in the oven coagulates the egg and sets the filling.

METHOD

1. Preheat oven to 200°C.
2. Trim crusts from bread. Roll each slice flat with a rolling pin to compress the slice and ensure there are no holes in the slices of bread.
3. Brush one side of each slice of bread with melted butter.
4. Carefully place the bread butter side down into small, greased muffin tins. Each slice will form a small cup with pleats in it.
5. Add bacon, spring onion and cheese to each bread case.
6. Lightly beat eggs and milk with a fork, then pour some of this mixture into each bread case.
7. Bake for approximately 15–20 minutes or until golden brown and the filling is puffed. Remove from muffin tray and serve.

EVALUATION

1. Identify the process that the eggs undergo during baking.
2. Outline the main role that egg plays in the structure of a quiche.
3. What are the nutritional benefits of using bread instead of pastry for the base of the quiche?
4. Suggest some other ingredients that could be used to flavour the Mini Quiches.
5. Outline two rules for using the oven safely.
6. If you were to repeat the production of Mini Quiches, what changes would you make?

4 EAT WELL, BE WELL

KEY KNOWLEDGE

- Food and me
- What is food?
- Nutrients in food
- Water
- Digestion
- Selecting food wisely
- The Healthy Eating Pyramid
  - Foundation layers
  - Middle layer
  - Top layer
- Nutrition throughout life
- Adolescent food needs
- Five healthy eating tips for teenagers

KEY TERMS

dietary fibre a nutrient found in the cell walls of all plant foods that improves the health of the digestive system by adding bulk to faeces

food any substance that we eat or drink that provides the body with chemical substances called nutrients

nutrients chemical substances in food that are broken down during digestion, including protein, carbohydrates, fat, vitamins and minerals

AUSTRALIAN CURRICULUM LINKS

DESIGN AND TECHNOLOGIES
- Knowledge and Understanding
- Processes and Production Skills

HEALTH AND PHYSICAL EDUCATION
- Personal, social and community health

GENERAL CAPABILITIES
- Critical and creative thinking

FOOD AND ME

People, like plants, come in a variety of shapes and sizes. To grow and remain healthy, plants need food, water, clean air and sunshine. People also need nutritious foods as well as plenty of water, fresh air, exercise and sunshine to maintain good health.

Every day, you eat a variety of foods at various times, and often in many different places. The reason you eat may be because you are hungry, because you always eat something at a particular time of the day, or because you are at a social occasion at which food is being served.

WHAT IS FOOD?

Food is any substance that we eat or drink that provides the body with chemical substances called nutrients. Nutrients from the food we eat are used to produce energy. They also enable us to grow and repair body tissue.

All of the food we eat comes from two sources: either plants or animals. Throughout the world, a wide variety of plants are eaten as food, including cereals, vegetables, fruits, nuts and legumes.

Animals that provide us with meat or dairy products include cattle, sheep, pigs, goats, kangaroos, rabbits, crocodiles, deer and wild birds. Chickens, ducks and turkeys also provide eggs in addition to meat.

Food from plant sources

Because plants and animals provide a range of different nutrients, health experts recommend that you include food from both of these sources as part of a healthy diet.

NUTRIENTS IN FOOD

The main nutrients found in food are protein, carbohydrates (including dietary fibre), fats, vitamins and minerals. The table below gives information about the different kinds of nutrients and their roles in the body. During digestion, these nutrients are broken down into tiny molecules that are absorbed into the bloodstream and then into the cells of the body. The water that is in the food we eat or drink is also absorbed into the bloodstream to provide the body with the essential fluid its cells need to function.

Each type of food is made up of a different combination of nutrients, meaning each food looks, tastes and feels different.

NUTRIENT	WHAT NUTRIENT DOES IN THE BODY	WHERE NUTRIENT IS FOUND
Protein	• Part of all cells in the body • Needed for growth • Helps to repair body cells after injury and allows the immune system to produce antibodies to fight illness	• Animal foods such as meat, milk and dairy foods, fish and eggs • Plant foods such as nuts, legumes and wholegrain cereals
Carbohydrates (including dietary fibre)	• Source of energy for all cells in the body • Keep essential systems operating (for example, the heart beating to pump blood) • Enable muscles to work so that we can move around • Dietary fibre is not absorbed during digestion. It improves the health of the digestive system by adding bulk to the faeces	• Plant foods such as cereals (bread, rice, pasta, noodles, dry biscuits), vegetables, fruit and legumes • Sugars in foods such as cakes, biscuits, confectionary, desserts and soft drinks
Fats	• Concentrated source of energy • Provide the body with almost twice as much energy per gram as carbohydrates or protein • Contain essential fatty acids that are needed in body cells and systems • Omega-3 fatty acids enable the brain and nervous system to develop and function properly and prevent heart disease	• Animal foods such as meat, poultry, cheese, cream and butter • Plant foods such as margarine, olive oil, canola oil, peanut oil and avocado • Omega-3 fatty acids in fish

NUTRIENT	WHAT NUTRIENT DOES IN THE BODY	WHERE NUTRIENT IS FOUND
Vitamins	• Essential to the functioning of body systems, such as creating new cells for growth and repair, producing energy, and absorbing other nutrients • Vitamins A, D, E and K are fat soluble • The B group vitamins are water-soluble and include thiamine, niacin and riboflavin. They help the body to utilise energy • Folate is a B-group vitamin that is essential for normal cell division during pregnancy • Vitamin A is necessary for healthy skin and eyes, and for the formation of bones	• Fat-soluble vitamins: fish and vegetable oils; milk and cheese; eggs; and nuts and butter • Vitamin A: spinach; carrots; broccoli; milk; cantaloupe; and salmon • B group vitamins: wholegrain cereals; green, leafy vegetables; legumes and fortified breakfast cereals • Vitamin C: oranges; kiwifruit; dried apricots; capsicum; parsley; and broccoli
Minerals	• Calcium, in conjunction with phosphorus, is essential for the formation of bones and teeth • Calcium helps to prevent the development of osteoporosis • Balance fluids in the body • Iron helps to form haemoglobin in the blood and assists in energy production	• Calcium: dairy foods • Iron: red meat; lentils; green, leafy vegetables; and dried apricots • Potassium: bananas; potatoes; and spinach • Magnesium: nuts; wholegrain cereals; and green, leafy vegetables • Zinc: wholegrain cereals; lean meat; and eggs

WATER

Water is an important substance that you need to keep your body healthy. Water is a major component of every cell in the body – approximately 70 per cent of your body is composed of water. Water is also important in assisting in the digestion of other nutrients, as well as in helping in the removal of waste products from your body.

Drink plenty of water

You lose water from your body every day through perspiration, tears, urine and faeces, as well as from the lungs when you breathe out. Therefore, you should drink at least six to eight glasses of water daily to replace the water you lose in these ways. Water is found in most foods, but is especially high in fruits and vegetables, as well as in milk and other liquids such as fruit juices.

If you do not drink enough water or other fluids, your body gives you signals to let you know that you need to drink more to prevent dehydration. One of these signals is urine colour – a dark or yellow colour means that the urine is too concentrated and that you need to drink more fluid. Light-coloured or clear urine means that your body is hydrated and your fluid intake is adequate.

Drinking lots of tap water is also important because in many countries, tap water contains fluoride, which helps to reduce the incidence of dental cavities. In recent years, there has been a significant increase in the popularity of drinking bottled water, and dentists have seen evidence of a large increase in the number of children who have tooth decay.

Testing knowledge

1. Define 'food'.
2. What are the two main sources of food? Give three examples of foods from each source.
3. Explain why the appearance, aroma, flavour and texture of an apple is different from that of an orange.
4. Why is it important to include protein in your diet? Give two examples of protein from animal sources and two examples of protein from plant sources.
5. Outline two main functions of carbohydrates in the body.
6. Explain why fats are said to be a concentrated source of energy.
7. Why is it important to include omega-3 fatty acids in the diet?
8. Explain why you need to include both vitamins and minerals in your daily food intake.
9. What are the three main uses of water in the body? List four of the best food sources of water.
10. Describe how you can tell if you are not drinking enough water.

DIGESTION

The digestive system begins in the mouth and finishes at the anus.

When you eat an apple or a ham and cheese sandwich, your teeth begin the process of digestion by grinding up the food. The saliva in your mouth contains the enzyme amylase, which begins to break down the carbohydrates in the food into sugars. When you swallow the food, it is pushed down the oesophagus through a series of muscular contractions called peristalsis.

As the food enters your stomach, it passes through a sphincter, or valve, that stops it from going back up into the oesophagus. The muscles in the wall of your stomach churn and mix the food with acid and enzymes to break down the food into a thick liquid.

From your stomach, the food passes through the small intestine, which is like a long tube about six metres long that is loosely coiled in your abdomen. The first section of the small intestine is called the duodenum. In the duodenum, more digestive juices are added to the food from the pancreas and liver. These enzymes break down the food into even smaller particles. As the food moves through the rest of your small intestine, its nutrients are absorbed through the villi, or the finger-like projections that line the wall of the intestine.

Any waste material that is not absorbed in your small intestine is moved into the large intestine, or bowel. Here, any water is removed, and the waste material, called faeces, is then stored in the rectum before it is passed out of the body through the anus.

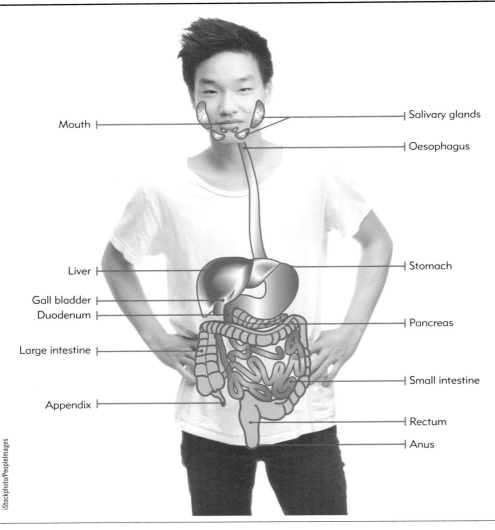

The digestive system

SELECTING FOOD WISELY

To ensure that you obtain all the nutrients you require for good health, you need to eat a wide variety of foods. Dietitian and health experts have developed food models such as the Healthy Eating Pyramid and the Australian Guide to Healthy Eating to help members of the community to make healthy food choices more easily. Using one of these food selection models will assist you to select foods for a healthy diet and enable you to maintain a healthy body weight.

THE HEALTHY EATING PYRAMID

Nutrition Australia developed the Healthy Eating Pyramid to help us select foods that will provide us with a balanced diet, and in recognition of the importance of exercise in achieving good health. The Healthy Eating Pyramid highlights the importance of including a variety of foods from the five core food groups, and limiting salt and added sugar. Based on these key ideas, the food in the pyramid is organised into layers: the foundation layers, made up of vegetables, legumes, fruit and grains; the middle layer, made up of lean meats, poultry, fish, eggs, dairy, non-dairy alternatives, nuts, seeds and legumes; and the top layer, which refers to healthy fats.

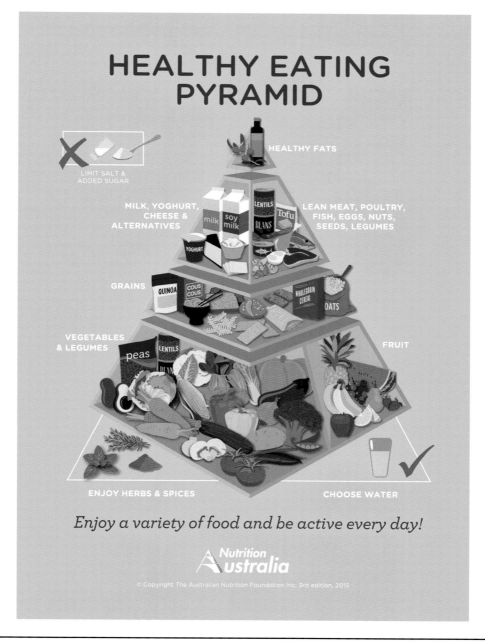

Foundation layers – eat most

The foods in the foundation layers of the Healthy Eating Pyramid are all plant based. They include grains, fruits, vegetables and legumes.

You should eat these foods most because they provide your body with its main sources of carbohydrates and fibre. Approximately 70 per cent of your energy needs should come from these foods.

Choose high-fibre, low-fat, low-sugar, low-salt breakfast cereals.

Grains

- Include breads, cereals, rice, pasta and noodles
- Provide carbohydrates for energy
- Wholemeal varieties provide dietary fibre, iron and vitamin B
- Contain small amounts of incomplete protein
- Low in fat

Vegetables and fruit

Vegetables and legumes

- Vegetables are approximately 70 to 97 per cent water
- Vegetables are low in fat and many are fat free
- High in vitamins, minerals and dietary fibre
- Legumes are rich in carbohydrates and fibre, but low in fat and cholesterol

Middle layer – eat moderately

These foods come mostly from animal sources with the exception of alternatives to dairy products and the inclusion of seeds, nuts and legumes. Legumes are a source of both protein and carbohydrate and so are included in both the foundation layers and the middle layer. These foods should be eaten in moderate amounts to provide the protein and minerals (especially calcium and iron) that are necessary for the growth and repair of body tissue. These nutrients are especially important during adolescence, when young people are growing rapidly.

Bread and cereals from grain sources

Fruit

- Contain approximately 70 to 97 per cent water
- High in dietary fibre
- Low in starch, which changes to sugar as fruit ripens
- Low in kilojoules
- Low in fat and many fruits are fat free
- Good source of a variety of vitamins and minerals
- Can be high in sugar when dried, but also higher in fibre, minerals and vitamins than fresh fruit

Nuts and legumes

Foods from animal sources contain high levels of saturated fat and cholesterol, so select lean cuts of meat, serve poultry without the skin and choose reduced fat options where possible.

Lean meat, poultry, fish, eggs, nuts, seeds and legumes

- Foods from animals have complete proteins, plants have incomplete proteins (there are exceptions)
- Meat is a good source of vitamin B, zinc and potassium
- Red meat is high in iron
- Fish is an excellent source of omega-3 fatty acids
- Soybeans contain complete protein; nuts and lentils contain incomplete protein
- Nuts are high in B group vitamins but also high in fat

Meat, fish, poultry and eggs

Milk, yoghurt, cheese and alternatives

- Dairy is high in protein and a rich source of calcium
- Contain vitamin D and phosphorus, which help the body to utilise calcium
- Milk is 87 per cent water
- Milk is also an important source of vitamin A
- Lactose, the carbohydrate in milk, cannot be digested by some people
- Soy, rice and cereal milks are often fortified with calcium and can contain 100 mg or more of calcium per 100 mL

Top layer — eat in small amounts

While we should limit intake of salt and added sugar, it is important to eat small amounts of healthy fats each day to support our brain function and heart health. Healthy fats include extra virgin olive oil and nut and seed oils, which are made from plants and contain either monounsaturated or polyunsaturated fats. These are better for your health than saturated fats. You will also have other healthy fats in your diet from eating the oils in other foods, such as avocadoes, nuts, seeds and fish, so you can limit your intake elsewhere.

Fats should make up 30 per cent of your total energy intake. Of this, less than 10 per cent should be saturated fat. Animal fats such as butter and cream contain saturated fats, which are linked to obesity, heart disease and certain cancers. They are not healthy fats. Only eat these foods in very small amounts. Snack foods high in fat, such as fried foods, pastries and potato crisps should only be eaten very occasionally. These foods are no longer included on the food pyramid.

Milk, yoghurt and cheese

Healthy and unhealthy fats

Additional messages

Enjoy herbs and spices

Herbs and spices offer flavour and aroma to food and many also have health benefits in larger amounts. Using dried, fresh or ground herbs and spices in cooking is an easy way to add flavour to your cooking without needing to add salt.

Choose water

Drinking water continuously throughout the day is encouraged so that you stay hydrated. Water also supports vital bodily functions. Avoid sugary options such as energy, soft and sports drinks.

Limit salt and added sugar

Sugar is found in sweet foods such as honey, jams, confectionery such as chocolate, ice-cream and carbonated beverages. A can of soft drink contains approximately 14 teaspoons of sugar! Health problems from obesity to tooth decay can result from overconsumption of sugar, and can increase your risk of developing type 2 diabetes. Check your foods for added sugars.

Although a small amount of salt is needed for good health, high-salt diets are linked to high blood pressure and strokes. Salt is contained in many snack foods and processed foods, such as potato crisps and salami. Choose foods with less than 120 mg salt per 100 g.

ACTIVITY 4.1

True or false: test your skill

Answer the following 20 questions as true or false to test your knowledge and understanding of nutrition.

1. Wholegrain cereals are a good source of vitamin B.
2. Plant foods such as breads, cereals and pasta are good sources of energy because they are high in fat.
3. Iron is important in the body because it assists in building healthy blood and producing energy.
4. We should gain most of our energy from plant foods.
5. Broccoli, dried apricots and kiwifruit are all very good sources of vitamin C.
6. Vegetables are made up of 20 per cent water.
7. The main use of sugar by the body is to help to build and repair new tissue.
8. Calcium and phosphorus work together to build strong bones and teeth.
9. Legumes, nuts, lean meat and fish are all found in the foundation layers of the Healthy Eating Pyramid.
10. Magnesium, zinc and calcium are all types of vitamins.
11. Fruit is low in sugar and dietary fibre.
12. Legumes are a good substitute for meat; as they contain large amounts of complete protein.
13. Meat contains only small amounts of cholesterol.
14. We should eat fish on a regular basis because it contains high amounts of omega-3, which is needed for the brain to function properly.
15. It is important to include lean meat in the diet because it contains lots of protein and iron.
16. Vitamin D helps the body to absorb calcium.
17. Fat is important in the body because it contains essential fatty acids, which are needed to build all body cells.
18. We should get 30 per cent of our energy needs from fat from animal sources.
19. One of the main causes of stroke is eating too many foods that are high in salt.
20. Canned soft drinks are good substitutes for water because they only contain small amounts of sugar.

How did you score? Check your answers with your class teacher.

RESULTS

- **20** Congratulations – what a star! You have learnt lots about the nutrients in food.
- **17–19** A fantastic effort! Check the answers you did not get correct so that you get a perfect result next time.
- **14–16** A very good try. You have obviously learnt a lot about the nutrients needed for good health.
- **11–13** You have learnt quite a lot about nutrition, but there is still a lot more to understand so that you can optimise your health.
- **8–10** You will need to work hard to learn more about the nutrients needed for a healthy diet and a healthy body.
- **5–7** You should re-check the information on the nutrients in food so that you are able to do better next time.

Less than 5 You have a lot of work to do to learn about the nutrients needed for a healthy diet and for good health.

ACTIVITY 4.2

Evaluating meals

1. Draw up two diagrams to represent the Healthy Eating Pyramid.
2. Evaluate each of the meals illustrated below. Use the diagrams you have drawn of the Healthy Eating Pyramid and note the main ingredients for each meal in the appropriate sections of the pyramid.
3. Rate each meal as very healthy, healthy, or not very healthy depending where the majority of the meal sits in the pyramid.
4. Suggest how each meal could be improved to enable it to be rated very healthy.
5. Prepare the recipes for Oriental Chicken Kebabs on Page 71. How does this meal rate according to the Healthy Eating Pyramid?

Evaluating meals

ACTIVITY 4.3

Evaluating snacks

1. Draw up two diagrams to represent the Healthy Eating Pyramid.
2. Evaluate each of the snacks shown below. Use the diagrams you have drawn of the Healthy Eating Pyramid and note the main ingredients for each snack in the appropriate sections of the pyramid.
3. Rate each snack as very healthy, healthy, or not very healthy depending on where the majority of the snack sits in the pyramid.
4. Suggest how each snack could be improved to enable it to be rated very healthy.
5. Prepare the recipes for Cupcakes on Page 289. How does this snack rate according to the Healthy Eating Pyramid?

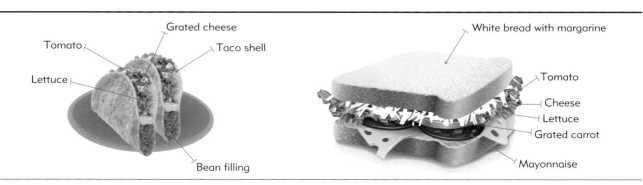

Evaluating snacks

Testing knowledge

11. Explain the first step in the process of digestion.
12. Describe what happens to food when it enters the stomach.
13. Why is the small intestine an important part of the digestive process? What happens to food once it reaches the large intestine?
14. Where do foods in the two foundation layers of the Healthy Eating Pyramid come from? Explain why foods in this group are described as being 'nutrient-dense'.
15. Draw a sketch of a carrot, and then annotate it to show the main nutrients found in vegetables.
16. Explain three reasons why you should include nuts and legumes in your diet.
17. Why should we eat some foods from the middle layer of the Healthy Eating Pyramid each day?
18. Which nutrients would you find in a glass of milk?
19. Why are butter, margarine and sugar meant to be eaten only in small amounts?
20. Explain why fats from plant sources are considered to be better for your health than those from animal sources.

NUTRITION THROUGHOUT LIFE

Regardless of our age, we all need the same basic nutrients to enable growth and repair of body tissues, to regulate all our life processes and to provide us with energy. However, our age and stage of life does affect the amount of each of the nutrients our bodies needs to meet its specific requirements for growth and development.

ADOLESCENT FOOD NEEDS

One of the key characteristics of adolescence is the rapid growth and development that takes place during the teenage years. In general, during adolescence, most girls grow 15 centimetres in height and, consequently, their weight increases by approximately 15 kilograms. Boys also experience significant growth, increasing in height by approximately 20 centimetres and gaining 20 kilograms in weight. This increase in weight for both boys and girls is a result of increased bone mass, additional muscle tissue and an increase in the size of all the internal organs, as well as additional requirements for skin and blood volume.

For many adolescents, this growth spurt is accompanied by an increase in appetite that reflects the fact that food needs are greater during adolescence than at most other periods in life. During adolescence, both males and females have an increased need for kilojoules, along with all the macronutrients and micronutrients that are needed to sustain the growth they are experiencing.

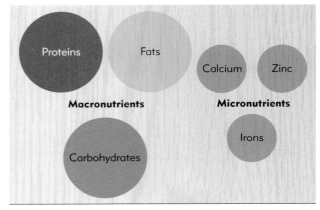

Macronutrients and micronutrients

However, recent studies indicate that, while the diets of most adolescents are able to supply the energy required for their rapid growth and development, the diets of some adolescents, particularly girls, may be deficient in the important micronutrients iron, calcium and zinc. Each of these nutrients is essential in ensuring optimum growth and development throughout life. It appears that the main cause of these deficiencies is that many young people skip meals, particularly breakfast, leading to a reduced intake of many essential nutrients. Another factor linked to the low intake of many important micronutrients is that few adolescents eat sufficient fruit and vegetables – almost 30 per cent of young people rarely or never eat fruit.

Regular snacking, or grazing, is another food habit associated with adolescence and is linked to adolescents' high energy requirements. Snacks can form a large part of adolescents' diets–accounting for approximately 25 per cent of their kilojoule intake. A negative aspect of constant snacking is that many young people frequently consume drinks that are high in sugar, such as carbonated cola drinks, sports drinks and energy drinks, instead of drinking water or milk.

Stages of a lifespan

Pregnancy
- During pregnancy, there is an increased need for protein, B group vitamins, vitamin C, folate, calcium and iron to meet the needs of the developing foetus.

Infancy and early childhood (0–2 years)
- This is a period of very rapid growth – infants double their birth weight in the first six months of life.
- Infants and young children require a diet that is high in energy, protein and calcium to support rapid growth of bones and soft tissue.

Childhood (approximately 2–11 years)
- Height and weight increases steadily.
- Childhood is often a period of intense physical activity.
- Only a gradual increase in nutrients is required.
- Nutrient-dense foods should be consumed.

Adolescence (12–18 years)
- There is an increased need for all nutrients to meet the needs of rapid growth.
- A growth spurt occurs, and continues until adult height is reached.
- Nutrient-dense, rather than energy-dense, foods are important.
- There is an increased need for calcium and exercise to maximise peak bone mass.
- Adolescence is often a period of intense physical activity.

Adulthood
- A well-balanced diet is needed to provide all essential nutrients for energy needs and growth and repair of body tissues.
- Activity levels may decrease, so energy-dense foods should be limited.
- Women continue to need a high intake of calcium as they reach menopause.

Late adulthood
- Energy needs reduce by 15 to 20 per cent in response to reduced levels of activity.
- Weight increase may occur, so it is important to consume nutrient-dense foods and restrict energy-dense foods.
- A well-balanced diet containing carbohydrates (from wholegrains and vegetables), protein, vitamins and minerals is important, although in smaller amounts.
- Smaller, regular meals are important.

FIVE HEALTHY EATING TIPS FOR TEENAGERS

1 Breakfast

Breakfast is probably the most important meal of the day because it provides energy for the day ahead and helps to improve concentration. It can also help to prevent the desire to snack on high-fat foods later in the morning.

2 Snacks

Limit the number of snacks you eat that are high in fat, salt or sugar. Choose fresh fruit, vegetable sticks, yoghurt, fruit smoothies, dried fruit and plain popcorn instead.

3 Variety of food

Each day, make sure you eat a wide variety of foods. You should try to eat 20 to 30 different types of food each day.

4 Eating out

Select foods in line with the guidelines of the Healthy Eating Pyramid. Select extra salad where possible, and fresh fruit for dessert. Avoid deep-fried foods.

5 Drinks

Drink plenty of water each day. Milk and pure fruit juices are great alternatives because they provide a wide variety of nutrients important for good health.

Shutterstock.com/ElenaGaak, Shutterstock.com/Kathie Nichols, Shutterstock.com/Bernabea Amalia Mendez, Shutterstock.com/Denphumi

ACTIVITY 4·4

Healthy food options in Australians' diets

Worksheet

1 Before you read the article below, 'Australians are including fewer healthy food options in their diets', record the amount of the following fruits and vegetables you would eat for:
 a breakfast
 b lunch
 c dinner
 d snacks.

 Now read the article. How does your fruit and vegetable consumption compare with the recommended five daily servings?

2 Identify the two age groups that were the focus of the survey.

3 Explain why these age groups were chosen as the focus of the survey.

4 How does the fruit and vegetable consumption of Australians today compare with that of 15 years ago?

5 Which foods have replaced fruits and vegetables in the diets of many Australians?

6 Outline the major health concerns linked to this change in eating habits of many teenagers and adults.

7 Why is this issue of concern to governments in Australia?

8 Outline the other factors that have led to an increase in the weight of many Australians.

9 Explain how the marketing campaigns of many fast-food outlets have influenced the eating habits of teenagers.

10 Do you agree that supermarkets have a responsibility to promote healthy food? Justify your response.

AUSTRALIANS ARE INCLUDING FEWER HEALTHY FOOD OPTIONS IN THEIR DIETS

THE SYDNEY MORNING HERALD, 9 MAY 2014, LUCY CARROLL

Australians are eating less fruit and vegetables than ever before, with teenagers leading the charge in unhealthy eating dominated by fast food, new data from the Bureau of Statistics shows.

In the first profile of Australia's eating habits available in 15 years, the *Australian Health Survey* of 12 000 people found we are eating 30 per cent less fruit and vegetables than 15 years ago, with one in four adults eating no vegetables on an average day and only 7 per cent eating the daily recommended five servings.

Professor of Health Policy at Curtin University Mike Daube said 'incredibly low' vegetable consumption reveals that fast food has eclipsed vegetables as a dietary staple.

'It is a major concern,' he said. 'Unless governments take the way we eat seriously, then there will be dire implications for health budgets and the cost of diabetes will blow out. The results are a triumph for the mass marketing of junk food.'

Australians eat about three kilograms of food and drink each day, with just over one-third of daily energy from foods high in saturated fat and sugar such as cake, biscuits, alcohol, soft drink and chips.

Professor Daube said most people would be 'horrified' to realise much of their energy intake is from food that is essentially useless.

But the survey results also show that despite Australians weighing about four kilograms more than 20 years ago, overall we are eating less.

Men are consuming about 9600 kilojoules each day – 1400 kJ less than 15 years ago – and women's average energy intake has dropped 1 per cent to 7400 kJ each day. The survey also shows people are eating about 226 grams of carbohydrates daily, the equivalent of about 12 to 14 pieces of bread, which is about 12 per cent less than 15 years ago.

National spokeswoman for the Heart Foundation Kellie-Ann Jolly said the 'dismal' daily intake of vegetables combined with fatty food diets means many people are 'unaware' of what they are eating.

'We are seeing a drop in physical activity combined with eating far too much saturated fat,' she said. 'Food like pastries and cake, which were once occasional, have become daily.'

Jane Martin, executive manager of the Obesity Policy Coalition, said the survey results highlight how pervasive unhealthy eating has become.

Ms Martin said the 'huge amounts' of highly processed foods eaten each day show the 'marketing success of McDonald's, Hungry Jack's and KFC, which are targeted to teenagers'.

'There is an increasing responsibility of supermarket chains to promote healthy food. The onus isn't on teenagers – it's with governments, parents and retail food outlets.'

Testing knowledge

21 List five nutrients that are required in increased amounts during pregnancy.

22 Why do infants have a need for nutrient-dense foods?

23 Describe the similarities and differences between the nutrients needed during infancy and childhood.

24 Why is an increased intake of calcium and exercise seen to be important during adolescence?

25 Outline two key issues associated with nutrient intake during adulthood.

26 Explain why men and women in late adulthood should try to decrease their intake of energy-dense foods.

27 Identify the main factors that account for the increased appetite that is common among adolescent girls and boys.

28 List the main nutrients that an adolescent diet is likely to be deficient in. Why are some adolescents likely to lack some of these important nutrients in their daily food intake?

29 Explain why many teenagers snack, or graze, throughout the day.

30 How many different foods should we try to eat each day? Why is this an important tip for teenagers?

HEALTHY EATING TIPS FOR TEENAGERS — THINKING SKILLS 4.1

Work with a partner to prepare an animation or video clip to highlight one of the five healthy eating tips for teenagers outlined on Page 66. Your animation or video clip should be appropriate to include in a health promotion campaign aimed at teenagers.

1 Write up a running sheet of the main points to be included in your animation or video clip.

2 Develop a list of all the scenes to be used in the animation, or props needed in the video.

3 Use an appropriate app or program to make the animation or video clip.

4 Edit the animation or video clip to approximately 30 seconds in total length.

5 Show your animation or video clip to the class. Obtain feedback on the positive aspects of the animation or video and any areas for improvement.

6 Write up an evaluation of your animation or video clip and its suitability to be included in a health campaign aimed at teenagers.

DESIGN ACTIVITY 4·1

Fried rice

Rice is a great ingredient to use as a base for family leftovers. A wide range of ingredients can be added to rice, to enhance its flavour and to make a quick and easy accompaniment or a meal in itself. With the addition of rice, a simple but delicious meal can be prepared from a range of ingredients in your refrigerator or pantry – the results will only be limited by your imagination!

Fried rice recipe leaflet

DESIGN BRIEF

A major supermarket is planning to run a campaign to highlight the importance of eating a variety of cereals and vegetables. As part of the campaign, they will provide a series of recipe leaflets for consumers to use when preparing recipes at home.

1. Write a design brief for a new fried rice-style recipe and an accompanying recipe leaflet. Include in your brief:
   - who the new recipe will be targeted towards
   - a description of when the fried rice-style recipe could be served
   - why the rice dish will help to promote the importance of eating cereals and vegetables
   - how the recipe leaflet will be a helpful guide to the preparation and cooking of the recipe.
2. Develop four evaluation criteria questions to judge the success of your end product.

INVESTIGATING

1. With a partner, research recipes for fried rice. Based on your research, develop a list of additional ingredients for fried rice. Fill in the gaps in the recipe map opposite.
2. Research information on recipe leaflets in major supermarkets, such as those prepared by Curtis Stone or Jamie Oliver.

GENERATING

1. Using the recipe map, design your own fried rice.
2. Based on your research, prepare a recipe leaflet for the recipe, incorporating the new ingredients you have chosen. Draw a recipe leaflet similar to the one on the left, or design your own leaflet.

Your recipe leaflet should include:

a. a title for your fried rice recipe
b. an illustration
c. a list of ingredients
d. the method to be used
e. the estimated preparation and cooking time
f. a list of equipment required to prepare the recipe
g. serving suggestions
h. a top tip for preparing, cooking, serving or storing the fried rice dish.

PLANNING AND MANAGING

1. Complete a food order for your product.

PRODUCING

1. Prepare the fried rice recipe you have designed.
2. Using a digital camera, photograph the completed dish to add to your recipe leaflet.

EVALUATING

1. Answer the four evaluation criteria questions you developed earlier.
2. Describe the sensory properties – appearance, aroma, flavour and texture – of the completed fried rice using the descriptors on Page 2.
3. Identify the safety precautions that you followed when cooking the rice.
4. Why is it important to store your fried rice in the refrigerator if you do not want to eat it immediately?
5. Annotate your photo, and suggest ways you could improve the appearance of the dish if you were to cook and serve it again.
6. Give your recipe leaflet to a friend and ask them to read through the recipe and method, and to cook the recipe at home, if possible. Ask them to comment on whether they find the recipe easy to follow and a helpful guide to the preparation and cooking of the recipe. Considering this feedback from your friend, what improvements could be made to your recipe leaflet?

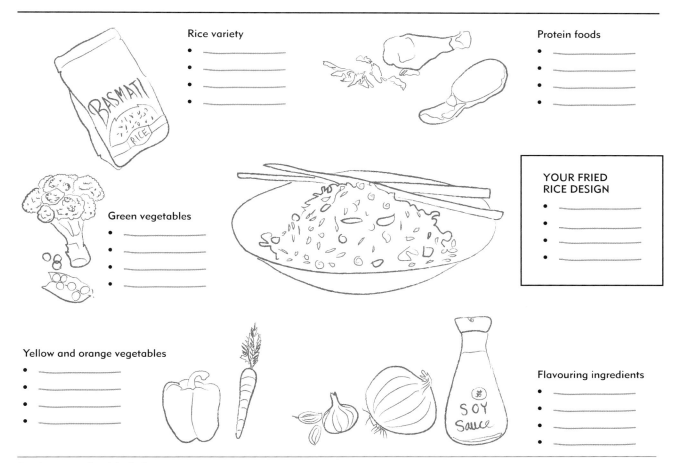

Recipe map for fried rice

Fried rice

DRESSED BAKED POTATOES

1 medium potato
1 rasher bacon, finely diced
1 spring onion, finely sliced
20 grams tasty cheese, grated
1 tablespoon milk
salt and pepper
1 tablespoon sour cream
1 teaspoon parsley, chopped

METHOD

Baking whole potatoes

1. Preheat the oven to 200°C.
2. Scrub the potato well and pierce the skin in several places with a skewer or fork.
3. Place on the oven rack for 1–1 ½ hours. Alternatively, microwave on high for 2 minutes per potato. Once finished, crisp the skins by placing in a hot oven for 20–30 minutes or until tender.

Preparing the filling and topping ingredients

1. Select the flattest side as the bottom of the potato. Cut off the top of the cooked potato.
2. Carefully scoop out the flesh of the potato with a metal spoon and place in a bowl. Leave the skin intact.
3. Pan-fry the bacon until lightly browned and beginning to crisp.
4. Combine the flesh with the cooked bacon, spring onion, grated cheese, milk, salt and pepper and spoon into potato skin.
5. Reheat before serving if necessary. Garnish with the sour cream and chopped parsley and serve.

EVALUATION

1. Describe the sensory properties – appearance, aroma, flavour and texture – of your dressed potato.
2. Write up the quantity of ingredients you would need if you had to make dressed potatoes for six people.
3. Explain two rules to observe to ensure you use the oven safely.
4. Evaluate the Dressed Baked Potatoes by plotting their ingredients on a diagram of the Healthy Eating Pyramid or the Australian Guide to Healthy Eating.
5. Describe an accompaniment you could serve with your Dressed Baked Potatoes to improve their health rating.

ORIENTAL CHICKEN KEBABS

4 bamboo skewers
½ chicken breast fillet
½ clove garlic
½ teaspoon fresh ginger
1 tablespoon light soy sauce
4 small mushrooms, halved
4 cherry tomatoes
⅙ green capsicum, diced
¼ cup pineapple pieces
1 teaspoon oil
1 quantity boiled rice

 SERVES ONE

METHOD

1. Soak the bamboo skewers in water.
2. Cook rice (see recipe on Page 72) and keep warm. (Take the saucepan off the heat, strain and cover with a lid or plate.)
3. Cut the chicken into 2-centimetre pieces.
4. Crush the garlic, grate the ginger and mix with the soy sauce.
5. Add the chicken pieces. Allow them to marinate in the refrigerator for as long as possible.
6. Cover the grill pan with foil.
7. Preheat the griller.
8. Thread the chicken, mushrooms, tomatoes, green capsicum and pineapple pieces onto the skewers, arranging the vegetables and chicken so that you vary the colours.
9. Brush the oil over the kebabs.
10. Gently grill the kebabs for approximately 3 minutes on each side. Remember to turn them several times while they are cooking.

Serve on a bed of the rice.

EVALUATION

1. Why does this recipe suggest soaking the bamboo skewers in water before using them?
2. Explain one important food safety rule to observe when preparing the chicken for the kebabs.
3. Explain the purpose of marinating the chicken in this recipe.
4. Why is it important to turn the kebabs several times while they are cooking?
5. Use your knowledge of nutrition to justify why this dish would be rated as a healthy meal.

BASIC FRIED RICE

4 cups water

⅔ cup rice

1 egg

1 clove garlic

1 cup selected vegetables

¼ cup protein food (for example, cooked chicken)

1 tablespoon oil

½ onion, diced

1 tablespoon soy sauce

SERVES TWO

METHOD

1. Bring the water to the boil.
2. Add the rice. Stir once or twice to separate the grains. Simmer for 12–15 minutes or until tender. Do not overcook.
3. Drain and spread the rice in a thin layer on a tray covered with absorbent paper. Refrigerate uncovered until required. (This process helps to dry out the rice so the grains remain separated.)
4. Beat the egg with a fork in a small cup to combine its white and yolk.
5. Crush the garlic and prepare the vegetables and protein food.
6. Gently heat the oil in a wok. Add the beaten egg and cook until set.
7. Remove the egg from the wok and finely dice the omelette.
8. Add the onion, garlic and any yellow, red or other hard vegetables to the wok. Stir-fry for 2–3 minutes or until the onion is transparent.
9. Add the cooled rice, green vegetables, protein food, diced egg and soy sauce. Toss over the heat until the ingredients are well combined and heated through.
10. Serve immediately garnished with sliced spring onions.

EVALUATION

1. Why is it important not to overcook the rice in Step 2?
2. Explain why the rice is refrigerated after boiling.
3. Suggest an alternative cooking utensil to a wok.
4. Identify the safe work practices you followed when boiling and stir-frying the rice.
5. Plot the ingredients you used on a diagram of the Healthy Eating Pyramid or the Australian Guide to Healthy Eating. Comment on the nutritional value of your Basic Fried Rice.

PRAWN AND VEGETABLE RICE PAPER ROLLS

20 grams rice vermicelli

1 spring onion, finely julienned

¼ carrot, grated

1 tablespoon coriander, chopped

1 tablespoon mint, chopped

⅙ red capsicum, finely julienned

20 grams bean shoots

1 tablespoon cashews, finely chopped

3 teaspoons sweet chilli sauce

8 rice paper sheets

8 prawns, cooked, peeled, de-veined and halved lengthwise

Dipping sauce

½ lime, juiced

1 tablespoon sweet chilli sauce

1 centimetre lemongrass, finely chopped

2 teaspoons coriander, finely chopped

2 teaspoons fish sauce

MAKES 8 RICE PAPER ROLLS

Rice paper is a delicate, edible wrapper used to package delicious ingredients in a tasty, low-fat snack. Roast pork, chicken or tofu could be used to replace the prawns for the protein component of this recipe.

METHOD

1. Soak the vermicelli in warm water for 10–15 minutes or until softened. Drain well.
2. Combine the vermicelli with the spring onion, carrot, coriander, mint, capsicum, bean shoots, cashews and sweet chilli sauce. Mix well.
3. Assemble the rice paper rolls one at a time. Soak a rice paper sheet in warm water for about 30 seconds, until just softened, and drain on a clean, dry tea towel.
4. Place a portion of the filling on one end of rice paper. Add prawn, then tuck in the ends and roll to enclose filling.
5. Cover with plastic wrap and repeat the process.
6. To make the dipping sauce, combine all ingredients and mix well.
7. Serve rolls on a platter accompanied by a bowl of the dipping sauce.

EVALUATION

1. Describe the sensory properties – appearance, aroma, flavour and texture – of your Prawn and Vegetable Rice Paper Rolls.
2. What are some other products that could be substituted for the rice paper?
3. Which cereal product is used in the recipe, and what is its function?
4. Discuss the nutritional advantages and disadvantages of including raw vegetables in snack foods.
5. Evaluate the Prawn and Vegetable Rice Paper Rolls by plotting their ingredients on a diagram of the Healthy Eating Pyramid or the Australian Guide to Healthy Eating. Write a brief paragraph to explain whether they would be suitable as a healthy snack.

RICEY LETTUCE PARCELS

- 2 slices lean bacon
- 1 cup white or brown rice, cooked
- 1 stick celery, finely sliced
- 1 carrot, grated
- 1 tablespoon light soy sauce
- lettuce leaves, washed, dried and chilled

 SERVES TWO

METHOD

1. Chop the bacon and place between sheets of absorbent paper. Microwave on high for 90 seconds to remove fat.
2. Mix the bacon with the rice, celery, carrot and soy sauce.
3. Heat the rice mixture in a covered container in a microwave on high for 1–2 minutes.
4. Place a heaped tablespoon of rice on the chilled lettuce leaf. Fold into an envelope to serve.

EVALUATION

1. Describe the sensory properties – appearance, aroma, flavour and texture – of your Ricey Lettuce Parcels.
2. What are some other products that could be substituted for the lettuce leaf?
3. Suggest reasons why soy sauce is used in the recipe. What section of the Healthy Eating Pyramid would soy sauce appear in? Why?
4. Suggest an alternative method for cooking the bacon if a microwave oven was not available.
5. Evaluate the Ricey Lettuce Parcels by plotting their ingredients on a diagram of the Healthy Eating Pyramid or the Australian Guide to Healthy Eating. Write a brief paragraph to explain whether they would be suitable as a healthy snack.

CREAMY CARROT AND TOMATO SOUP

½ white onion, finely chopped
250 grams carrots, thinly sliced
30 grams butter
225 grams canned tomatoes
2 cups chicken stock
pinch of salt, pepper and sugar
¼ cup milk
1 teaspoon parsley or chives, finely chopped

SERVES TWO

METHOD

1. Sauté the onion and carrot in the butter until lightly coloured.
2. Add the tomatoes, chicken stock, salt, pepper and sugar.
3. Cover and gently cook for 20 minutes or until carrot is tender.
4. Remove from heat. Blend until smooth.
5. Reheat. Add the milk, taking care not to reboil.
6. Garnish with the parsley or chives.

EVALUATION

1. Describe the sensory properties – appearance, aroma, flavour and texture – of your Creamy Carrot and Tomato Soup.
2. Explain two important food safety rules to observe when preparing this recipe.
3. Make a list of other ingredients that could be used to vary this recipe.
4. Discuss the advantages and disadvantages of preparing your own soup rather than purchasing a commercial product.
5. Evaluate the Creamy Carrot and Tomato Soup by plotting the ingredients on a diagram of the Healthy Eating Pyramid or the Australian Guide to Healthy Eating. Write a brief paragraph to explain whether this recipe would be suitable to serve as a healthy snack or light meal.

SATAY VEGETABLES AND TUNA

¾ cup rice or noodles

3 tablespoons coconut milk powder

1 cup water

½ vegetable stock cube, crushed

1 small onion, diced

1 clove garlic, crushed

1 teaspoon curry powder

2 teaspoons oil

1 small carrot, peeled and cut into rings

1 stick celery, sliced

5 small florets of broccoli

2 tablespoons crunchy peanut butter

185 grams canned tuna, in brine, drained and flaked

SERVES TWO

This recipe is an example of a dinner recipe that is nutrient-dense and a good source of omega-3 and calcium.

METHOD

1. Cook rice (see recipe on Page 72) or noodles and keep warm. (Take the saucepan off the heat, strain and cover with a lid or plate.)
2. Blend the coconut milk powder and water. Add the stock cube.
3. Fry the onion, garlic and curry in oil until tender.
4. Add the carrot and celery and stir-fry for 3–5 minutes.
5. Add the broccoli florets and continue to stir-fry for 2–3 minutes.
6. Add the blended coconut powder mixture and peanut butter and stir well. Simmer until vegetables are slightly crunchy.
7. Add the flaked tuna, gently stir through and heat through.
8. Serve on top of rice or noodles.

EVALUATION

1. What other methods could you use to keep the rice warm?
2. Why is it important to blend the coconut milk powder with water?
3. Why are the carrot and celery added before the broccoli?
4. Why is it important to gently stir through the flaked tuna?
5. Evaluate the Satay Vegetables and Tuna by plotting its ingredients on a diagram of the Healthy Eating Pyramid. Write a brief paragraph to explain whether this recipe would be suitable to serve as a healthy dinner.

Mark Fergus Photography

CHICKEN AND VEGETABLE STIR-FRY

1 chicken fillet, sliced in strips
2 teaspoons lemon juice
1 tablespoon soy sauce
1 tablespoon oil
½ onion, cut in wedges
1 garlic, sliced
1 centimetre fresh ginger, diced
¼ red capsicum, sliced
4 baby sweetcorns, halved lengthwise
6 green beans, sliced
1 bok choy, coarsely sliced
1 tablespoon vegetable oil
¼ cup chicken stock
1 tablespoon sweet chilli sauce
1 tablespoon soy sauce (extra)
160 grams egg noodles

SERVES TWO

METHOD

1. In a small bowl, combine the lemon juice, soy sauce and oil to make a marinade.
2. Slice the chicken fillet across the grain into thin strips and mix into the marinade. Cover and refrigerate.
3. Prepare the vegetables. (Remember to keep them separate, since they have different cooking times and will be added at different times.)
4. Boil water in a medium saucepan and pour over noodles to soak while the chicken and vegetables are cooking.
5. Heat the oil in a wok. Add onion, garlic and ginger. Stir-fry for 30 seconds.
6. Drain the chicken and add to the wok. Stir-fry for 2 minutes or until the chicken changes colour.
7. Add the vegetables separately to the wok, starting with the vegetable that takes the longest to cook.
8. Pour in the chicken stock and cook for a further 2 minutes.
9. Drain the noodles and toss through the chicken and vegetables. Add the sweet chilli sauce and extra soy sauce.

EVALUATION

1. Describe the sensory properties – appearance, aroma, flavour and texture – of your Chicken and Vegetable Stir-Fry. Rate the success of your product with a hedonic face (see Page 4).
2. List two reasons why the chicken was marinated in Step 2.
3. Describe the role of the following ingredients in the marinade:
   - lemon juice
   - soy sauce
   - oil.
4. Explain why the marinating chicken was covered and then stored in the refrigerator while the vegetables were prepared.
5. What is cross-contamination?
6. How did you prevent cross-contamination from occurring when preparing the ingredients for this recipe?
7. How did you decide on the order in which vegetables were added to the stir-fry during the cooking process?
8. Explain why the noodles were soaked in boiling water rather than cooked on a stovetop before being added to the stir-fry.
9. Describe how you worked safely when cooking the stir-fry in the wok.
10. If you were to repeat the recipe, what changes would you make to improve the end product?

CHEESE AND SPINACH PASTRIES

2 spring onions

60 grams frozen spinach, defrosted

50 grams ricotta cheese

50 grams feta cheese

1 egg

pinch of nutmeg

4 shakes black pepper

8 sheets filo pastry

¼ cup olive oil

 MAKES 8

METHOD

1. Preheat oven to 200°C. Brush an oven tray with melted butter.
2. Finely slice the spring onions and mix with the spinach, ricotta cheese, feta cheese, egg, nutmeg and black pepper.
3. In a bowl, divide the mixture into 8 portions.
4. Lay one piece of filo pastry on the bench and lightly brush with oil. Fold into thirds, lengthways.
5. Place one portion of the filling on the lower edge of the pastry. Fold up the pastry to form a right-angled triangle, then continue to fold the triangles to the length of the pastry.
6. Place the finished pastry on the oven tray and brush with oil. Repeat the folding process to make 8 pastries.
7. Bake for approximately 10 minutes or until a pale-golden colour.

EVALUATION

1. Describe the sensory properties – appearance, aroma, flavour and texture – of your Cheese and Spinach Pastries.
2. How were you able to prevent the filo pastry from drying out while you were preparing the pastry filling?
3. Explain the function of the egg in the filling.
4. Discuss three safe work practices you followed when using the oven to bake the filo triangles.
5. If you were to repeat this recipe, what changes would you make to either the ingredients or the method?

5 GET UP AND GO!

KEY KNOWLEDGE

- The role of breakfast
- Breakfast eating habits
- Processed breakfast cereals
- Milk
  - Nutrients in milk
  - Milk production
  - Types of milk
  - Primary and secondary processing of milk
- Breakfast drinks
- Eggs
  - Egg production
  - Testing for freshness of eggs
  - Nutrients in eggs
  - Cooking eggs

KEY TERMS

breakfast the first meal you eat soon after waking up from your night's sleep

coagulation the permanent change of the physical and chemical structure of protein

fast a period of time during which we eat nothing

processed breakfast cereals grains such as corn, wheat and rice that have been softened by precooking and then dried; most become fortified or have had vitamins and minerals added during processing

rolled oats wholegrains of oats that have been cooked in water to soften them and make them easier to digest, and then rolled

AUSTRALIAN CURRICULUM LINKS

DESIGN AND TECHNOLOGIES
- Knowledge and Understanding
- Processes and Production Skills

HEALTH AND PHYSICAL EDUCATION
- Personal, social and community health

GENERAL CAPABILITIES
- Critical and creative thinking

CROSS-CURRICULUM PRIORITIES
- Sustainability

THE ROLE OF BREAKFAST

Breakfast is the first meal you eat soon after waking up from your night's sleep. According to nutritionists, it is the most important meal of the day, and it should supply your body with one-third of your daily food intake. When you eat breakfast, you are 'breaking the fast' by supplying your body with its first food for the day after eight to 12 hours of sleep. A **fast** is a period of time during which we eat nothing. Eight to 12 hours is a long time to go without food – equivalent to at least a full day at school – and people of all ages and occupations need a healthy, adequate breakfast to refuel the body.

When you are asleep, your body is resting and does not require as much energy or kilojoules as when you are active during the day. The body uses the glucose produced from the carbohydrates you eat during the day, and stores it in the muscle tissue and liver as glycogen. During the night, these glycogen stores are slowly released into the bloodstream to keep your blood sugar levels stable. When you eat breakfast, your body is able to replenish these stores of glycogen and provide a store of energy for the day's activities, as well as boost your metabolism.

Research has shown that eating breakfast helps to improve mental and physical performance and contributes important nutrients to the diet. It improves concentration, problem-solving ability, memory and mood, and enables people to think more quickly and clearly. Eating breakfast before you go to school helps to prevent hunger, snacking at recess time and the temptation to eat foods that are high in fat and sugar. Skipping breakfast does not help people to lose weight, since it often leads to overeating later in the day. Nutritionists recommend that breakfast be based on carbohydrates, preferably from fruits, wholegrain breads, cereals and grains or vegetables.

BREAKFAST EATING HABITS

Many teenagers go to school in the morning having eaten nothing for breakfast or having drunk only fluids such as cordial, soft drink, tea or coffee, all of which have very little nutritional value. Nutritional studies suggest that people who eat breakfast every day are less stressed and depressed than those who do not. Skipping breakfast can lead to poor physical growth, poor health and a reduced ability to concentrate and learn. Research shows that during the school or working week, cereals, milk and bread are the most popular breakfast foods, and only a small number of people eat a cooked breakfast before going to school or work.

Preparing breakfast

ACTIVITY 5·1

Breakfast eating habits

The following graphs summarise the key findings from the Australian *Children's Nutrition and Physical Activity Study*.

Cereal and milk	48.4%	
Nothing	19.9%	
Toast/bread	17.8%	
Other	7.4%	
Juice only	1.3%	
Water only	1.9%	
Milk drink	1.5%	
Coffee/tea only	1.2%	
Soft drink/Cordial	0.6%	

Boys' breakfast eating habits

Cereal and milk	37.8%	
Toast/bread	24.2%	
Nothing	19.5%	
Other	8.8%	
Juice only	3.0%	
Water only	2.6%	
Milk drink	2.2%	
Coffee/tea only	1.5%	
Soft drink/cordial	0.4%	

Girls' breakfast eating habits

1. Which group consumed the most breakfast cereals?
2. What percentage of boys and what percentage of girls ate nothing for breakfast?
3. Give two reasons why girls and boys are likely to eat nothing for breakfast.
4. What was the most popular breakfast drink for boys and for girls?
5. Which group was most likely to eat toast or bread for breakfast?
6. What factors are likely to influence breakfast food choices for boys and girls?
7. If parents laid out fruit, cereal, milk and juice, do you think boys and girls who presently eat nothing would be encouraged to eat a healthy breakfast? Why or why not?
8. What other strategies do you think could be used to encourage children to eat a healthy breakfast?
9. Explain the effect that missing breakfast is likely to have on a student's performance at school.
10. What are the health concerns for those who drink soft drink and cordial for breakfast?

ACTIVITY 5·2

What do you eat for breakfast?

1. Did you eat breakfast before coming to school this morning? If so, what foods did you eat?
2. How many times do you eat breakfast in a week?
3. Do you prepare your own breakfast?
4. If you skip breakfast, what is the reason for this?
5. Are the breakfast foods you eat on the weekend different from those you eat during the week?
6. If someone else was going to prepare your breakfast, what would you order?
7. Suggest two ways of improving your breakfast.
8. What foods are available at your school cafeteria that would be suitable for a healthy breakfast?
9. Make a list of other suitable breakfast foods for teenagers that could be added to your school's cafeteria menu.
10. Access the Victorian Government's Better Health Channel website. Search the site using the keyword 'breakfast' to respond to the following questions.
    a. Explain why it is important for children to develop good breakfast eating habits when they are young.
    b. Suggest three reasons why breakfast is often skipped.
    c. Explain how eating a healthy breakfast can reduce the risk of illness.
    d. Identify why skipping breakfast can lead to snacking later in the day, and how this can affect your nutrient intake and health.
    e. Many breakfast cereals are high in fibre. Explain why this nutrient is important in a healthy diet.

PROCESSED BREAKFAST CEREALS

Processed breakfast cereals are grains such as corn, wheat and rice that have been softened by precooking and then dried. This enables the cereals to be eaten with cold milk. Most of these cereals are fortified or have had vitamins and minerals (for example, iron) added during processing.

Australia's most popular breakfast cereal is Sanitarium's Weet-Bix. It is said that if the total number of Weet-Bix eaten in Australia each year was laid end to end, would circle the equator 2.8 times!

Wholegrains such as oats are cooked in water to soften them and make them easier to digest. The softened grain is then rolled and made into a versatile product called rolled oats, which can be used in muesli, cooked in porridge or used in biscuits. These cereals are less processed and contain more of the wholegrain, including the valuable vitamins and minerals and the bran layer, which is high in fibre.

The best choice of breakfast cereal is one high in carbohydrates and fibre and low in fat, salt and sugar.

ACTIVITY 5·3
Top-selling processed breakfast cereals

A survey by *Choice* magazine identified the top-selling breakfast cereals in Australia. The nutritional information of each of the four top-selling processed breakfast cereals are shown below. Carefully analyse this information and then answer the questions that follow.

Top-selling processed breakfast cereals

Sanitarium Weet-Bix
Nutrition information per 100 grams of major ingredients
- Energy: 1490 kilojoules
- Fibre: 11 grams
- Fat: 1.4 grams
- Sugar: 3.3 grams
- Salt: 2.9 milligrams

Kellogg's Nutri-Grain
Nutrition information per 100 grams of major ingredients
- Energy: 1600 kilojoules
- Fibre: 2.7 grams
- Fat: 0.6 grams
- Sugar: 32 grams
- Salt: 480 milligrams

Kellogg's Corn Flakes
Nutrition information per 100 grams of major ingredients
- Energy: 1580 kilojoules
- Fibre: 3.3 grams
- Fat: <0.1 grams
- Sugar: 7.9 grams
- Salt: 550 milligrams

Kellogg's Coco Pops
Nutrition information per 100 grams of major ingredients
- Energy: 1610 kilojoules
- Fibre: 1.2 grams
- Fat: 0.4 grams
- Sugar: 36.5 grams
- Salt: 465 milligrams

1. Which cereal has the most fibre?
2. Explain why this cereal is likely to contain the most fibre.
3. Why should we try to consume foods with a high fibre content?
4. Which cereal would people who wish to reduce the fat in their diet be advised to choose?
5. Which cereal is highest in sugar? What health concerns are associated with consuming cereals that are high in sugar?
6. Which cereal is lowest in salt?
7. Why is it important to read the labels when making cereal choices?
8. Which cereal would nutritionists be most likely to recommend, and for what reasons?
9. Identify a television advertisement for a breakfast cereal. Describe the way that the advertisement promotes the cereal product. Which age group is it aimed at?
10. For further information about cereals and good health, visit the websites of companies such as Kellogg's, Uncle Toby's and Sanitarium, or the Go Grains website.

ACTIVITY 5.4

Taste testing breakfast cereals

AIM

To determine which cereal is most suitable as a breakfast food for adolescents

METHOD

Worksheet

1. Conduct a taste test of the four breakfast cereals discussed in Activity 5.3. You may eat them either with milk or dry. Whichever method of tasting you choose, you must taste each cereal in the same way.
2. Wash and dry your bowl between samples.
3. Record your impressions of each cereal in the following table.

RESULTS

Cereal type	Flavour	Texture	Appearance
Sanitarium Weet-Bix			
Kellogg's Nutri-Grain			
Kellogg's Corn Flakes			
Kellogg's Coco Pops			

ANALYSIS

1. Which cereal had the sweetest flavour?
2. Does your sensory test for sweetness agree with the information gained from the label analysis in Activity 5.3?
3. Which cereal took the longest to chew and swallow?
4. Suggest a reason why the textures of each cereal are different.
5. Was the cereal with the highest amount of dietary fibre the hardest to chew?
6. Which cereal had the most appealing appearance, and which the least appealing? Why?

CONCLUSION

Based on the information from the taste test you have just completed and the information in Activity 5.3, which cereal would you recommend as a healthy breakfast food for a teenager? Justify your answer.

Testing knowledge

1. What percentage of the day's food intake should breakfast provide?
2. Define 'fast' in relation to eating habits.
3. Why is it important to eat breakfast?
4. List three advantages to school students of eating breakfast.
5. What do nutritionists recommend that breakfast foods should be based on?
6. Define 'processed breakfast cereal'.
7. Why are processed breakfast cereals a popular food for breakfast?
8. What are rolled oats and how are they made?
9. Discuss why wholegrain cereals are recommended as good breakfast foods.
10. List three recommendations you would give to a person trying to choose a healthy breakfast cereal.

MILK

For many people, milk plays a key part in their morning breakfast menu. A bowl of muesli, porridge or low-GI cereal with milk provides many important nutrients that will sustain you throughout the morning.

Nutrients in milk

Mammals, such as cows, goats and sheep, produce milk for their young that is suitable for humans to consume. Milk is made up of 87 per cent water, and contains a wide range of important nutrients including protein, fat, carbohydrates (lactose) and vitamins – especially vitamins A D, and riboflavin. Milk is also a valuable source of the mineral calcium, which is essential for bone development.

Milk production

Before we buy milk, it is pasteurised to kill pathogenic or disease-causing bacteria. When milk is pasteurised, it is heated to 72°C for 15 seconds and then rapidly cooled to 4°C. This process, called high-temperature, short-time pasteurisation, kills all disease-causing bacteria and extends the milk's shelf life.

Milk is also homogenised to break down into tiny particles the globules of fat that are present in it. This prevents the cream from rising to the milk's surface.

When milk is homogenised, high pressure is used to force it through a very fine membrane, which breaks down the fat molecules into tiny particles.

Types of milk

A wide variety of milk is available to consumers.

- Full cream milk contains approximately 4 per cent fat.
- Reduced-fat milk contains about half the fat (less than 2 per cent) of full cream milk. Reduced-fat milk has vitamins A and D added to it to replace the nutrients that are lost when the fat is removed during processing.
- Skim milk has the lowest proportion of fat (between 0 and 1 per cent). Skim milk, too, has vitamins A and D added to it to replace the nutrients that are lost when the fat is removed during processing.

Primary and secondary processing of milk

Primary processing of milk:
1. Cows are milked using automated milking machines
2. Milk is pumped to refrigerated storage vats and cooled to 4°C
3. Milk is tested for freshness and quality at the dairy
4. Refrigerated trucks transport the milk to the processing plant
5. At the processing plant, milk is again tested for freshness, quality and protein content
6. Milk is pasteurised and homogenised ready for packaging or secondary processing
7. Milk is packaged into cartons or bottles and stamped with a use-by date
8. Packaged milk is sent to shops and supermarkets for sale to consumers

Primary processing of milk

Secondary processing of milk – making cheese:
1. The enzyme rennet is added to milk to enable the protein in the milk to form a clot and create the curd
2. The curd is gently heated, causing it to shrink, become tougher and expel the whey
3. The curd is cut or milled into small, even pieces
4. Salt is added to the milled curd to reduce the moisture content, add flavour and prevent it from becoming too acidic
5. The curds are packed into moulds and pressed to form blocks of cheddar
6. The cheddar is aged – from one to two months to up to one year

Secondary processing of milk – making cheese

BREAKFAST DRINKS

Nutritious drinks have become a popular breakfast food. Fruit drinks are often included on breakfast menus, and dairy drinks such as flavoured milk or breakfast replacement drinks are also popular.

Freshly squeezed orange juice is one of the most nutritious breakfast drinks

One of the most nutritious breakfast drinks is freshly squeezed orange juice, which has a delicious flavour and the benefit of being full of vitamin C. You can also make a beautiful breakfast drink by blending together oranges, fresh strawberries, a banana and even some pineapple. Alternatively, you can purchase juices in a wide range of flavours. It is important to remember, though, that many of these juices are high in sugar and contain less fibre than a piece of fresh fruit. Consequently, fruit juices are less filling than whole fruit, and it is very easy to drink more of them than we really need as part of a healthy diet. Milk-based breakfast drinks, too, can be high in sugar, and so should only be consumed occasionally.

ACTIVITY 5·5
Comparing the nutrient content of breakfast drinks

AIM
To compare the nutrient content of three breakfast drinks

METHOD
1. Select three different breakfast drinks from each of the following categories.
   - Fruit juice
   - Breakfast replacement drink
   - Flavoured milk drink
2. Read the nutrition panels provided with each drink and record the information in a table similar to the one at the top of the next column.

Nutritive value per 100mLs	Pure pineapple juice	Drinking yoghurt	Strawberry-flavoured milk
Protein			
Fat			
Sugar			
Vitamin C			
Calcium			

3. Sample each drink. Using the sensory descriptors on Page 2, record the results in the table below.

Appearance			
Aroma			
Flavour			
Texture/ mouth feel			

ANALYSIS
1. Which drinks contained vitamin C?
2. Why is vitamin C an important nutrient for the body?
3. List the drinks in order of their fat content, from highest to lowest.
4. Why is it important to have some fat in the diet?
5. Which drink contained the most protein?
6. What role does protein play in a healthy diet?
7. Which drink contained the most sugar? Compare this with your sensory test for the sweetest-tasting drink.
8. Why is it important to include calcium in the diet?

CONCLUSION
Which drink would you choose as the most nutritious breakfast drink? Why?

EGGS

Eggs can form the basis of a cooked breakfast or a quick, easy family meal. A special breakfast on a Sunday morning can be a time when families relax and enjoy being together.

Egg production

On average, Australians eat 192 eggs each per year. Approximately 61.6 per cent of these are eggs produced using intensive cage production systems in which the hens are kept in large numbers in small cages. Barn-laid eggs account for about 8.2 per cent of sales, and come from chickens kept in large barns; these hens have space to spread their wings and stretch. Free-range eggs are the fastest-growing sector of the egg market in Australia, and now make up approximately 29 per cent of all egg sales. In a free-range system, hens are free to move around in open spaces and scratch in the dirt for worms and seeds. Organic eggs come from farms that are free range and do not use artificial fertilisers and pesticides. Organic eggs make up only about 1.2 per cent of all egg sales in Australia.

When purchasing eggs, always check the best before date to ensure that the eggs you purchase are fresh. Eggs are usually purchased in a carton, where they sit in individual spaces, point-down so that the yolk remains in the centre of the white. Remember that eggs are a perishable food, so it is best to store them in the refrigerator to maximise their keeping qualities.

A 'Best before' label usually shows the day and month

Always check the 'Best before' date on an egg carton to ensure you buy eggs that are fresh.

Testing the freshness of eggs

Eggs can be tested for freshness by filling a bowl with cold water and gently lowering the egg into the water. The egg:
- is fresh if it stays on the bottom of the bowl
- is stale if it floats to the surface of the bowl.

Testing the freshness of eggs: fresh eggs sink and stale eggs float.

ACTIVITY 5·6
Investigating the structure of eggs

1. Collect an egg, a small plate and a skewer.
2. Break the egg onto the plate, making sure not to break the yolk.
3. Carefully examine both the shell and the egg.
4. See if you can locate all of the components of the egg shown in the diagram below. Place a tick next to each component as you identify it.
5. What do you think is the purpose of the shell membrane?
6. Carefully pierce the egg yolk with your skewer. What happens to the yolk?
7. Explain how the yolk keeps its round shape.
8. What do you think is the role of the chalaza in an egg?
9. What is the purpose of the germinal disc in an egg?

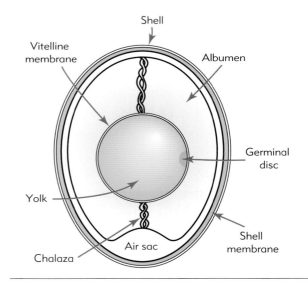

Components of an egg

Nutrients in eggs

Eggs contain most of the important nutrients that are needed for growth and repair of body tissue. However, they do not contain any vitamin C, and the amount of carbohydrate in eggs is also very minimal. The fat and cholesterol that are present in eggs are only found in the egg yolk. Because eggs are such a nutritious food, the Heart Foundation states that you can eat up to six eggs per week as part of a healthy, balanced diet.

APPROXIMATE NUTRIENT CONTENT OF A 55-GRAM EGG	
Energy	297 kilojoules
Water	37.7 kilojoules
Protein	6.35 grams
Fat	5.05 grams
Carbohydrate	0.15 grams
Cholesterol	187.5 milligrams
Sodium	66.5 milligrams
Potassium	57 milligrams
Calcium	19.5 milligrams
Iron	0.8 milligrams
Zinc	0.45 milligrams

Cooking eggs

Protein is one of the major nutrients in eggs, and it can be changed by heat or mechanical methods. Coagulation is the permanent change of the physical and chemical structure of protein. The texture of the egg protein is permanently changed during the cooking process. As the egg is heated, the protein begins to set. When eggs are cooked, the white cooks first and sets, or coagulates. In other words, the egg white forms a soft gel that no longer flows. The yolk takes slightly longer to set to a pale-yellow, slightly firmer gel. If eggs are overcooked, they become tough and rubbery and the yolks become dry.

Eggs are a very versatile ingredient and can be prepared in a variety of ways. One of the simpler methods of serving an egg is to have it either soft- or hard-boiled. Soft-boiled eggs are delicious when served with fingers, or 'soldiers', of toast that can be used to dip into a runny yolk. Fried eggs and bacon are another popular breakfast or brunch food; eggs can also be served scrambled or poached. Making an omelette is slightly more challenging, but allows you to add a range of other ingredients to the eggs to make a more substantial dish.

Another delicious way to serve eggs for breakfast or brunch is to separate the egg yolk from the egg white and then sit the yolk in a nest of beaten egg white to make a shirred egg. Once the egg is baked, the white puffs up to make a mini soufflé, producing a very special dish. French toast, too, is made using eggs, but because the bread is fried in butter, it should only be eaten occasionally.

You can find recipes for all of these dishes on Pages 91–8.

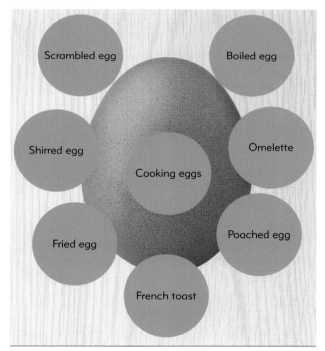

Methods of cooking eggs

Testing knowledge

11 Outline the process used to pasteurise milk, and explain why pasteurisation is an important process in milk production.

12 Describe how milk is homogenised, and explain why milk undergoes homogenisation before being sold to consumers.

13 List four key steps in the primary processing of milk.

14 What is the best method for storing eggs?

15 Explain how organic eggs differ from other eggs.

16 Explain how you would test an egg for freshness.

17 Draw a Venn diagram that demonstrates the similarities and differences in the nutrient content of milk and eggs.

18 What is the name of the part of an egg that holds the yolk in place?

19 Why do health experts suggest you should eat up to six eggs a week as part of a healthy, balanced diet?

20 Explain what happens when an egg coagulates.

DESIGN ACTIVITY 5.1

A special breakfast

Worksheet

DESIGN BRIEF

Design a healthy breakfast suitable for a special occasion that includes eggs in the menu and serves one person. You should also present a 'Welcome to breakfast' menu card to give to your guest when you serve the breakfast.

1 Write a breakfast design brief that includes the following information:
- Who – who is the breakfast for?
- Why – why is the special breakfast being served?
- What – what will the menu include (for example, an egg, fruit and a wholegrain cereal product)?
- When – when will you be serving the special breakfast?
- Where – where will you prepare and serve the breakfast?

2 Using the design brief, develop five criteria questions to evaluate the success of your special breakfast.

THINKING SKILLS 5.1

1 Draw up a bar graph that demonstrates the fat content of full cream, reduced-fat and skim milk.

PREPARING A BREAKFAST FOLDABLE

1 Using an A4 sheet of paper, prepare a tri-fold book foldable (see the diagram opposite) that summarises the key information about breakfast.

Prepare the foldable using the following headings:
- 'Breakfast heading and definition'
- 'Why breakfast is important and what the research says about breakfast'
- 'Breakfast eating habits: boys compared to girls'
- 'My dream breakfast' (divide this section into half). List the menu, including drinks for your dream breakfast, on the left-hand side. On the right-hand side, plot the food items you have included on a diagram of the Healthy Eating Pyramid. Write a comment on the nutritional value of your dream breakfast.

2 Glue the completed foldable into your workbook.

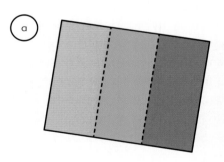

Breakfast foldable

DESIGNING

1. Research a food selection model such as the Healthy Eating Pyramid or the Australian Guide to Healthy Eating and list the guidelines that will be important to consider in designing a healthy breakfast.
2. Use the recipe index at the back of this book or search the internet to assist you with your research for:
   - two recipes that contain fruit and are suitable to serve for breakfast
   - two ideas for using wholegrain cereal products in the menu
   - two methods of cooking eggs for breakfast
   - two suggestions for a suitable breakfast beverage; this could include some of the key ingredients identified in the design brief.

GENERATING

1. Use the ideas from your research to develop two menu options for your breakfast.
2. Select your preferred option, and justify your decision in relation to the specifications in your design brief.
3. Design a creative 'Welcome to breakfast' menu card that includes the food items that are to be served at your special breakfast.

PLANNING AND MANAGING

1. Prepare a food order for your breakfast.
2. Prepare a production plan for the breakfast menu. See Page 11 for an example of how to prepare a production plan.

PRODUCING

1. Produce the breakfast menu you have planned.

EVALUATING

1. Evaluate your breakfast for a special occasion using your evaluation criteria and feedback from your guest.
2. Describe the sensory properties – appearance, aroma, flavour, texture – of each dish on your breakfast menu.
3. If you prepared a hot dish, were you able to prepare the breakfast so that it was still warm when it was served? How did you present the food so that it encouraged a cheerful start to the day?
4. Explain two safety rules you needed to consider when preparing the breakfast.
5. Analyse your breakfast menu by plotting all of its ingredients onto a diagram of the Healthy Eating Pyramid or the Australian Guide to Healthy Eating. Comment on the nutritional value of your breakfast.
6. If you were to prepare the breakfast again, would you make any changes to the menu or to your production plan? Consider how well your menu met the requirements of the design brief.

A special breakfast of a vegetable omelette, english muffins, fresh fruit and a glass of orange juice

QUICK APPLE MUESLI

- 1 teaspoon lemon juice
- 1 large, crisp apple, grated
- 2 tablespoons rolled oats
- 1 tablespoon sultanas
- 2 tablespoons yoghurt
- 2–3 teaspoons honey

SERVES ONE

METHOD

1. Pour lemon juice over the grated apple to prevent it from going brown.
2. Mix in rolled oats, sultanas, yoghurt and honey.
3. Place in serving bowl.
4. Serve immediately.

For added flavour, garnish with chopped nuts and cinnamon.

EVALUATION

1. Explain the best way of safely using the grater to prepare the apple for this recipe.
2. Why should lemon juice be poured over the grated apple?
3. How are the rolled oats softened during the preparation of the muesli?
4. List other fresh and dried fruits that could be added to the muesli.
5. Based on the guidelines of the Healthy Eating Pyramid or the Australian Guide to Healthy Eating, evaluate whether or not the Quick Apple Muesli is a healthy breakfast.

BOILED EGGS

1 egg

🍴 SERVES ONE

If the egg is taken straight from the refrigerator and placed into boiling water, its shell can crack as a result of the sudden, drastic change in temperature. Placing the egg into cold or warm water can overcome this problem.

METHOD

1. Cover the egg with cold or warm water.
2. Bring water to the boil.
3. Turn the heat to low.
4. If you like your egg soft, cook for 2–3 minutes. For hard-boiled eggs, cook for 8 minutes once the water has boiled. (An egg is hard-boiled if its shell dries quickly when it is lifted from the water.)
5. Carefully remove the egg from the boiling water.

To serve a soft-boiled egg, place in an egg cup, cut the top off and serve with fingers of warm toast. To serve a hard-boiled egg, drain the water from the saucepan and immediately run cold water over the egg. Crack the shell against the edge of the sink, or a bench, and remove the shell once it is cool enough to handle. Cracking the egg shell will prevent a green-grey ring from forming around the yolk. This ring is formed when eggs that are not fresh are heated; the iron in the yolk and the sulphur in the egg white react to form ferrous sulphide on the surface of the yolk.

EVALUATION

1. Why is it important to put eggs in cold or warm water before boiling them?
2. Describe how you can tell if water is boiling.
3. Explain the differences in the textural properties of the yolk of a soft- and a hard-boiled egg.
4. Explain the reason for cracking the shell of a boiled egg immediately after removing it from the boiling water and rinsing it under cold water.
5. List one safety rule to consider when cooking a boiled egg.

POACHED EGG

- 1 egg
- 1 teaspoon vinegar
- ½ avocado, smashed (optional)
- Squeeze of lemon juice (optional)
- Hot, buttered toast (optional)
- Salt and pepper to serve (optional)

SERVES ONE

For a delicious breakfast, mix half a smashed avocado with a squeeze of lemon, salt and black pepper. Spoon the avocado mixture on a slice of hot, buttered toast and top with the poached egg.

METHOD

1. Crack the egg into a jug or cup.
2. Fill a frying pan with water to about 2 centimetres deep. Add a teaspoon of vinegar. The vinegar will lower the temperature at which the protein in the egg coagulates and help to set the egg.
3. Bring the water to the boil. Once boiling, stir briskly with a wooden spoon to create a whirlpool.
4. Gently pour the egg into the centre of the swirling water. (The swirling water helps to keep the egg in a round shape.)
5. Turn down the heat so that the water is just simmering.
6. Gently cook the egg for 2–3 minutes or until just set.
7. Carefully lift the egg from the water using an egg lifter.

EVALUATION

1. Why is it important to crack the egg into a jug or cup before pouring it into the boiling water?
2. What is the purpose of adding vinegar to the water when poaching the egg?
3. What does the swirling water do to the egg during poaching?
4. Describe how you can tell if water is simmering.
5. List one safety rule to consider when cooking a poached egg.
6. Explain why poaching eggs is a better method of cooking than frying eggs.

FRIED EGG

1 tablespoon vegetable oil
cube of bread (to test the oil)
1 egg

SERVES ONE

METHOD

1. Add the oil to a small frying pan and place over medium heat for approximately 30 seconds.
2. Add the cube of bread to the oil. Once it becomes golden brown, the oil is hot enough to cook the egg. Remove the cube of bread from the fry pan.
3. Carefully crack the egg into the hot oil.
4. Cook for 2—3 minutes or until the white is set and the yolk slightly set.
5. Use an egg slice to remove the egg from the frying pan. Drain it on paper towel before serving.
6. Allow the frying pan to cool before washing.

EVALUATION

1. Discuss why it is important to use a small frying pan rather than a large one if you are cooking only one egg.
2. Why is it important to preheat the oil before adding the egg?
3. Why is it desirable to drain the egg on paper towel before serving?
4. List one safety rule to consider when cooking a fried egg.
5. Suggest some accompaniments to serve with the Fried Egg to make a delicious meal.

SCRAMBLED EGGS

2 eggs

4 tablespoons milk

pinch of salt

pepper

1 teaspoon butter

1 teaspoon parsley, chopped

SERVES ONE

METHOD

1. Whisk the eggs, milk, salt and pepper until well combined.
2. Melt the butter in a small frying pan.
3. Pour in the egg mixture and gently stir over a low heat until the mixture thickens but is still soft. Do not overcook the eggs.
4. Serve with warm toast and sprinkle with the chopped parsley.

EVALUATION

1. Why is it important to stir the eggs while they are cooking?
2. Explain how the scrambled egg mixture thickens during cooking.
3. Describe the sensory properties of the Scrambled Eggs.
4. What effect would overcooking have on the texture of the eggs?
5. List one safety rule to consider when cooking scrambled eggs.
6. Discuss why serving scrambled eggs on toast is considered a healthy breakfast.

SHIRRED EGG

1 teaspoon butter

1 egg

30 grams cheese

SERVES ONE

Mark Fergus Photography

METHOD

1. Grease a small ramekin or ovenproof dish with some of the butter.
2. Preheat the oven to 200°C.
3. Separate the egg yolk from its white.
4. Grate the cheese.
5. Beat the egg white until stiff peaks form.
6. Carefully fold half the grated cheese through the egg white.
7. Place the egg white into the ramekin or ovenproof dish. Use a spoon to make a small nest in the centre of the white.
8. Carefully drop the yolk into the centre of the egg white.
9. Sprinkle remaining grated cheese over the egg.
10. Bake in the preheated oven for approximately 8 minutes or until the egg white has set.

EVALUATION

1. Why is it important to preheat the oven before use?
2. Explain how you would separate the egg yolk from the white when making the Shirred Egg.
3. Describe how to tell when egg whites are beaten to stiff peaks.
4. What is the purpose of covering the egg yolk with grated cheese before baking?
5. List one safety rule to follow when cooking the Shirred Egg in the oven.

CHEESE OMELETTE WITH BACON

2 eggs
1 tablespoon water
2 shakes pepper
1 teaspoon butter
1 tablespoon parsley, finely chopped
¼ cup cheddar cheese, grated

SERVES ONE

METHOD

1. Mix eggs with water and gently beat with a fork until well combined.
2. Season with the pepper.
3. Melt the butter in a non-stick frying pan over medium heat until it is foaming, but not brown.
4. Pour the egg mixture into the frying pan. Gently shake to ensure the omelette mixture covers the base of the pan.
5. Carefully lift the edges of the omelette with a spatula so that the mixture can run to the edges and set.
6. When the mixture is almost set, sprinkle the grated cheese and chopped parsley over the front half of the omelette.
7. Carefully lift the back half of the omelette over the top of the cheese and parsley.
8. Turn onto a warmed plate and serve immediately. Serve with bacon (see below).

Cooking bacon in the microwave

Dice two rashers of bacon. Place a sheet of absorbent paper on a plate. Spread the diced bacon over the paper and cover with another sheet of absorbent paper. Microwave on high for 1–2 minutes or until crisp.

Fried bacon

Place bacon in a frying pan and fry over medium heat until the bacon is crisp and lightly browned. Drain on absorbent paper.

EVALUATION

1. Why is it important to beat the eggs and water until they are well combined?
2. Explain what happens to the texture of the egg as it cooks in the hot pan.
3. Discuss why is it important to lift the edges of the omelette and allow the uncooked mixture to cover all of the pan's base.
4. Outline two advantages of cooking bacon in the microwave.
5. Why is it important to drain bacon on absorbent paper after cooking?

SAVOURY EGG ROLL

1 round wholemeal bread roll

1 spring onion

1 rasher bacon

10 grams butter

1 egg

1 tablespoon cheese, grated

 SERVES ONE

METHOD

1 Preheat the oven to 200°C.

2 Cut the top from the bread roll and keep aside – this will form the lid for the Savoury Egg Roll.

3 Using a teaspoon, scoop out a small amount of bread from the inside of the bread roll to make a hollow case. Take care not to make holes in the outside of the roll.

4 Finely slice the spring onion. Dice the bacon.

5 Melt the butter in a small frying pan and sauté the spring onion and bacon until just beginning to brown.

6 Spoon the bacon and spring onion mixture into the bread roll case.

7 Break the egg into the bread roll.

8 Sprinkle with the grated cheese. Place the lid on the top.

9 Wrap the roll in aluminium foil and place on an oven tray.

10 Bake in the preheated oven for approximately 15 minutes or until the egg is almost set.

11 Remove the roll from the oven and turn back the foil. Return to the oven for a further 5 minutes to allow to crisp.

EVALUATION

1 Why is it important to cook the bacon and spring onion before adding them to the bread roll case?

2 List one safety rule to consider when cooking the spring onion and bacon in the frying pan.

3 Explain how the physical properties of the egg change once it is baked.

4 What is the purpose of wrapping the Savoury Egg Roll in aluminium foil before it is baked?

5 Describe the sensory properties of the cooked Savoury Egg Roll.

FABULOUS FRENCH TOAST

1 egg
½ cup milk
rind of half an orange, grated
2 thick slices stale bread
1 teaspoon caster sugar
½ teaspoon cinnamon
2 teaspoons butter
½ banana, peeled and sliced
1 tablespoon maple syrup

SERVES ONE

METHOD

1. Beat together the egg, milk and orange rind.
2. Cut each slice of bread into two triangles.
3. Mix the caster sugar and cinnamon together in a small container.
4. Heat the butter in a frying pan.
5. Dip each triangle of bread in to the egg mixture, then place in the hot pan.
6. Cook on both sides until golden brown. Drain on absorbent paper.
7. Serve on a warmed plate. Top with the sliced banana and drizzle with maple syrup.
8. Sprinkle with the cinnamon and sugar mixture.

EVALUATION

1. Why would it be important that the bread is at least two days old?
2. What would happen if the butter was not hot enough when the battered bread was placed in the frying pan?
3. Why is it important to drain the cooked toast on absorbent paper before serving?
4. List three other fruits that could be used to top the cooked toast instead of banana.
5. Discuss why you should limit your intake of ingredients such as caster sugar and maple syrup.

6
THE GREENGROCER

KEY KNOWLEDGE
- Fruit
  - Buying fruit in season
  - Classification of fruit
  - Fruit for good health
  - How much fruit should I eat?
- Apples
- Bananas
- Citrus fruits
- Packaging and labelling fresh fruit
  - Packaging fruit
  - Labelling fruit
- Processing fruit
  - Primary processing of fruit
  - Secondary processing of fruit
- Vegetables – the colours of the rainbow
  - Vegetables for good health
  - How many vegetables should I eat?
  - Top tips for cooking vegetables
- Orange vegetables: sweet potatoes
- Green vegetables: green beans
- White vegetables: potatoes
- Preparing fruit and vegetables safely

KEY TERMS
anthocyanins substances that produce red to purple to blue colours in fruits and vegetables
carotenoid a substance that produces the orange colour in fruit and vegetables
chlorophyll a substance that produces the green colour in fruit and vegetables
enzymatic browning a process that occurs when the enzymes in cut or peeled fruits cause browning when exposed to oxygen in the air
in season the time of year when a fruit or vegetable has its best sensory properties
modified atmosphere packaging (MAP) a system that changes or modifies the atmosphere or gas inside a package, in order to extend the shelf life of the food
pomme fruit fruit with crisp, juicy flesh surrounding a core that contains seeds; for example, apples and pears
solanine a toxin found in green potatoes
sustainable farming farming practices that maintain the land's productivity so that it will be available for future generations
tuber a vegetable that grows under the ground and has a high starch content; for example, potato

AUSTRALIAN CURRICULUM LINKS
DESIGN AND TECHNOLOGIES
- Knowledge and Understanding
- Processes and Production Skills

HEALTH AND PHYSICAL EDUCATION
- Personal, social and community health

GENERAL CAPABILITIES
- Critical and creative thinking

CROSS-CURRICULUM PRIORITIES
- Sustainability

FRUIT

Fruit is often described as nature's perfect snack food. It comes in its own biodegradable package; it is convenient to eat; there are many varieties; it is a good source of vitamins, minerals, carbohydrates and dietary fibre; and it tastes great.

Buying fruit in season

Fruit matures with the seasons; as the weather changes – for example, from summer to autumn – some fruits are at their peak. Because they are *in season*, they have their best flavour, texture and aroma, and, because they are plentiful, they are available at the best price.

Fresh fruit can be served whole, cut into interesting pieces, canned, juiced and dried, and be included in a wide range of desserts and cakes.

Classification of fruit

Fruits are classified according to how they grow and their physical properties.

Fruit

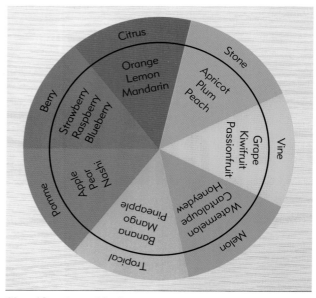

Classification of fruit

Fruit for good health

Fruit offers the best nutrient package when it is eaten raw or lightly cooked. It is a good source of carbohydrates, is high in fibre, has a wide range of vitamins and minerals, and most fruits are low in fat. Most raw fruit is a good source of vitamin C, especially citrus and berry fruits. Mangoes, cantaloupes and apricots, the orange-coloured fruits, are good sources of vitamin A. Some fruits, such as such as bananas and dried figs, are good sources of phosphorus, and dried fruits are a good source of iron. Fruit eaten with the skin on is a better source of dietary fibre than peeled fruit, and whole fruit has more dietary fibre than fruit juice.

How much fruit should I eat?

The Australian Guide to Healthy Eating recommends eating at least two serves of fruit per day and trying to include a wide variety of fruits throughout the week.

 =  =  =

1 piece of medium-sized fruit – 1 apple, 1 orange, 1 pear

2 small pieces of fruit – plums, peaches, apricots, kiwifruit

1 metric cup of canned or chopped fruit

Some dried fruit – 4 dried apricots or 1 ½ tablespoons of sultanas

One serve of fruit equals …

APPLES

There are more than 7000 varieties of apple grown throughout the world. Along with pears, apples are a member of the pomme fruit family, which means they have a compartmentalised core.

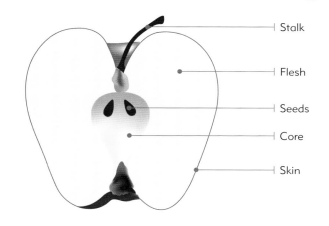

Cross-section of an apple

Apples growing

Australia is one of the world's healthiest fruit-growing environments. It is protected from acid rain and industrial pollution, which cause problems in Europe, where apples are a significant crop. Australia is also free from a number of the world's pests and diseases because of its isolation. Apples are grown in all states of Australia, with Victoria a leading producer.

Apple varieties

In Australia, two major varieties of apples are grown: Red Delicious and Granny Smith. In the past, these two varieties accounted for over 75 per cent of the total volume of apples grown in Australia. However, consumer demand for more choices has led to an increase in plantings of new varieties, such as Australian Fuji, Gala, Sundowner and Pink Lady.

The Gala originated in New Zealand and is an early-season apple, usually available from February onwards. It has a dense, crisp texture and is sweeter than Red Delicious. The Australian Fuji was bred in Japan and is a late-season apple, available from April onwards. It has a dense flesh and a sweet, distinctive flavour, and its skin is a blushed pink-and-red colour. The Pink Lady was bred in Western Australia and is available late in the season. Its flesh is crisp and firm and has a flavour similar to that of Golden Delicious.

The Australian Granny Smith apple was first grown in New South Wales by Maria Ann Smith in 1860, and is now popular all over the world. It has a green, 'greasy' skin and a tart flavour. It is popular for cooking and for processing into apple products.

ACTIVITY 6·1

Varieties of apples

AIM
To compare the sensory properties of different varieties of apples

Worksheet

METHOD
1. Select four different varieties of apples. Wash and dry them before the comparison.
2. Draw up the following table in your workbook and record the appearance of each variety of apple before slicing it for the taste testing.

Apple varieties grown in Australia

	Royal Gala	Jonathan	Jonagold	Golden Delicious
FLESH	Firm, crisp, sweet, juicy, excellent texture	Crisp, sweet, juicy, pleasantly tart	Flavoursome, crisp, subacid to sweet, juicy	Crisp, creamy, flavoursome, sweet, juicy
COLOUR	Red blush to solid red over golden background	Scattered red stripes and blushed over greenish background	Bright red striped over yellow with a tinge of green background	Greenish to golden yellow, occasionally with a slight pink blush
SIZE	Medium	Medium, can be small	Large to very large	Medium
AVAILABILITY	Early February–late August	Late February–mid-July	Mid-March–August	Late March–early December
COMMENTS	Very attractive apple of excellent eating quality	Older variety with a loyal following	Universally popular apple showing both Jonathan and Golden Delicious attributes	Internationally popular apple, always in demand

	Jazz®	Red Delicious	Granny Smith	Fuji
FLESH	Firm, dense, crunchy with a tangy, sweet flavour, similar to peaches and melon	Medium sweetness, crisp, juicy, white	Hard, crisp, tart flavour, white	Very sweet, firm, crisp, juicy
COLOUR	Pink/red flush over a light green background	Solid bright red with slight stripe, attractive	Green to greenish yellow, occasionally with a slight pink blush	Predominantly blushed dull pink/red, some russet evident
SIZE	Small to medium	Medium to large	Medium round	Medium, but can be small
AVAILABILITY	All year	All year	Harvest April, available all year	Mid-April–mid-December
COMMENTS	Distinct new variety, a cross between Royal Gala and Braeburn	Internationally popular apple, always in demand	Australia's own and world-renowned green apple	Very sweet 'honey core' is preferred by some customers

	Braeburn	Pink Lady™	Sundowner™	Lady Williams
FLESH	Good, sweet, subacid, very juicy	Firm, sweet, crisp, juicy, flavoursome	Sweet, flavoursome, firm, juicy	Sweet-tart flavour, improves with storage
COLOUR	Red stripe and blush over yellowish background	Pink or light red blush over yellow-green background	Some red to full red with a little russet	Usually deep red with a coverage of at least 75 per cent
SIZE	Medium to large	Large uniform	Medium	Medium
AVAILABILITY	Late April–mid-October	Late April–February	May–February	Late May–February
COMMENTS	Unique combination of sweet and tart	Unique and very popular apple that eats exceptionally well	Good, long-keeping, late-season apple	Very firm, long-keeping apple with Granny Smith-like characteristics

Apple and Pear Australia Ltd.

RESULTS

Sensory properties of apple varieties

Apple variety	Appearance		Flavour	Texture	Rating
	Colour	Shape			
	• Red • Yellow-green • Pink • Mottled • Striped • Speckled	• Round • Oval • Flat • Big • Small • Uneven	• Sweet • Sour • Sharp • Tart • Bland	• Crisp • Crunchy • Soft • Floury • Firm • Juicy • Dry	5 = like a lot 4 = like 3 = OK 2 = dislike 1 = dislike a lot

ANALYSIS

1. Why is it always advisable to wash fruit before eating?
2. Which variety of apple did you like best? Why?
3. Which variety of apple did you like least? Why?
4. Which sensory property is the most important to you when selecting an apple to eat raw?
5. Which qualities would be the most important if you wanted to cook with an apple? Why?
6. Why have apple growers begun to supply the market with a greater variety of apples?
7. What is meant by the term 'in season' when discussing fruit?
8. Record the months when the apple varieties you compared are in season.
9. How may the sensory properties of a fruit change when it is not in season?

CONCLUSION

Which variety of apple do you prefer to eat raw? Justify your answer.

Nutrition pack

Apples are one of nature's best snacks. The firm skin keeps the crisp, juicy flesh in good condition for days and ensures that the apple is easily transported in a lunch box or bag without refrigeration. The saying 'an apple a day keeps the doctor away' reflects the fact that apples are a very healthy food, as they are high in dietary fibre in the form of soluble dietary fibre called pectin. There are also a small amount of vitamins and minerals available in apples.

A medium-sized eating apple contains about 270 kilojoules and is a filling snack. In an apple:

- 10 per cent is carbohydrate
- more than 80 per cent is water
- approximately 4 per cent is vitamins and minerals
- approximately 6 per cent is dietary fibre.

Dietary fibre is important for good health because it helps to keep the digestive system working. Apples contain some dietary fibre in their skin.

Selecting the best

Apples are at their best during autumn and winter. Choose firm fruit with brightly coloured skin and no blemishes.

Storage

Apples lose their crispness when they are stored at room temperature. For a longer life, store apples in a sealed plastic bag in the crisper of a refrigerator for up to three weeks.

Cooking with apples

When apples are peeled and prepared for cooking, they can sometimes turn brown. This is called enzymatic browning. Enzymatic browning is a process that occurs when the enzymes in cut or peeled fruits cause browning when exposed to oxygen in the air. This browning can be prevented by covering the cut fruit with water or sprinkling it with lemon juice.

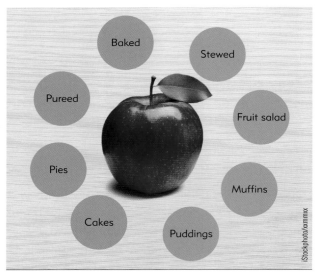

Cooking with apples

BANANAS

Bananas are another of nature's cleverly packaged, healthy snacks. In fact, they are sometimes described as 'brain food'. That is because bananas are high in potassium, which is a major nutrient in the process of carrying the trillions of messages that nerves move around the human body. Eating foods that are good sources of potassium helps your concentration at school and home, as well as assisting your performance in physical activity.

Growing bananas using sustainable practices

Bananas grow in tropical regions on large, palm-like plants. The banana palm is not a tree, because it does not have a woody trunk. The banana palm develops a flower stem containing female flowers, from which the bananas grow. A single banana is called a 'finger'; a number of bananas – about 12 – makes a 'hand'; and several hands form a bunch of bananas. The most popular variety of bananas grown in Australia is the Cavendish, which is long and thin. Lady Finger bananas are short and plump, and have a very distinctive flavour and texture.

Some farmers are now producing bananas using sustainable farming practices. These are practices that maintain the land's productivity so it will be available for future generations. When growing bananas in a sustainable manner, farmers use reduced amounts of fertilisers and pesticides on their crops, which has considerable benefits for the environment. Instead of using chemicals to increase production, sustainable methods use a more natural growing cycle, which in turn improves and maintains the quality and fertility of the soil. Banana crops are also watered by efficient drip systems that prevent run-off and protect water quality.

Red wax-tipped bananas

While these sustainable farming practices slow down the growing process, producers believe that bananas grown in this way have a firmer texture, a sweeter flavour and an extended shelf life. Before the bananas are sent to market, farmers dip their ends in red wax to indicate that they have been grown using sustainable farming practices.

Bananas growing

Nutrition pack

Bananas are low in fat and a good source of kilojoules, and so are a great high-energy food. As bananas ripen, their starches are converted to sugar, which the body can turn into energy quickly. Bananas are an excellent source of the minerals potassium, calcium, magnesium, phosphorus and iron. They are also high in vitamins A, B and C. Bananas are rich in dietary fibre, which means you feel full after eating them and so don't feel like eating unhealthy, sweet, fatty food. The other benefit of dietary fibre is that it helps to absorb water in the intestines, and so makes the digestive process more efficient.

Selecting the best

Bananas are at their best in autumn and winter. Select fruit that is firm, bright yellow in colour and free from bruises. As soon as bananas are peeled or cut, they begin to discolour as a result of enzymatic browning. This can be prevented by a squeeze of lemon juice.

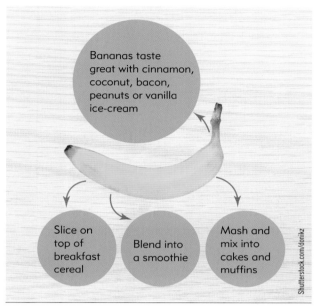

Banana recipe ideas

Storage

Allow bananas to ripen at room temperature out of direct sunlight.

Once ripe, bananas can be wrapped in a brown paper bag and stored in the crisper of a refrigerator for two to three days. Even though this turns the skin black, the flesh inside will remain delicious to eat.

CITRUS FRUITS

The citrus fruit family includes oranges, lemons, mandarins, grapefruit, limes, cumquats and pomelos. Citrus fruit trees are evergreen trees that grow in subtropical climates. They were introduced into Australia in the late 18th century. To grow, they require rich, well-drained soil with a good water supply. The most suitable citrus-growing areas in Australia are in the irrigation regions of the Murray and Murrumbidgee rivers, on the plains in New South Wales and on the outskirts of Perth in Western Australia.

ORANGES

The two most commonly available varieties of orange in Australia are the valencia and the navel. Valencia oranges are considered best for juicing, and are available throughout summer and autumn. They have a smooth skin, which often has a green tinge. The navel orange is the most popular orange to eat fresh because it is almost seedless and is easy to peel and break into segments. It is available from May to December.

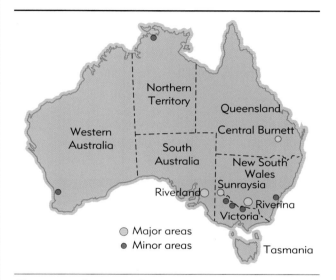

Citrus-growing areas in Australia

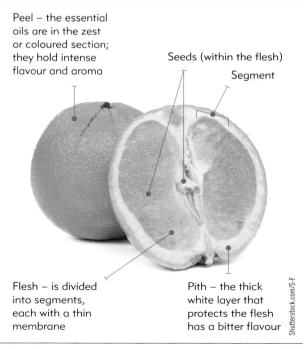

Parts of an orange

Nutrition pack

Oranges are naturally high in vitamin C as well as being a good source of folic acid, potassium and dietary fibre. Vitamin C helps in the healing process if you have been ill or injured, helps to strengthen body tissues and bones and assists in the absorption of iron.

Selecting the best

Unlike some other fruits, oranges do not continue to ripen once they have been picked, so colour, juice content, sugar and acid levels are measured carefully before harvesting begins. An orange should feel heavy for its size when held in the hand. The fruit should have a fresh, sweet, citrus aroma with no mouldy smell. The skin should not be wrinkled or have any soft patches.

Storage

Oranges can be stored for two weeks at room temperature, and for longer in the refrigerator.

Koala Easy Peel navel oranges

ACTIVITY 6·2

Citrus cordial comparison

AIM

To compare the sensory properties of homemade and commercial citrus cordial

METHOD

1. Using the recipe on Page 128, prepare the ingredients for a quantity of citrus cordial.
2. Make a glass of homemade citrus cordial and a glass of commercial citrus cordial, using the ratio of one tablespoon of cordial to 100 millilitres of water.
3. Compare the ingredients and sensory properties of the homemade and the commercial citrus cordials.

ANALYSIS

1. Which cordial had the most intense colour?
2. How do you account for the difference in colour between the two cordials?
3. Which cordial had the strongest flavour?
4. Why was the homemade cordial cooled for 15 minutes before straining?
5. How do you ensure the bitter flavour from the pith does not flavour the cordial?
6. Why is a pastry brush, not a knife, the best tool to use to remove the zest from the grater?
7. Identify the ingredients that are not in both products.

CONCLUSION

Which cordial did you like best? Explain why.

RESULTS

	Ingredients	Colour	Flavour	Texture	Rating 5 = like a lot 4 = like 3 = OK 2 = dislike 1 = dislike a lot
Homemade cordial					
Commercial cordial					

Testing knowledge

1. Explain how fruit is classified in family groups. List the groups.
2. Why should fruit be part of everyone's daily diet?
3. State how much fruit is needed each day to help maintain a balanced diet. List examples of fruit you would eat to meet this recommendation.
4. Describe the characteristics of fruits that belong to the pomme family.
5. Identify two varieties of apples and describe the differences in their sensory properties.
6. Explain how the browning of cut apple can be prevented.
7. Explain the benefits to the environment of farmers using sustainable methods to grow bananas.
8. Outline how bananas should be stored to maintain top quality.
9. Identify the major citrus-growing areas in Australia, and explain why they are the most suitable citrus-growing environments.
10. Outline the three main parts of an orange, and describe the features you would look for to select good-quality citrus fruit.

ACTIVITY 6·3
Fruity snacks

1. Develop a list of foods that students in your class eat as snacks.
2. Position each item on the list on a diagram of the Healthy Eating Pyramid or the Australian Guide to Healthy Eating. How many items are fruit-based and in the foundation layer of the Healthy Eating Pyramid or the fruit section of the Australian Guide to Healthy Eating? Explain why it is a good thing if many of the snacks are in these sections.
3. Explain why fruit is an excellent snack food in terms of nutrition and convenience.
4. List four examples of processed drinks and foods that contain fruit and could be consumed as snacks.
5. Are any of the items you selected in Question 4 high in fat or sugar? How do you know this?
6. Crunching and chewing on an apple helps to reduce dental plaque. What other health benefits are there to eating an apple?

PACKAGING AND LABELLING FRESH FRUIT

Packaging fruit

Transporting fresh fruit without damage, from orchards in rural areas, to supermarkets, and then to the consumer's home, involves many factors. Some aspects of packaging design that are taken into account are:
- how effectively the container will hold the fruit during transport
- whether the container protects the fruit from damage
- whether the container maintains the sensory properties of the fruit
- how much the packaging material costs
- whether the materials can be recycled
- whether incorrect disposal may injure wildlife
- whether consumers can see the fruit through the packaging material
- whether the portion size will meet the needs of consumers.

Labelling fruit

The introduction by growers of a large range of apple and orange varieties led to the development of a system of cataloguing the produce, so that retailers would know exactly what they were selling and shoppers could easily identify each variety. Supermarket chains also influenced this process, because they required a labelling system that could be read by laser scanners at the check-out. Small, stick-on labels are now attached to the fruit in packing sheds after it has been weighed and classified as large, medium or small. The information on the stickers is very useful for supermarkets, because it means the correct weight and price of the fruit can easily be calculated. It also enables sellers to keep their data files up to date with information on the most popular varieties, the amounts sold and when to order new stock.

The stickers are non-toxic; some are even made of edible plastic.

Stickers on fruit

Corrugated cardboard cartons

- Prevent damage; strong and lightweight
- Can withstand a wide range of temperatures
- Waxed surface helps to resist moisture during transportation
- Easy to stack and carry
- Communication space – flat surface is excellent for printing
- Environmentally friendly – made from recycled paper

Plastic bags

- Fruit is visible
- Small holes allow fruit to breathe as it continues to ripen
- Suitable for storing in a controlled atmosphere to extend shelf life
- Fruit can be graded into the same size
- Convenient for consumers to carry home
- Easy to open
- Suitable for use on automatic packaging machines

Shrink-wrap

- Suitable for individual pieces of fruit
- Protects fruit from damage and contamination
- Easy to transport from the supermarket
- Good surface on which to stick pricing information

Mesh bags

- Low-cost
- Lightweight
- Good ventilation
- Easy to see the fruit
- Do not damage the fruit
- Make an attractive display in stores
- Convenient for consumers to transport home

Rigid plastic packs

- Strong and lightweight
- Prevent damage from crushing
- Fruit is visible
- Inexpensive, recyclable plastic
- Unaffected by cold temperatures
- Stack easily whether empty or full

ACTIVITY 6·4

Packaging and labelling of fruit

1. List the information that is included on the stickers on fresh fruit.
2. Explain why labelling fresh fruit with stickers is useful to supermarket managers.
3. Why are some bananas dipped in red wax?
4. What would be the advantages and disadvantages of using wax instead of stickers as a means of identification?
5. Using the information on Page 108, compare the methods of packaging fresh fruit.

Worksheet

Style of packaging fresh fruit	Example/s of fruit packaged	Characteristic/s that prevent damage to fruit	Benefit/s to consumers when purchasing fruit	Environmental issue/s associated with its use
Corrugated cardboard	Apples	Absorbs bumps during transport	Apples without bruises	Cardboard has been made from recycled paper
Plastic bags				
Mesh bags				
Shrink-wrap				
Rigid plastic packs				

PROCESSING FRUIT

Primary processing of fruit

Primary processing of fruit involves the stages that farmers use to make fruit safe to eat.

Primary processing begins with harvest and ends with fruit that is ready either for the fresh fruit market or to undergo secondary processing. The physical form of the fruit changes very little as a result of primary processing.

Secondary processing of fruit

Not all the fruit that is grown is destined for the fresh fruit market. Some will undergo a range of secondary processes to create a wide variety of fruit products.

Some of the reasons for the secondary processing of fruit are:
- convenience
- adding variety to the diet
- improving the eating properties of a fruit
- increasing shelf life
- avoiding wastage of fresh fruit.

Secondary processing significantly changes the physical and sensory properties of the fruit. Other ingredients can also be added to the fruit to create

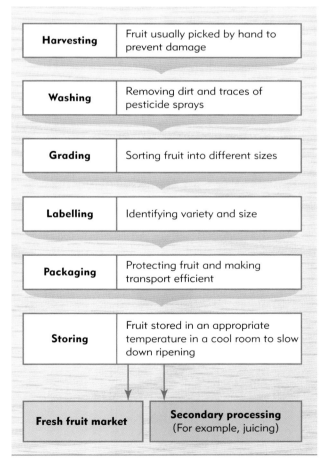

Stages in primary processing of fruit

new food products. Fruit can be canned, juiced, dried, chilled and frozen, as well as combined with other ingredients to make jam, cordial, soft drink and confectionary.

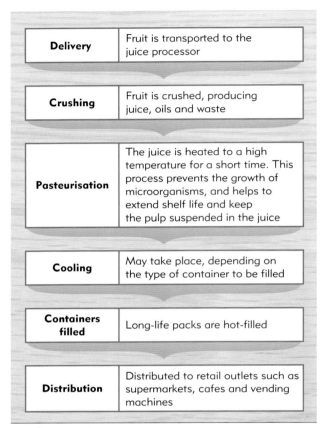

Stages in the secondary processing of fruit juice

Apple juice

Testing knowledge

11 Outline four reasons for packaging fruit.

12 Why is labelling fruit with stickers useful for supermarkets?

13 Describe the advantages and disadvantages of packaging fruit in mesh bags.

14 List two types of fruit that are packaged in rigid plastic containers.

15 Identify the stages in the primary processing of fruit.

16 Explain why fresh fruit is usually picked by hand.

17 Why is the washing step important during the primary processing of apples?

18 Why is the secondary processing of fruit extremely useful for our food supply?

19 Explain how the production of juice helps to prevent waste in the fresh fruit industry.

20 Why are fruit juices pasteurised?

VEGETABLES: THE COLOURS OF THE RAINBOW

Vegetables are nutrient-dense plants that can be eaten raw or cooked and which help to make our daily diet interesting. Vegetables can be classified according to either the way they grow or the part of the plant that is eaten. If you choose vegetables in all the colours of the rainbow, you will consume a wide range of nutrients that are essential for good health. Vegetables make meals more interesting because they contribute a variety of colours, flavours and textures. The varying colours in vegetables are due to different pigments; for example:

- chlorophyll produces a green colour
- carotenoids create orange and yellow colours
- anthocyanins produce red to purple to blue colours.

Vegetable rainbow

Vegetables for good health

Vegetables are a valuable source of dietary fibre, which is necessary for a healthy digestive system. They are low in fat and high in vitamins and minerals. In particular, they are a good source of vitamins such as vitamins A and C, and folate. Dark-green, leafy vegetables and orange and red vegetables supply vitamin A, which helps to repair and maintain tissue and assists with night vision. Dark-green vegetables, such as silverbeet and broccoli, provide some of the iron that the body needs to help transport oxygen. Green vegetables and dried peas and beans are good sources of folate. Vegetables also contain a range of other minerals, including magnesium, zinc and calcium. Vegetables are 70 to 95 per cent water, so most do not contribute many kilojoules to the daily energy intake. Vegetables such as potatoes and sweet potatoes are high in starch and are a good source of energy. However, when potatoes are cooked in a lot of fat, they soak it up like a sponge. Potato crisps and chips are concentrated sources of energy, and should be eaten only occasionally.

How many vegetables should I eat?

Vegetables are in the foundation layers of the Healthy Eating Pyramid, and make up almost one-third of the 'plate' in the Australian Guide to Healthy Eating. Health experts recommend that we eat at least five to seven serves of different vegetables every day. This could mean starting the day with a drink of fresh vegetable juice, nibbling on a celery or carrot stick during the day, having a super stack of salad vegetables in a sandwich for lunch, and then enjoying a selection of vegetables with some grilled meat for a main meal in the evening.

 =  =  =

1 medium potato | ½ cup cooked peas, lentils or dried beans | ½ cup cooked vegetables | 1 cup salad vegetables

One serve of vegetables equals …

ACTIVITY 6·5

Thinking about vegetables

1. Sketch and colour a food rainbow. List at least two vegetables for each of the following colours:
   - red
   - yellow
   - orange
   - brown
   - purple
   - green
   - white.

2. Share your list of vegetables with other class members, and add to your list after the discussion. Which colour had the greatest variety of vegetables?

3. Identify the two vegetables you like most and the two you like least. Compare your choices with those of other class members, and then explain why there are similarities and differences in your choices.

4 Which vegetables can be eaten raw and which are more enjoyable cooked?

5 Which colour of vegetables would be a good source of iron in the diet?

6 Suggest some vegetables you could eat to include vitamin A in your diet.

Look at the graph below to answer the following questions.

7 Why do you think potato crisps have such a high energy value?

8 Considering this, why should potatoes cooked this way be eaten only occasionally?

9 Baked potatoes have a lower energy value than roasted potatoes. Explain how the differences in the cooking methods of baking and roasting may account for this.

10 Why do mashed potatoes have a higher energy value than boiled potatoes?

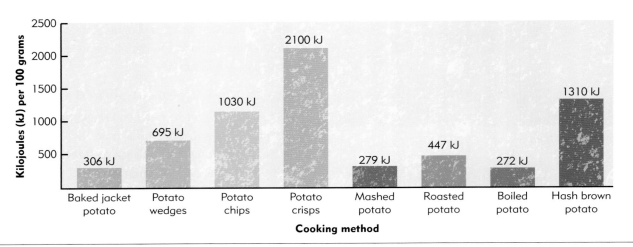

Energy value for 100-gram portions of potatoes

Top tips for cooking vegetables

1 Preparing and cooking vegetables changes their appearance, aroma, flavour, texture and, sometimes, their nutrients.

2 After washing, vegetables should be peeled and trimmed lightly, because many nutrients are found just under the skin.

3 The heat applied during cooking can intensify flavours and soften the texture of vegetables so that they are more enjoyable to eat.

4 The processes of baking and frying cause some vegetables, such as onions and potatoes, to brown, making them more appealing to many people.

5 The processes of boiling and steaming should be carried out until the vegetables are 'just' cooked to avoid loss of water-soluble nutrients, such as vitamin C, and to prevent vegetables, particularly 'greens', from losing their colour.

6 Microwaving is a very quick cooking process that minimises nutrient loss and allows vegetables to maintain their colour.

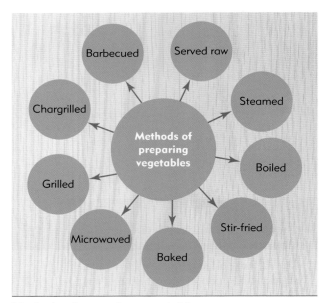

Methods of preparing vegetables

Testing knowledge

21. Identify two reasons for including vegetables in the diet.
22. What creates different colours in vegetables? Provide two examples.
23. Which coloured vegetables are a good source of vitamin A?
24. Which vegetables supply the mineral iron?
25. Explain how iron is used in the body.
26. Why don't most vegetables contribute much to energy intake?
27. How many serves of vegetables is it recommended that you eat every day?
28. Discuss the reasons for cooking some vegetables.
29. Explain how baking and frying can improve the appeal of some vegetables.
30. Why is microwaving a good method of cooking vegetables?

ACTIVITY 6·6
What's in the box

1. Search the internet to find out at least 12 facts about a vegetable of your choice.
2. Your research should cover:
   - growing the vegetable – how, where, season
   - nutrient value – highlight the main nutrients
   - sensory properties – raw and/or cooked
   - suitable dishes/recipes using the vegetable
   - other interesting facts.
3. Present your information on the six sides of a small box. Write at least two facts about your vegetable – without naming it – on each side of the box. Place your vegetable inside the box, where it cannot be seen.
4. With your class members as the audience, read out the information on your vegetable so they can guess what it is.

ORANGE VEGETABLES: SWEET POTATOES

Sweet potatoes are native to Central America and were among the foods brought back to Europe by Christopher Columbus in the 15th century. They are now grown around the world. The sweet potato is a large, edible root or tuber with a lumpy, pale, pinky-yellow skin that hides delicious orange flesh. It can be baked, fried, mashed or combined with other ingredients to make interesting vegetable dishes.

Nutrition pack
Sweet potatoes are high in carbohydrates and vitamins A and C.

Selecting the best
When buying sweet potatoes, select those that are small to medium in size and with smooth, unbruised skin.

Storage
Sweet potatoes should be stored in a similar way to potatoes – in a cool, dry, dark place. They will keep for three to four weeks.

GREEN VEGETABLES: GREEN BEANS

The most common green bean varieties in supermarkets and markets are green beans, broad beans and runner beans. Snake beans, which are very long with dark-brown tips, are the most popular in Asian-style cooking. Some of the green bean harvest undergoes secondary processing so that consumers can purchase beans frozen, canned or dried.

Nutrition pack
Fresh green beans should be cooked quickly to retain their vitamins A and C.

Selecting the best
Fresh green beans should have a bright-green, slightly glossy appearance and snap cleanly when broken in half. Beans that bend, are wrinkled or have some discoloured spots have poor flavour and texture.

PROCESSING OF VEGETABLES

Primary processing of green beans

Locally grown vegetables, including green beans, are picked and delivered to fresh food markets and supermarkets within 48 hours of leaving the farm. The speed of this primary process ensures that consumers can purchase vegetables with excellent sensory properties and maximum nutritional value.

Harvesting	Beans are picked early in the morning when they are in best condition
Washing	Beans are washed under high pressure to remove traces of dirt and chemicals
Grading	Checking for damage, size and freshness
Transporting	Beans are transported in refrigerated trucks to fresh food markets and supermarkets

Fresh food market or supermarket

Secondary processing (For example, freezing)

Stages in primary processing of green beans

Growing green beans

Secondary processing of green beans

Many people are very busy and do not have time to prepare fresh green beans. They find frozen beans very convenient because they have already been washed, peeled and sliced, and are ready to cook.

Green beans and other fresh vegetables can undergo a wide range of secondary processes, such as freezing, drying and canning, as well as be combined with other ingredients to make stir-fried meals and frozen meals. These processes provide products that are very convenient for the home cook.

Snipping	Beans are topped and tailed
Blanching	Beans are plunged into boiling water and then cooled quickly. This process makes the colour of the beans brighter and deactivates enzymes that cause spoilage during storage
Cutting to size	Beans are either sliced or left whole
Freezing	Beans are blasted with cold air while on a continuously moving belt, so they are frozen quickly to retain their shape and size
Packaging	Beans are packed in different-sized packets ready for transport to retail outlets

Stages in freezing green beans

Frozen beans

Storage

Fresh green beans can be stored in a plastic bag in a refrigerator for up to four days.

Freezing

Freezing is a method of food preservation in which the temperature of the water in food is lowered to below 0°C and the food is stored at −18°C. Lowering the temperature of food reduces the chemical reactions and microbial growth that cause food to deteriorate. The rate at which freezing takes place affects the quality of frozen food. When food freezes quickly, its texture is retained because there is minimal damage to its cell structure. This is because small ice crystals that form within the food do not move much, and so do not ruin the cell wall. Slow freezing allows large ice crystals to form, damaging the cells. After defrosting, the food becomes soft and flabby and excretes a lot of excess water.

ACTIVITY 6·7 Cooking green beans

Worksheet

AIM
To compare the flavour, texture and colour of fresh green beans cooked by different methods

INGREDIENTS
6 fresh green beans per test
1 teaspoon of oil

EQUIPMENT
Study the results table and determine the equipment list required for the test.

METHOD
1. Wash and top and tail all the beans.
2. Use a vegetable knife to slice each bean on a 45° angle.
3. Prepare the cooking equipment for each test.

RESULTS

	Flavour	Texture	Colour
Boil beans in 2 cm water, lid on. Cook 6 minutes.			
Boil beans in 2 cm water, lid off. Cook 6 minutes.			
Steam beans, lid on. Cook 6 minutes.			
Microwave beans with 1 tablespoon water in covered container. Cook 1 minute 30 seconds.			
Stir-fry beans in 1 teaspoon oil. Cook for 2 minutes.			

ANALYSIS
1. Which cooking method produced beans with the best flavour?
2. Which cooking method produced beans with the best texture?
3. Which cooking method produced beans with the best colour?
4. Which cooking method would retain the most nutrients in the fresh beans?
5. Which cooking method would be the least healthy way of cooking fresh beans?

CONCLUSION
What was the best cooking method to achieve good flavour, texture and colour in fresh green beans?

PREPACKED AND READY-TO-EAT SALAD MIXES

Consumers are changing the way they purchase and serve salads. Until recently, if a consumer wanted to make a salad at home, they would purchase a variety of individual salad ingredients. However, because many people today lead very busy lives, they are now often choosing prepacked and ready-to-eat salads from the supermarket, seeing this as a far more convenient and easy method of purchasing salads. Vegetables, such as a mix of salad leaves, are washed, cut and assembled in a bag containing sufficient salad for a family. The bag is ready to open and simply tip into a salad bowl – all the cook needs to do is to add a splash of salad dressing.

These salad mixes are packaged using a system called **modified atmosphere packaging (MAP)**. MAP modifies the atmosphere or gas inside a package in order to extend the shelf life of a food. This packaging system alters the gas inside the package by adding carbon dioxide and reducing the level of oxygen. The bag that contains the salad is made from a breathable film that also controls the flow of gases in and out of the bag.

Salad mix

The MAP process benefits the consumer because it ensures there is little food wastage, since the salad has a longer shelf life and stays fresh for longer. The producer also benefits, because there are fewer deliveries needed and the ingredients are protected from contamination by insects and dust during delivery.

WHITE VEGETABLES: POTATOES

Potatoes were grown by the ancient Incas in South America and introduced into Europe during the 16th century. They became the staple food of Ireland during the 17th century, and since that time have been grown and consumed in large quantities throughout the world. In Australia, potatoes have been one of the staple vegetables since the arrival of the Europeans. They are a nutritious food, a good source of carbohydrate and are fat-free. The potato is a tuber that grows underground on the roots of the plant. The leaves of the potato plant are poisonous. In Australia, potatoes come in many different skin colours, but the flesh inside the potato is usually white or cream in colour. In Europe, potatoes with a yellow flesh are more popular than the white potato.

Nutrition pack

Potatoes are a good source of carbohydrates, vitamin C, potassium and dietary fibre. They also provide some magnesium, niacin and thiamine. Potatoes are fat-free, but the methods used to cook them can increase their energy value.

Selecting the best

All potatoes should be firm, well-shaped for their variety and blemish-free. Avoid potatoes that are wrinkled and beginning to shoot. Greening of potatoes is caused by the development of a toxin called solanine, which develops when potatoes are exposed to light, either natural or fluorescent. Light causes chemical changes in the pigment chlorophyll, resulting in a green colour on all exposed surfaces. Potatoes that have a green tinge should be thrown away, because the toxin solanine can cause illness.

Storage

Potatoes stored in a cool, dark, well-ventilated cupboard will keep for two to three weeks. Keeping them away from light will prevent them from greening, and storing them separately from onions will maximise their storage life.

Varieties and their uses

Coliban
- Round, white potato with smooth skin; floury flesh
- Suited to baking and mashing, but sometimes disintegrates with boiling

Desiree
- Waxy, long, oval; pink skin with creamy yellow flesh
- Suited to baking, boiling and mashing, but not deep-frying

Dutch Cream
- Large, oval; yellow, rich, waxy flesh
- Roasts and mashes well

Kennebec
- Irregular, white, floury potato
- Can be baked, boiled, fried and mashed
- Used for processed chips in Australia

Kipfler
- Small, long, oval; sometimes called the 'finger potato'; yellow skin and yellow, waxy flesh
- Best steamed

Pontiac
- Round, red/pink skin with waxy, white flesh
- Denser than other varieties, so requires longer cooking
- All-purpose potato, good for baking, boiling and grating

Sebago
- Round/oval with dry flesh
- Ideal for baking, roasting, boiling, frying and mashing

Toolangi Delight
- Round, with white flesh and purple skin
- Suitable for all purposes except frying; mashes well

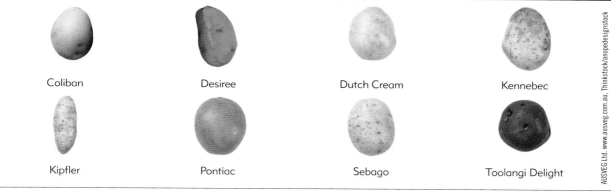

Potato varieties

Characteristics of potatoes

New potatoes
- Dug early in the season; thin, papery skin and moist flesh

Mature potatoes
- Dug three to four weeks later than new potatoes
- Skins have set, forming a protective coating, ensuring the potato's extended shelf life

Floury
- Low in moisture and sugar; high in starch
- Mash and bake well
- When fried, produce golden chips
- Tend to collapse when boiled because of low sugar content

Waxy
- High in moisture; low in starch
- When boiled, hold their shape and remain firm
- Not suited to mashing or making chips

PREPARING FRUIT AND VEGETABLES SAFELY

- Select fruit and vegetables that are not damaged. Discard any produce that is mouldy.
- Always wash fruit and vegetables before eating – do not eat from a display.
- Use serving equipment when selecting produce from cut fruit and vegetable displays.
- Packaged salads should be refrigerated.
- Refrigerate all cut and peeled fruit and vegetables within two hours of preparing.
- Wash leafy vegetables in plenty of clean, cold water.
- Store fruit and vegetables separately from meat, poultry and fish.

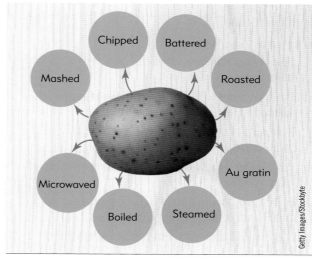

Methods of cooking potatoes

Testing knowledge

31 How should sweet potatoes be stored to maintain best quality?

32 What features should you look for when selecting green beans to buy?

33 Describe the best method of storing green beans.

34 Why is it important that the primary processing of green beans takes only 48 hours?

35 Why are green beans frozen very quickly during the freezing process?

36 Explain how modified atmosphere packaging (MAP) extends the shelf life of salad mix.

37 Why do some people think potatoes are fattening?

38 How should potatoes be stored? Why is this important?

39 Explain the differences between floury and waxy potatoes.

40 List four safe work practices that should be followed when preparing fruit and vegetables.

ACTIVITY 6·8
Investigating the best cooking method for varieties of potatoes

Worksheet

AIM
To determine the cooking method that produces the best eating properties in waxy and floury potatoes

EQUIPMENT
4 varieties of potatoes
1 steamer
1 saucepan with lid
1 container suitable for use in microwave
chopping board
vegetable knife

METHOD
1. Select four varieties of potatoes – two with waxy flesh and two with floury flesh.
2. Preheat oven to 220°C and set up steamer; fill saucepan with water and set to boil.
3. Peel and cut each of the varieties of potatoes into 1-cm cubes.
4. Divide the potato cubes of each variety between each method of cooking.
5. Brush the potato cubes to be baked with oil and bake until tender and golden brown.
6. Steam, boil and microwave the remaining potato cubes.
7. When cooked, present samples on plates for evaluation.
8. Record your results on the table.

RESULTS
Copy the table below and record your observations of the sensory properties – appearance, aroma, flavour and texture – for each variety of potato.

Potato variety	Baking	Steaming	Boiling	Microwaving

ANALYSIS
1. Which cooking method allowed the most varieties of potato to retain their shape?
2. Identify the varieties of potato that performed best for:
   - baking
   - steaming
   - boiling
   - microwaving.
3. Consider the details about varieties of potatoes on Page 116. Compare them with your results and identify those that were significantly different. If there were differences in your results, suggest reasons for this.
4. How does each cooking method affect the nutrient value of potatoes?
5. Suggest other foods that would complement each method of cooking potatoes to make up a main meal.
6. List a safety consideration for each method of cooking potatoes.

CONCLUSION
1. Make a decision about which cooking method produces the best eating properties for each variety of potato.
2. 'The sensory properties of cooked potatoes will vary depending on whether the flesh is floury or waxy, and on the method used to cook them.' Do your results support this statement? Explain your answer.
3. Make recommendations about the selection of varieties of potatoes for future recipes you might prepare.

THINKING SKILLS 6·1
COMPARE THE NUTRIENT CONTENT OF FRUIT AND VEGETABLES

1. Select two fruits and two vegetables to compare. Each of the fruits and vegetables must be from a different classification.
2. Identify the key vitamins and minerals each of the fruits and vegetables contains.
3. Identify the amount of fibre present in each fruit and vegetable.
4. Explain how the fruits and vegetables are similar and different with respect to the nutrients they contain.
5. Make recommendations about how frequently people should eat the fruits and vegetables you have compared, and explain why they are important to good health.

DESIGN ACTIVITY 6·1

A baked fruit dessert

DESIGN BRIEF

1 Design and produce a high-fibre baked fruit dessert that is suitable to serve in individual portions. Your dessert should have a fruit base and a crunchy topping, and be quick and easy to prepare.

   In your design brief, you should include information about:
   - who will eat the dessert
   - what type of dessert you will be designing
   - when the dessert will be served
   - where the dessert will be served
   - why it is important to design a dessert that is based on fruit and high in fibre.

2 After writing the design brief, use the specifications to develop five criteria for evaluating the success of the finished dessert.

INVESTIGATING

Work in small teams to complete the investigation.

1 Research the fruit that is currently in season.
2 Investigate preserved fruits, such as frozen, canned or dried fruits, that are suitable for a baked dessert.

GENERATING

1 Read the recipe for the Fruit Crumble (Page 123). You could use the quantities in that recipe as a guide for your baked fruit dessert.
2 Copy and complete the recipe map (below) for your baked fruit dessert.

PLANNING AND MANAGING

1 Write up your new recipe.
2 Prepare a food order for your baked fruit dessert design. Remember to include extras such as ice-cream, yoghurt or cream.

Recipe map for baked fruit dessert

3 Write up a production plan.
4 Make a list of the aspects of the production task that will rely on you and your partner sharing and working collaboratively.

PRODUCING

1 Make your baked fruit dessert.
2 Record any modifications you made during production on your work plan.

EVALUATING

1 Evaluate the success of your baked fruit dessert using the criteria you have developed.
2 Describe the sensory properties – appearance, aroma, flavour and texture – of your baked fruit dessert using the sensory wheel on Page 2.
3 Plot the ingredients of your baked fruit dessert on a diagram of the Healthy Eating Pyramid or the Australian Guide to Healthy Eating and comment on the health rating you would give your product.
4 Discuss the benefits of using preserved fruit as an alternative to fresh fruit.
5 Identify two safe work practices you followed when using the oven to bake your dessert.
6 What changes would you make to either the ingredients or the method if you were to make the dessert again?
7 How well did you manage your time according to your production plan? Justify your answer.
8 Explain how you worked cooperatively with your bench partner/s in:
- using the stove top
- using the oven
- washing up
- leaving your work area clean and tidy.

A vegetable parcel

DESIGN BRIEF

Wrapping vegetables in pastry is a great way to present a delicious snack or light meal. Using filo pastry keeps the fat content of the snack low and allows the designer to wrap the vegetables in a creative way.

1 Write a design brief for a parcel containing deliciously flavoured vegetables wrapped in filo pastry. It should serve one person and be able to be reheated. The brief should include information about:
- who the parcel will be served to
- what is to be prepared
- when the parcel will be served
- where the parcel will be served
- why the parcel will be served.

2 After writing the design brief, use the specifications to develop five criteria suitable for evaluating the success of the finished product.

INVESTIGATING

1 Use recipe books, magazines and the internet to research vegetables and flavourings that would be suitable to be wrapped in pastry.
2 Practise your skills by preparing the recipe for Sweet Potato Parcels on Page 130.

GENERATING

1 Sketch four ideas of ways to wrap vegetables in filo pastry.
2 Complete the recipe map on Page 121 to design your vegetable parcel recipe.
3 Sketch and annotate your two ideas for the vegetable parcel.
4 Explain why you selected the ingredient combination for the parcel.

PLANNING AND MANAGING

1 Write out your new recipe so that it is ready for production.
2 Write up a production plan.
3 Make a list of the aspects of the production task that will rely on you and your bench partner sharing and working collaboratively.

PRODUCING

1 Prepare the product.
2 Record any changes you made during production.

EVALUATING

1 Evaluate the success of your vegetable parcel using the criteria you have developed.
2 Describe the sensory properties – appearance, aroma, flavour and texture – of your vegetable parcel.

3. Did the pastry hold the filling ingredients successfully?
4. Describe any changes you would make to either the ingredients or the method if you were to make your vegetable parcel again.
5. Plot the ingredients of your vegetable parcel on a diagram of the Healthy Eating Pyramid or the Australian Guide to Healthy Eating and comment on the health rating you would give your product.
6. How effective was the timing you outlined in your production plan? Justify your answer.
7. Explain how you worked cooperatively with your bench partner/s in:
   - using the stove top
   - using the oven
   - washing up
   - leaving your work area clean and tidy.

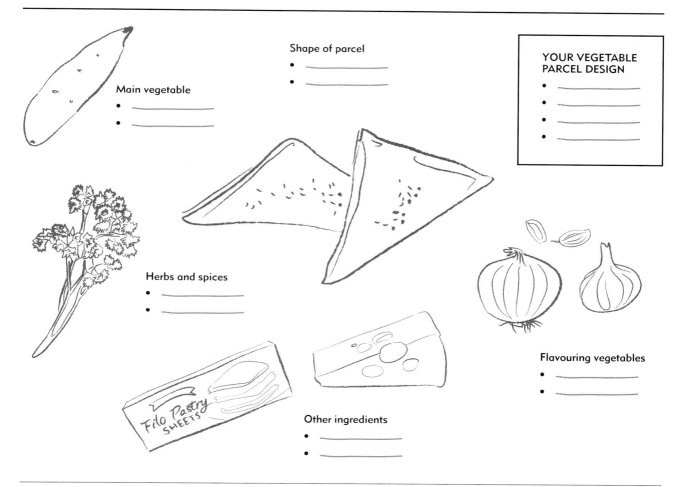

Recipe map for vegetable parcel

STEWED APPLES

2 Granny Smith apples
¼–⅓ cup water
1–2 tablespoons sugar
1 clove

SERVES TWO

METHOD

1. Wash and peel apples and cut into quarters.
2. Remove cores and slice apples thinly.
3. Place all ingredients in a small saucepan. Stir to coat the apple with all the ingredients.
4. Bring to boil, place lid on saucepan and reduce heat to simmer. Stir occasionally.
5. Cook until apple is tender.

EVALUATION

1. Define 'stewing'.
2. Why is it important to wash the apples before cooking?
3. Which knife would be most suitable for cutting and slicing the fruit? Describe the safest way to slice the apples.
4. Why is it important to keep the lid on when stewing apples?
5. List some other fruits that could be stewed successfully.

FRUIT CRUMBLE

1 cup fruit – stewed, canned or freshly sliced

Crumble topping

2 tablespoons sugar or brown sugar

2 tablespoons coconut or almond slivers

2 tablespoons rolled oats or fresh breadcrumbs

pinch of nutmeg and cinnamon

2 tablespoons self-raising flour or wholemeal self-raising flour

20 grams butter

SERVES TWO

METHOD

1. Preheat oven to 180°C. Lightly butter two ovenproof ramekins.
2. Divide the fruit evenly between the two ovenproof dishes.
3. Mix all dry ingredients in a bowl.
4. Either melt butter in microwave and stir into dry ingredients or, using fingertips, rub butter into dry ingredients until the mix resembles fresh breadcrumbs.
5. Sprinkle the fruit with crumble topping. It should have a rough surface, not smooth.
6. Bake for 15–20 minutes until golden brown.

EVALUATION

1. List four examples of fruits that would be suitable to use in the crumble recipe.
2. Identify the dry ingredients you used for your crumble and explain why they were not sifted into the bowl, just mixed together.
3. Describe the sensory properties – appearance, aroma, flavour and texture – of your finished crumble.
4. Identify two safe work practices you followed when using the oven.
5. If you were to repeat this recipe, what changes would you make to either the ingredients or the production method?

CORNISH PASTIES

Shortcrust pastry

1 cup plain flour

1 cup self-raising flour

pinch of salt

125 grams butter

¼ teaspoon lemon juice

⅓–½ cup water

Filling

250 grams lean minced beef

1 medium onion, finely chopped

1 medium potato, grated

½ carrot, grated

2 tablespoons frozen peas

2 teaspoons parsley, chopped

salt, pepper

Glaze

1 small egg

2 tablespoons milk

MAKES SIX PASTIES

METHOD

1. Sift flours and salt.
2. Chop the butter into small pieces and rub into the flour, using fingertips, until the mixture resembles breadcrumbs.
3. Make a well in the centre of the mixture and add lemon juice and sufficient water to make a firm dough. Use a spatula or knife to carry out this process.
4. Place the dough on a bench lightly sprinkled with flour and knead lightly until smooth.
5. Press the dough into the shape of a disc and cover with plastic wrap until the filling is completed.
6. Preheat oven to 200°C.
7. Prepare the glaze by beating the egg and milk together. This mixture can also be shaken in a lidded jar.
8. Break up the meat with a fork and add the prepared vegetables, parsley, salt and pepper.
9. Divide the pastry into six equal portions and roll out each to the size of a saucer.
10. Divide the filling into six portions and place one portion in the centre of each circle of pastry.
11. Using a pastry brush, brush the edges of one half of each circle of pastry with the egg and milk glaze.

2

3

4

10

12

12. Pick up the edges of each circle and join by pressing together to form a half-circle. Pinch the edges of the half-circle and place on an oven tray.

13. Prick the side of the pasty with a fork to form a small vent. Glaze with the egg and milk, avoiding the frilled edge.

14. Bake at 200°C for 10 minutes, then reduce the oven to 180°C for 30–35 minutes.

15. Serve with tomato sauce.

EVALUATION

1. Why is it important to sift the flours and salt together?
2. How will you know when the butter is rubbed in sufficiently?
3. Explain what could happen if too much water was added to the flour.
4. Why is it important to knead the pastry lightly?
5. How could you check to see if the six quantities of pastry and filling were equal?
6. Explain how you would test the pasties to tell if they were cooked.
7. Plot the ingredients of your Cornish Pasties on a diagram of the Healthy Eating Pyramid or the Australian Guide to Healthy Eating and comment on the health rating you would give the pasties.

PUMPKIN NOODLE SOUP WITH HERB AND GARLIC BREAD

Pumpkin Noodle Soup

400 grams butternut pumpkin (approx. 300 grams after peeling and seeding)

1 onion, diced

1 medium potato, peeled and diced

2 cups water

½ packet regular chicken noodle soup

ground pepper

½ tablespoon chopped parsley

SERVES TWO

METHOD

1. Carefully peel the pumpkin and remove the seeds. Cut into small pieces.
2. Place the pumpkin, onion and potato in a saucepan with the water and bring to the boil.
3. Add the chicken noodle soup mix.
4. Simmer for approximately 30 minutes or until the pumpkin is tender.
5. Puree the soup in a blender or using a stick mixer.
6. Season to taste with pepper.
7. Sprinkle with a little chopped parsley to garnish.
8. Serve hot with herb and garlic bread.

Herb and Garlic Bread

1 bread roll

1 teaspoon parsley

1 teaspoon chives

1 clove garlic

15 grams butter, softened

SERVES ONE

METHOD

1. Preheat oven to 200°C.
2. Cut the bread roll into slices nearly all the way through.
3. Finely chop the parsley and chives.
4. Crush the garlic.
5. Mix the softened butter with the herbs and garlic.
6. Spread one side of each bread slice with the herb and garlic butter.
7. Wrap the bread in aluminium foil.
8. Heat through in the oven for approximately 10 minutes until warm.

EVALUATION

1. Describe how to peel the pumpkin safely.
2. Describe an alternative method of cooking the pumpkin noodle soup other than cooking it on the top of the stove.
3. How else could you puree the soup if you didn't have a blender or stick mixer?
4. Describe the sensory properties – appearance, aroma, flavour and texture – of the Pumpkin Noodle Soup.
5. Describe two other accompaniments other than the herb and garlic bread that could be served with the soup.
6. Use your knowledge of nutrition to analyse the Pumpkin Noodle Soup with Herb and Garlic Bread as a light meal. Outline some suggestions of other foods you could add to make it a more substantial meal.

Mark Fergus Photography

SHEPHERD'S PIE

Filling

2 teaspoons oil

½ onion, finely diced

125 grams minced steak

⅓ carrot, grated

1 teaspoon parsley, finely chopped

1 tablespoon tomato paste

1 tablespoon tomato sauce

¼ cup beef stock

½ teaspoon Worcestershire sauce

¼ teaspoon mixed herbs

pepper

MAKES TWO PIES

Topping

2 large potatoes

2 tablespoons milk

2 teaspoons butter

pinch salt

40 grams cheese, grated

Mashed potato makes a delicious topping on this pie, and is much lower in fat than traditional pastry.

METHOD

1. Peel potatoes and cut into even-sized pieces. Place in a saucepan with sufficient water to cover. Cook with lid on for approximately 15 minutes until tender.
2. Heat oil in a saucepan. Add onions and fry lightly.
3. Add the minced steak and cook until brown. Stir constantly to prevent the meat clumping together.
4. Add carrot, parsley, tomato paste, tomato sauce, stock, Worcester sauce, herbs and pepper. Simmer for approximately five minutes until the mixture thickens slightly.
5. Preheat oven to 200°C.
6. Drain the potatoes. Mash and add milk, butter and a pinch of salt.
7. Place meat mixture in foil containers and spread mashed potatoes on top. Sprinkle with grated cheese.
8. Place on a tray and bake for 10 minutes or until golden brown.

EVALUATION

1. Why is it important to cut the potato into even-sized pieces when preparing it for mashed potatoes?
2. Why are milk and butter added to potato mash?
3. What are the health benefits of not using pastry in this pie?
4. Describe the sensory properties – appearance, aroma, flavour and texture – of your finished pies.
5. Identify two safe work practices you followed during production.

CITRUS CORDIAL

1 orange
1 lemon
½ cup sugar
300 millilitres water

METHOD

1. Wash and dry the orange and lemon.
2. Remove the zest from the orange and lemon using the smallest section of the grater. Use a pastry brush to remove the zest, not a knife or metal spatula.
3. Juice the fruit and mix with the zest and sugar.
4. Bring the water to the boil and pour over the fruit mixture. Stir until the sugar has dissolved.
5. Cool for 15 minutes, then strain through a fine sieve. Store in the refrigerator for one week.

FRUIT PUNCHES

Raspberry Punch

100 grams raspberries (fresh, frozen or canned)
200 millilitres raspberry cordial
½ lemon, juiced
1 litre soda water

SERVES FOUR

METHOD

1. Combine raspberries, cordial and lemon juice and chill.
2. Just before serving, place soda water and other ingredients in a large jug.

Sparkling Pineapple Punch

1 orange
1 lemon
2 tablespoons crushed pineapple
2 cups lemonade
1 cup pineapple juice

Garnish

½ lemon, sliced
1 tablespoon sugar coloured with 1 drop of green colouring
4 ice cubes
2 glacé cherries
2 toothpicks

SERVES TWO

METHOD

1. Rub the rims of two tall glasses with a slice of lemon and dip in the coloured sugar. Allow to dry.
2. Juice the orange and half the lemon and place juice in a jug.
3. Add crushed pineapple, lemonade and pineapple juice to the jug.
4. Place two ice cubes in each glass and pour in the mixture from the jug. Take care not to dissolve the frosted rim of the glass.
5. Spear a cherry with the toothpick, attach it to a slice of lemon and hang it on the side of the glass.

EVALUATION

1. Why is raspberry cordial added to the Raspberry Punch?
2. What would be the function of frozen raspberries in the punch?
3. What effect does the lemon juice have on the flavour of the punches?
4. How would crushed ice change the texture of the Sparkling Pineapple Punch?
5. Suggest other fruits that could be used in place of the pineapple in the Sparkling Pineapple Punch.

SWEET POTATO PARCEL

150 grams sweet potato
¼ leek
10 grams butter
2 teaspoons flour
¼ teaspoon mustard powder
pinch of cinnamon
½ cup milk
4 sheets filo pastry
2 teaspoons oil
1 teaspoon sesame seeds

🍴 MAKES TWO PARCELS

METHOD

Steaming the sweet potato

1. Half-fill the base of the steamer with hot water. Place steamer on top of the base and put on lid. Bring to the boil.
2. Wash and peel the sweet potato. Cut out any blemishes and cut into 2 cm chunks.
3. Place in the steamer, cover with the lid and steam for 10–15 minutes or until just tender, so the potato holds its shape.
4. Cool.

Making the sweet potato parcel

5. Preheat oven to 200°C.
6. Wash and trim the leek, discarding the green leaves, as they have a bitter flavour. Finely dice.
7. In a small saucepan, melt the butter, then sauté the leek until it is transparent. Do not brown.
8. Add flour, mustard powder and cinnamon, stirring over heat for 30 seconds.
9. Remove from heat and stir in milk gradually.
10. Return to heat. Bring to boil, stirring constantly. Cook for 1 minute or until the mixture thickens. Remove from heat and add cooked sweet potato. Cool.
11. Carefully brush one sheet of filo pastry with the oil, then lay the next filo pastry over the top of the first. Repeat with the other two sheets of pastry.
12. Divide the sweet potato mixture into two.
13. Spread one half of the sweet potato mixture along the one short end of the pastry.
14. Fold in the two long edges and roll into a parcel. Brush with oil and sprinkle with sesame seeds.
15. Repeat with the other two sheets of pastry and the remaining filling.
16. Bake in the preheated oven for 15 minutes or until golden brown. Serve with green vegetables or a salad.

EVALUATION

1. Identify an alternative method of cooking the sweet potatoes so they retain their shape.
2. Identify two safety considerations when using a steamer.
3. Why was it important not to brown the ingredients in step 7 of the recipe?
4. Why do you have to work quickly when using filo pastry?
5. Discuss the success of the production of your Sweet Potato Parcel.

7 GRAINS ARE GREAT

KEY KNOWLEDGE

- Cereals
- The structure of a wheat grain
- Wheat: from paddock to plate
  - Conventional wheat production
  - Sustainable wheat production: no-till farming
- Cereals for good health
- Gluten: the protein in wheat flour
- Preparing yeast doughs
  - Types of yeast
  - Processes in yeast baking
  - Top tips for baking yeast doughs
- Bread
  - Flatbreads
  - Labelling of bread
  - Storage of bread
- Couscous
  - Couscous for good health
- Rice
- Rice: from paddock to plate
  - Sustainable rice production
  - Rice for good health
  - Varieties of rice
  - Top tips for cooking rice

KEY TERMS

cereals edible seeds of certain grasses, including millet, oats, barley, wheat, rye, rice and corn

dextrinisation the process that occurs when the starch in flour is exposed to dry heat and is broken down into dextrin, resulting in a change in colour to golden brown

fermentation the process of yeast growing and reproducing by budding, then converting carbohydrates into carbon dioxide, alcohol and water

gelatinisation the process that occurs when starch granules in the endosperm of cereals absorb liquid in the presence of heat and thicken the liquid

gluten the main protein in wheat flour

kneading a process in which air bubbles are evenly distributed and the gluten strengthened in a yeast dough

no-till farming involves leaving the stubble from last year's crop to enrich and stabilise the soil. The new crop is planted by direct drilling in between the rows of the previous crop without tilling the soil

proving a process in which a yeast dough is rested to allow time for fermentation to take place – the gas bubbles are trapped in the structure, so the volume of the dough increases

yeast a single-celled, microscopic fungus

AUSTRALIAN CURRICULUM LINKS

DESIGN AND TECHNOLOGIES
- Knowledge and Understanding
- Processes and Production Skills

HEALTH AND PHYSICAL EDUCATION
- Personal, social and community health

GENERAL CAPABILITIES
- Critical and creative thinking
- Intercultural understanding

CROSS-CURRICULUM PRIORITIES
- Sustainability

CEREALS

Cereals, which are the edible grains or seeds of grasses, have been important as a source of food throughout the world since the earliest origins of humankind.

Wheat, rice, barley, oats, rye and maize (corn) are all cereals that are processed in some way for us to eat. Wheat covers more of the earth's surface than any other grain crop, and it is the staple grain food for much of the earth's population. However, specific varieties of cereals are cultivated in each of the world's regions due to being better suited to the climate and soil of specific areas. Due to their wide availability, cereals are the staple foods of many regions, and are often eaten several times a day.

Australia has traditionally been well known throughout the world for producing high-quality wheat. Today, Australian farmers are also competing on the world market with small quantities of top-quality rice.

Wheat
Used for breads, pasta, noodles, cakes, biscuits, extruded snack foods, bulgur and thickening agents

Rice
Used as cooked grains in savoury dishes, puddings, breakfast foods, biscuits, rice cakes, and extruded snack foods

Maize (corn)
Used as a fresh vegetable; the kernels can also be canned, frozen or dried. Used in breakfast cereal, polenta, tortillas, corn oil, corn syrup and popcorn

Barley
Used in breakfast cereal and as a thickening agent in soups and casseroles

Oats
Used as porridge (rolled oats, oatmeal) and in muesli and oatcakes

Rye
Used to make leavened bread and crispbreads

Cereals and their uses

THE STRUCTURE OF A WHEAT GRAIN

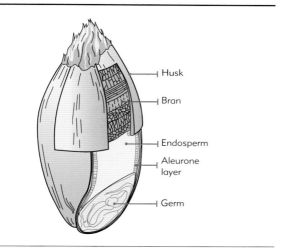

All cereal grains are made up of three main parts.
- The bran, the cover or outer layer of the grain, which is made up mainly of dietary fibre
- The endosperm, the main body of the cereal grain, which is composed almost entirely of starch
- The germ, which contains the nutrients needed for a new plant to grow – protein, fat, and some vitamins, and minerals and carbohydrates

Cross-section of a cereal grain

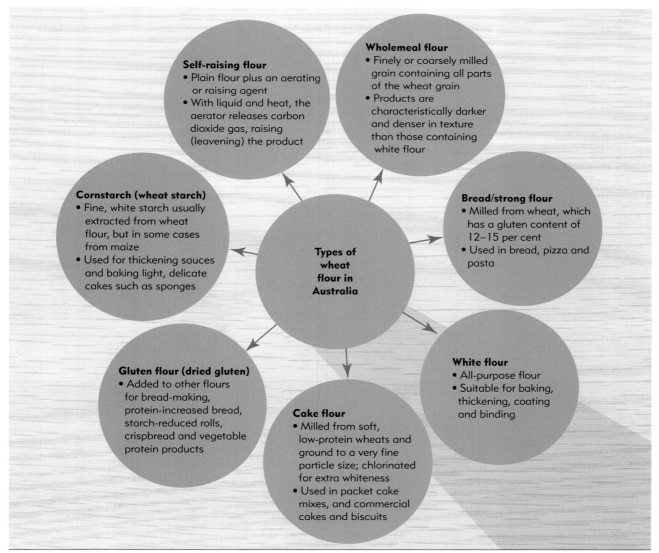

Types of wheat flour in Australia

WHEAT: FROM PADDOCK TO PLATE

Conventional wheat production

Ploughing – the soil in the paddock is ploughed in preparation for sowing the crop in autumn.

Sowing – the grains are sown evenly and efficiently using large machinery. Sowing usually takes place in autumn after rain.

Spraying – the crop is sprayed to reduce the growth of weeds and fungi that reduce the final yield.

Harvesting – the growing season is completed by early summer, when farmers harvest the crop. The harvester strips the head of the wheat stalks and separates the grains from the chaff.

Augering – the wheat is augured into trucks ready for transporting to the silo.

Storage – the grain is trucked to large silos or country receival points, where it is weighed and the quality of the grain is measured for moisture content and protein quality. The amount of broken grain is also calculated.

Milling – in flour mills, the grains are crushed and processed to different degrees to produce ingredients such as white flour, wholemeal flour and semolina.

Products made from wheat flour include breads, cakes, biscuits, pastry, pasta, scones, noodles, breakfast cereals, pizza crusts, crumpets, and snack foods such as pretzels.

Sustainable wheat production: no-till farming

Australian wheat farmers have an international reputation for producing high-quality wheat that is in high demand throughout the world. In recent years, many wheat farmers have worked towards developing strategies to make their cereal production more sustainable and environmentally friendly. Sustainable farming involves farming practices such as no-till farming that maintain the land's productivity so that it will be available for future generations. No-till farming is a method of producing cereal crops such as wheat and canola that has significant environmental advantages. This method of farming also provides economic advantages for farmers, who state that when using it, their crops provide higher yields, especially during dry years.

No-till farming is a revolutionary method of farming that involves leaving the stubble from the previous year's crop to enrich and stabilise the soil. The new crop is planted by direct drilling in between the rows of the previous crop, without tilling the soil. This differs from conventional farming methods that involve ploughing or tilling the land, which are the first stages in preparing the soil before the crop is planted.

Direct drilling seed

Retaining stubble after the harvest

Leaving the stubble from the previous crop in the soil after harvest holds significant benefits for soil health. 'Stubble' refers to the stalks of the cereal crop, which are left once the heads have been cut off during harvesting. As the remaining stubble breaks down, the nutrients it contains are returned to the soil. The stubble also provides a layer of mulch on the surface of the soil, ensuring that moisture is retained, especially in dry years. A further advantage of retaining stubble on the land is that the roots of the stubble hold the soil in place, minimising wind erosion.

Along with leaving stubble in the soil, no-till farmers use GPS technology to establish a system of controlled traffic lanes, or 'tramlines', that their large machinery follow when sowing and harvesting crops. Using designated traffic lanes for machinery means that the soil is not crushed or compacted, but remains moist. GPS also enables farmers to use precision sowing; that is, to sow crops in between rows from the previous year's crop.

CEREALS FOR GOOD HEALTH

The message from nutritionists is to eat more breads and cereals. Cereals are a plant food, are found in the foundation layers of the Healthy Eating Pyramid and are recommended by the Australian Guide to Healthy Eating. Nutritionists recommend that you eat six serves of cereal each day.

All the cereal grains are similar in nutrient value, so it doesn't matter whether it is wheat, rice or corn – they all provide a similar range of important nutrients. The nutrient value of the end product will largely depend on the amount of processing involved. For example, the removal of the bran and germ layer during milling to produce white flour reduces the end product's vitamin, mineral and fibre content.

The most nutritious cereals are wholegrain cereals, which are an excellent source of carbohydrates and dietary fibre, yet are low in fat. Wholegrain cereals also provide your body with small amounts of protein, B group vitamins and minerals. The bran of cereal grains is made up almost entirely of dietary fibre. Dietary fibre is important in the diet because it absorbs water, and therefore adds bulk and helps food to move through the digestive tract.

GLUTEN: THE PROTEIN IN WHEAT FLOUR

Gluten was originally a Latin word meaning 'glue'. Wheat flour has two major components, starch and protein, and its main protein is gluten. Gluten in wheat flour is made up of two component proteins, glutenin and gliadin.

The protein gluten is primarily found in wheat, although there is also some gluten in other cereal grains such as rye, barley and oats. Gluten is the component of flour that helps to form the structure of bread and other cereal products. In the bread-making process, the gluten in the dough stretches to form a skin that holds the bubbles of carbon dioxide (just like bubblegum) that are formed by the action of yeast. Without gluten, the carbon dioxide would escape, and the dough would not rise.

Hard or strong wheats contain the most gluten, and so are best for making bread. Gluten flour can be purchased at health food stores and added to other flours to make a strong dough. It is a creamy, greyish powder that has been separated from wheat endosperm and is made up of strands formed by about 20 different amino acids. Each strand has a helix formation, which makes it look like a tiny spring, and after water is added, a complex three-dimensional network is formed.

When bakers mix and knead a bread dough, they are said to be 'developing the gluten', because the strands of gluten are being straightened and overlapped. This means the 'springs' of gluten must be uncoiled, and the resulting straightened strands must then be recombined into a continuous, overlapping mesh. The new structure forms a three-dimensional network capable of retaining the carbon dioxide produced by the yeast during fermentation.

Random gluten in flour before hydration

Three-dimensional network of hydrated gluten in dough

PREPARING YEAST DOUGHS

Yeast is used as a leavening agent in many food products. To leaven a mixture means to add an ingredient such as yeast, baking powder, bicarbonate of soda or beaten egg white to make it rise. Yeast is used in dough to make bread, buns, pizzas and some batters to make them rise, so that the end product has a light, cellular texture.

Yeast is neither plant nor animal, but a single-celled microscopic fungus. As yeast grows, it reproduces by budding; during this fermentation, the yeast converts the carbohydrates in the dough to a gas called 'carbon dioxide', plus alcohol and water.

Three important conditions are needed to allow the fermentation of a yeast dough: warmth, moisture and food. It is also important to cover the dough with plastic wrap or a damp tea towel as the dough is proving. This will prevent the dough from drying out and forming a crust as it ferments.

The carbon dioxide aerates the dough by forming bubbles that stretch the gluten in the flour to create the structure of the dough. The alcohol and water evaporate during baking.

Fermentation is utilised in bread-making to produce carbon dioxide, which raises bread.

$$C_6H_{12}O_6 \text{ (glucose)} + \text{yeasts} \rightarrow 2CO_2 \text{ (carbon dioxide)} + 2C_2H_5OH \text{ (alcohol)} + \text{energy}$$

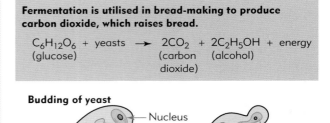

Reproduction of yeast cells

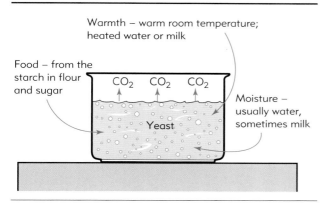

Ideal environment for the growth of yeast

Yeast fermenting

Types of yeast

Yeast can be purchased fresh as a compressed block, or dried in granules. Fresh yeast should be creamy in colour and smell sweet. It should not be dry, grey or crumbly. It will store in the refrigerator for about two weeks. In recipes using fresh yeast, the yeast is mixed with a small amount of flour and water and left in a warm place for the 'sponge' to set. The sponge is later incorporated into the dry ingredients. Active dried yeast is purchased in the form of tiny, dehydrated granules. The yeast cells are alive but dormant, due to the lack of moisture. This form of yeast comes in small sachets and is usually dissolved in lukewarm water before being added to the dry ingredients; it can also be added directly to the dry ingredients if time is limited. The yeast granules should be stored in the refrigerator or freezer.

ACTIVE DRY YEAST

- Looks like tiny, dehydrated granules
- Available in sachets or large packets
- Stored in the refrigerator or freezer
- The yeast cells are alive but dormant, because of the lack of moisture
- When mixed with a warm liquid, the cells are activated and begin to grow and produce alcohol and carbon dioxide

COMPRESSED FRESH YEAST

- Has a creamy colour, is firm but with a moist texture, and has a sweet aroma
- Purchased in blocks
- Must be stored in an airtight container in the refrigerator because it becomes stale quickly
- Loses its ability to leaven a dough as it becomes stale

ACTIVITY 7.1

The growth of yeast

Try these tests to determine the ingredients and the environment that yeast requires to grow and produce good bread. You will need:

- 8 test tubes
- 8 balloons
- 8 × 1 teaspoon yeast
- 6 × ½ teaspoon sugar
- 4 × 1 teaspoon flour
- 2 × 1 teaspoon bread improver
- 4 × 50 millilitres warm water
- 6 × 50 millilitres iced water.

METHOD

1. Put 1 teaspoon of yeast, then 100 millilitres of iced or warm water, into each test tube, according to the diagram opposite.
2. Follow the diagram opposite to add the other ingredients to each of the test tubes. Stir each test tube well.
3. Stretch a balloon over the neck of each test tube and watch what happens after five, 10, 15 and 30 minutes.

RESULTS
Record your results.

CONCLUSION

1. After observing each test, determine the ingredients and environments in which yeast grows best.
2. What are the results' implications about the temperature of the environment required for making bread?

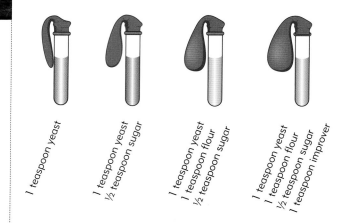

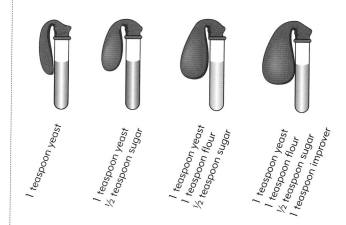

The growth of yeast

Processes in yeast baking

- **Fermentation** – sugars and some of the starch are converted to alcohol and carbon dioxide as the yeast ferments.
- **Kneading** – makes the gluten, the protein in wheat flour, more elastic and strong so that it can and begin to stretch and capture the carbon dioxide bubbles.
- **Proving** – gas bubbles are trapped in the structure of the dough. Gluten from the flour stretches as the yeast grows and the size of the dough increases.
- **Kneading and shaping** – evenly distributes the air cells to make an even-textured end product.
- **Proving** – the dough increases in size as the yeast continues to grow.
- **Baking** – the high temperature kills the yeast and evaporates the alcohol. The structure of the dough is set and the outside becomes golden brown.

Top tips for baking yeast doughs

1. Use flour with a high gluten content so that the dough can capture all the bubbles of carbon dioxide produced by the yeast during fermentation.
2. Prove the dough in a warm environment that is free from draughts. Cover with oiled plastic wrap or a damp tea towel to prevent the surface of the dough drying out and forming a crust.
3. Prove shaped rolls close to each other on the tray so that they can support one another and not collapse during proving.

4 Spray the dough with water just before it goes into the oven – this helps the dough to stay moist so it can expand at the start of baking.

5 Cook yeast doughs in a hot oven.

6 The dough is cooked if it is golden brown and sounds hollow when tapped. The browning on the outside of the loaf occurs as a result of **dextrinisation**. This is the process that occurs when the starch in the flour is exposed to dry heat and is broken down into dextrin, resulting in a change in colour to golden brown.

1 Water is added to a mixture of flour, salt, sugar, fat, yeast and bread improver.

2 Water is absorbed by the starch and protein in the flour.

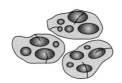

3 Gluten in the flour gives the dough elasticity by forming elastic films around tiny gas pockets, stopping them from combining.

4 The leavening (fermenting) of dough occurs when carbon dioxide is released from the sugars by yeast.

5 Kneading, stretching and folding develop the gluten.

6 Proving allows the yeast to ferment. The carbon dioxide increases and the dough rises as the gluten allows the dough to stretch.

7 The heat of the oven causes the expanded dough to become self-supporting and for dextrinisation to occur.

Stages in the preparation of a yeast dough

Testing knowledge

1 Define 'cereals' and list some examples of cereals.

2 Sketch and label the main sections of a grain of cereal.

3 Draw a flow chart that illustrates the stages in the primary and secondary processing of wheat grown using conventional farming methods.

4 Define 'no-till farming'.

5 Outline three reasons why farmers may use this method of producing cereal crops.

6 Explain why cereals are important in the diet.

7 Discuss the function of dietary fibre found in cereals in the human body.

8 Define 'gluten' and explain why gluten is an important component of the flour that is used to make bread.

9 Outline the three aspects of the environment that need to be considered when fermenting yeast.

10 Develop a simple concept map that provides some tips for making a successful yeast dough.

BREAD

Bread is one of the oldest and most diverse foods. Throughout the ages, bread has been an important staple food for many people, and has often been referred to as 'the staff of life'. Many types of flour have been used to make bread throughout the centuries. However, only the flour from wheat and, to a lesser degree, from rye, can produce dough that is capable of holding the leavening gases produced by yeast sufficiently well to yield well-risen loaves with a fine, soft cellular structure.

The physical properties – that is, the shape, size and texture – of bread made from wheat flour are the result of the presence of gluten in the flour. All the processes of bread-making – fermenting, kneading and proving – involve changing and improving the natural properties of gluten to make the dough strong enough to hold the bubbles of carbon dioxide that are produced during fermentation.

In traditional bread-making methods, this gluten modification occurs over several hours during fermentation of the dough. A 12-hour fermentation period may be required when yeast activity is low. Modern bread-making processes rely on other means of modifying the gluten and enable good-quality bread to be made in two hours.

Dough development can be hastened by the presence of very small quantities of oxidising agents known as bread improvers. Some flours contain gluten that is too tough and strong. If this is the case, improvers are added to soften the flours in order to make a loaf with soft texture.

Flatbreads

Some of the world's oldest and simplest breads are flatbreads. Flatbreads are usually quick to make, and are cooked either on a stovetop, under a griller or in a very hot oven. They may not use yeast as a raising agent.

Many different baking techniques are used to make flatbreads, although the basic steps are the same as for loaf bread. The formula is very basic, using flour, water and salt.

Flatbreads can be crisp or chewy, plain or rich. Traditional flatbreads come in many shapes, sizes and flavours. In some cultures, they are used to wrap around food and replace plates and cutlery. For example, in Greece, pitta bread is used as a wrap for meat kebabs, and in India, naan and chapatti are designed so that people can scoop food from their plates to their mouths.

Various types of bread

Types of flatbreads

NAME	COUNTRY/REGION	FLOUR (OR SIMILAR)
Pitta	Middle East	Wheat
Lavash	Middle East	Wheat
Naan	India	Wheat
Chapatti	India	Wheat
Puri	India	Wheat
Pappadam	India	Lentils; rice; potato
Tortillas	Mexico	Corn
Branch bread	Scandinavia	Wheat; rye
Griddle oatcakes	Scotland	Oats
Roti	Malaysia	Wheat

Flatbreads

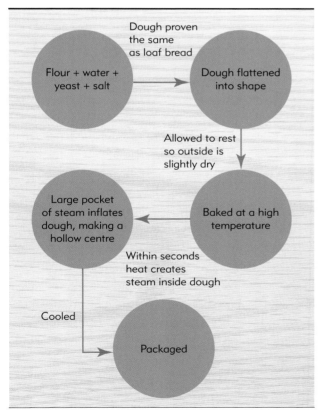

Production of pocket bread

ACTIVITY 7.2

Taste testing different types of bread

1. Taste test a range of different types of bread, including:
   - white
   - wholemeal
   - multigrain
   - rye blend
   - sourdough
   - flatbread.

   Focus on the sensory properties – appearance, aroma, flavour and texture – of each type of bread. Record your results in a table.

2. Rate the breads from the one you liked the most to the one you liked the least. Justify your ratings.

3. Predict which bread has the highest fibre content. Explain the criteria you used to make your decision.

4. With a partner, brainstorm and record why there is such a large variety of breads available today.

5. If you were responsible for buying bread for your family, which type would you choose? When writing your response, consider sensory properties, nutritional value and the way that bread is used in your home.

Labelling of bread

All food manufacturers are required by Food Standards Australia New Zealand (FSANZ) to comply with food labelling laws for bread. There are two main methods of food date-marking that food manufacturers are required to use:

1. use-by date – for foods that should not be consumed after a certain date for health and safety reasons

2. best before date – for foods that have a shelf life of less than two years, but may still be safe to eat after this date, even though they may have lost some quality or nutrient value.

Because the freshness of bread is important to consumers, FSANZ has developed additional standards for marking the date on bread labels. As well a use-by or best before date, bread and other baked products can also be labelled with two other dates if the bread has a shelf life of less than seven days:

- 'baked for' date – for bread that has been baked up to 12 hours before the marked date
- 'baked on' date – for bread that has been baked on the marked date.

Storage of bread

To ensure that bread will keep fresh for as long as possible, it is best to store it in the pantry in a well-ventilated bread crock or breadbin. Heavier or denser breads such as sourdough breads will usually keep fresh for longer.

To prevent bread from going mouldy, it is very important to keep the bread container clean and free of stale crumbs and bread scraps. Today, many bread manufacturers use additives such as mould inhibitors to prevent their bread from going mouldy.

Some types of bread can be kept in a refrigerator in a well sealed plastic bag or container. Bread is also suitable to freeze, and can be frozen for up to four months. The bread should be placed in a freezer bag; as much air as possible should be removed from the bag before sealing. To thaw, the seal should be removed and the bread left to defrost naturally.

Flatbreads freeze very well and are easy to warm up by sprinkling with a little water before heating in the oven.

COUSCOUS

Couscous is a grain product made from semolina, the coarsely ground endosperm of wheat, with the addition of some wheat flour. It has been used as a staple food in North Africa and some Middle Eastern countries since the earliest times in recorded human history. Today, couscous is becoming more popular because it is a very versatile food that makes a great alternative to pasta or rice. It is a light and fluffy grain that is perfect for serving with spicy foods or vegetable dishes, and is equally delicious served as a pilaf.

The precooked, instant form of couscous that is available in supermarket is much simpler to prepare than the traditional method, which involves hours of preparation.

Couscous

Couscous for good health

Couscous is made from the endosperm of wheat; therefore it is high in carbohydrate and is a good source of energy. Couscous is also high in vitamin B1 (thiamin) as well as vitamin B3 (niacin). Another advantage of incorporating couscous into the diet is that it is naturally low in fat.

ACTIVITY 7.3

Getting to know couscous

1. Work with a partner to prepare one of the instant couscous packaged mixes available in supermarkets.
2. While the couscous is cooking, examine the packaging and record the ingredients that are included in the product.
   a. What hints for serving have been included on the packaging?
   b. In which country was the couscous made?
   c. Why are the cooking instructions given in several languages?
3. Once your couscous is cooked, taste test the product. Rate the taste as excellent, very good, good, fair, or unacceptable.
4. List the other ingredients that have to be added to the couscous so that it cooks successfully.
5. Do you think the product would be improved by the addition of any other ingredients? What else could you add to improve the flavour of the couscous?

RICE

Rice is one of the most important staple foods eaten throughout the world. More than half the world's population eats rice on a daily basis. Rice has been harvested for thousands of years, and recent archaeological finds indicate that rice was present in China from around 3000 BCE. Rice is particularly important in many Asian cultures, where it often forms the focal point of every meal. In Japan, many shrines to the rice god, Inari, dot the countryside, and in India and China, when young couples are married they are showered with rice to ensure fertility. The throwing of confetti at weddings stems from this tradition.

Approximately 40 000 varieties of rice are grown throughout the world. The main rice-growing areas are found in South-East Asia, Africa, the Middle East, Australia, North America, South America and

Europe. In Australia, the main rice-growing areas are in the Riverina region of New South Wales and in the Murrumbidgee area near Leeton because of the suitable clay-based soils and the availability of water in these regions. Recently, rice-growing has also been introduced in the Ord Valley in Western Australia, as well as in parts of the Northern Territory.

RICE: FROM PADDOCK TO PLATE

Sustainable rice production

One of the key issues facing sustainable rice production is that rice farmers use large volumes of water to irrigate their rice paddies. In recent years, Australian rice growers have become aware of the importance of developing environmentally sustainable systems to produce rice. To overcome many of the environmental problems they face, rice growers have developed a variety of strategies to improve water efficiency on their farms. The latest research shows that Australian-grown rice now requires 50 per cent less water than rice grown by other major rice producers. Some of the strategies rice producers have introduced to improve the sustainability of their farming practices are to:

- move towards irrigation systems that use less water than flood irrigation. The reduction in flood irrigation also has the effect of lowering water tables and minimising salinity, as less water is used. Greenhouse gas emissions that are linked to flood irrigation are also reduced through the use of more efficient watering systems
- introduce laser levelling technology to level the ground where rice is to be grown, ensuring that water is evenly distributed
- use new technology to measure the amount of water that the rice crop requires, so that water is only released to the crop as it is needed. This gives farmers precise control of water both on and off the paddock, and has led to a 60 per cent reduction in the amount of water used in rice production
- develop closed rice production systems that recycle water and keep the water and nutrients on the property
- use high-yielding, shorter-season rice varieties that require less water to grow.

Rice for good health

Rice is high in carbohydrates, low in fat, sugar and salt, is a good source of B group vitamins and is gluten-free. Rice is found in the cereals section of the Australian Guide to Healthy Eating.

Rice growing in Western Australia

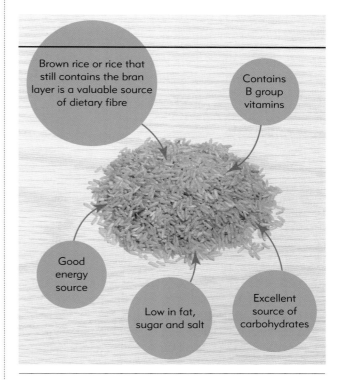

Nutrient value of rice

As well as being part of a variety of meals, rice is also used in a wide range of processed foods such as breakfast cereals, noodles, snack foods and crisp biscuits.

Varieties of rice

Rice is sometimes classified according to its colour. Brown rice has had its outer husk removed, but retains the bran, while white rice has undergone more processing stages, has been polished during milling, and has had its bran and germ removed.

There are thousands of different varieties of rice, but these are usually classified into three main groups.

1 Long-grain rice: the grains remain separate when cooked and are light and fluffy in texture. Long-grain rices such as basmati and jasmine are generally used for savoury dishes such as pilafs, and as accompaniments to curries. They may be either polished (white) or unpolished (brown or wholemeal)

2 Medium-grain rice: the grains are slightly rounder in shape than long-grain rices, stick together when cooked and are moist in texture. The grains of rice cling together more than long-grain rices, but are still separate when cooked. Chinese people generally prefer to use medium-grain rices in their cooking. Arborio rice is a medium-grain rice that is soft in texture; it is ideal for preparing risotto because it can absorb a lot of liquid, such as stock, without becoming gluey

3 Short-grain rice: the grains are much rounder in shape than long- and medium-grain rices. Short-grain rice is moist, tender and sticky when cooked. Traditionally, short-grain rice is used by Japanese people for making sushi, and by Spanish people for making paella. Classic rice puddings are also prepared using short-grain rice

Wild rice is not a true rice, but is the seed of an aquatic North American grass. It is not cultivated in a similar way to traditional rices, and is expensive to harvest. It ranges in colour from dark brown to black and has a very nutty flavour. Wild rice is often combined with other rice varieties to add colour and flavour to a dish.

Three types of rice (left to right): brown rice, arborio rice and basmati rice

ACTIVITY 7·4

Rice varieties

Visit the websites of the Rice Growers Association of Australia and Sunrice and complete the following research about rice varieties.

1 Outline the nutrient value of rice, and explain why rice is an important food.

2 Copy and complete the following table, giving an example of each rice variety and a summary of its properties.

Rice variety	Properties of the grain	Uses in food preparation
Long-grain		
Jasmine	Tender texture; fragrant aroma	Asian dishes: stir-fries; fried rice
Medium-grain		
Short-grain		

3 Using information from the websites, list examples of rice that have been processed so that they cook more quickly.

4 Where is rice grown in Australia? Draw a thumbnail sketch and explain why these regions are the most suitable for rice growing.

5 Outline the major environmental issues associated with rice-growing.

Comparison of uncooked and cooked rice

Top tips for cooking rice

Rice is a very versatile food and can be cooked in a variety of ways. In South-East Asia, India and China, rice forms the main part of most meals, and is frequently served as an accompaniment to other, much spicier foods.

When rice cooks, it undergoes gelatinisation. This is the process that occurs when starch granules in the endosperm of rice absorb liquid in the presence of heat, thickening the liquid and softening and swelling the grain.

During cooking, rice increases in volume; one cup of uncooked rice will produce approximately three cups of cooked rice. Allow approximately one-third of a cup, or 70 grams of raw rice, per person.

Care must be taken when using cooked rice, because bacteria quickly multiply in warm rice if it is not stored in the refrigerator. Sushi and rice salads must always be refrigerated after preparation and not left in a warm atmosphere. When using cooked rice in a recipe such as fried rice, always reheat the cooked rice until it is very hot to prevent bacterial growth.

Rice is usually cooked in one of four ways.

1. Rapid boil – Bring a large saucepan of water to the boil. Add ⅓ cup rice per person. Stir once or twice only. Boil uncovered for 15 minutes. Drain through a sieve. Rinse under running water. Allow to drain completely.

2. Absorption method – Allow 1 cup of rice to 2 cups of stock or water. Place the rice and stock into a saucepan with a well-fitting lid. Bring to the boil. Reduce heat to very low and allow to gently simmer for 15 minutes. Remove from the heat and allow to rest for 10 minutes.

3. Microwave – Read the instructions on the packet of rice, as cooking times vary for different varieties. Also remember to check the instructions in your microwave manual, since power levels vary considerably between models.

4. Rice cooker – Use equal quantities of rice to water. Rinse the rice and place in the rice cooker. Add water and switch on. Allow to rest for 10 minutes when the rice cooker has switched off and the cooking is complete. Some rice cookers have a warming element that will keep the rice warm for several hours.

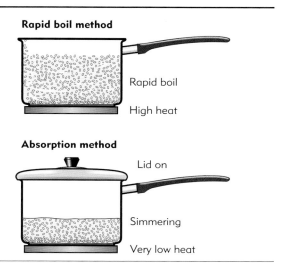

Two methods of cooking rice

ACTIVITY 7.5
Comparing methods of cooking rice

AIM
To compare the sensory properties, cooking time and volume of rice cooked by a range of methods

EQUIPMENT
- 2 saucepans (one with a tight-fitting lid)
- Rice cooker
- Microwave-safe bowl
- 4 × ¼ cup Calrose rice.

METHOD
1. Using the information on methods of cooking rice on this page, cook a quarter of a cup of rice by each of the four methods.
2. After you have cooked and drained each batch of rice, record the volume produced.
3. Record your results in a table similar to the one below.

	Rapid boil	Absorption	Microwave	Rice cooker
Cooking time				
Volume				
Texture				
Flavour				
Separation of grains				

ANALYSIS
1. Were there differences in the times required to cook the rice until it was tender using the four methods? Which method had the shortest cooking time? Which method took the longest?
2. How would you account for the increases in volumes when cooked?
3. Was there a significant difference in the volumes of rice produced using each cooking method? Explain.
4. Which cooking method produced the rice with the best texture and flavour?
5. Identify the method/s of cooking that produced grains of rice that were clearly separated.

CONCLUSION
If you were going to cook rice for a meal, which method of cooking would you recommend? Why?

Testing knowledge

11 What is bread improver, and why do many commercial bread manufacturers use it in their production process?

12 Explain why flatbreads are a popular food throughout the world.

13 Identify the type of information that is required to be included on the label of a loaf of bread to help consumers determine if it is fresh.

14 What is couscous? Identify the regions in the world where couscous is a staple food.

15 Outline three strategies rice growers are using to improve the sustainability of their farms.

16 Explain why rice is a valuable food to include in the diet.

17 Develop a concept map that includes the different varieties of rice, as well as dishes made from each variety.

18 Identify the process that describes the changes in starch when rice is cooked. Explain how this process makes rice more palatable to eat.

19 Why is it important to store cooked rice in the refrigerator? How can rice be reheated to ensure it is safe to eat?

20 List the different methods that can be used to cook rice. Which method would require the most accurate timing to prevent the rice from becoming overcooked?

THINKING SKILLS 7·1

1 SWOT ANALYSIS OF NO-TILL FARMING

Complete a SWOT analysis of no-till farming – strengths, weaknesses, opportunities, threats.

Strengths	Weaknesses
Opportunities	Threats

2 STAGES IN PRODUCING A PIZZA DOUGH

Draw a flow chart of the key steps in producing a pizza dough, including the process of fermentation.

DESIGN ACTIVITY 7·1

Pizza

DESIGN BRIEF

The school council has approached your Food Technology class to support their latest fundraising program by designing a pizza that can be sold in single portions. The school council wishes to sell the pizzas as a takeaway snack during the recess break in a few weeks' time. It has also requested that the pizza be packaged in an environmentally friendly container in support of the school's environment policy.

1 Use the information below to write a design brief for a pizza that can be served in single portions as a takeaway food item that is packaged in an environmentally friendly manner. Use the five Ws – who, what, when, where, why – as the basis of your design brief.
   - Who: explain who will be purchasing the new pizza
   - What: a single-portion pizza suitable to serve as a snack
   - When: describe when the pizza will be served
   - Where: outline where the pizza will be sold
   - Why: comment on why the pizza will help promote the school's environment policy

2 Based on the specifications in your design brief, develop five criteria to evaluate the success of your pizza.

INVESTIGATING

1 With a partner, research recipes for pizza.

2 Investigate ideas for suitable environmentally friendly packaging.

GENERATING

1 Complete the recipe map that follows to design your pizza.

2 Sketch and annotate two ideas for the pizza – the toppings and the shape.

3 Select your preferred option.

4 Explain why each of your designs is appropriate to serve in single portions as a takeaway food.

5 Select an interesting name for your new pizza, and write out your recipe so it is ready for production.

PLANNING AND MANAGING

1 Write up a production plan.

2 Make a list of the aspects of the production task that rely on you and your bench partner sharing and working collaboratively.

PRODUCING

1. Prepare the product.
2. Record any modifications you made during production.

EVALUATING

1. Answer your five evaluation criteria questions in detail.
2. Describe the sensory properties – appearance, aroma, flavour and texture – of your pizza. Share your pizza with two other people and record their comments.
3. Was the dough base firm enough for you to eat a single serve of pizza without the topping spilling off it?
4. After the pizza was baked, was the arrangement of the topping ingredients successful, and did it look appetising?
5. Plot the ingredients of your pizza on a diagram of the Healthy Eating Pyramid or the Australian Guide to Healthy Eating. Comment on the nutritional value of your pizza.
6. Explain how the packaging for your pizza met the requirement to be environmentally friendly.
7. Based on your own experience and the comments of your two taste testers, discuss any improvements you would make to either the ingredients or the production process if you were to make the pizza again.

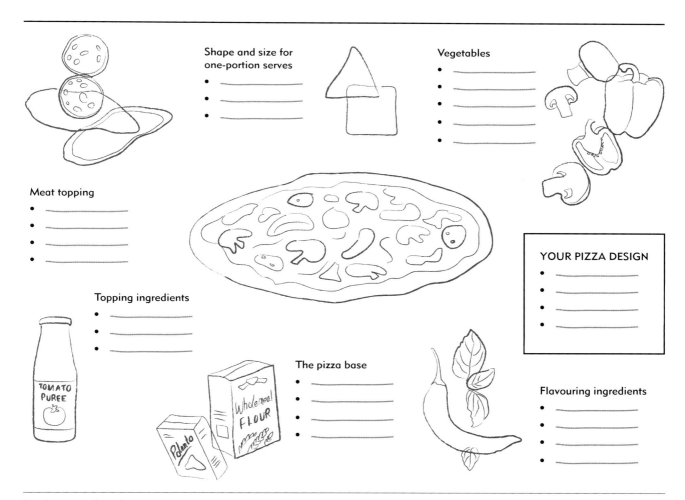

Recipe map for pizza

DESIGN ACTIVITY 7.2

Tear-and-share bread

DESIGN BRIEF

A hand-shaped loaf of bread is often served as part of a lunch. People often like having a pull-apart loaf to share because it is easy to pull off a portion to eat.

Worksheet

1 Write a design brief based on the five Ws – who, what, when, where, why.
   - Who: explain who will be eating the bread
   - What: a bread with a high fibre content, an interesting flavour and a garnish
   - When: describe when the pull-apart loaf of bread will be served
   - Where: outline where the bread will be served
   - Why: comment on why it is important that the shape of the loaf indicates where each portion of the loaf can be torn apart

Tear-and-share bread

2 After writing your design brief, develop five evaluation criteria questions from the specifications to evaluate the success of your tear-and-share loaf of bread.

INVESTIGATING

1 Research a variety of flours, grains, vegetables, nuts and seeds that are high in dietary fibre and would be suitable to include in the bread mix.

2 Develop a list of the most desirable sensory properties of bread to help you evaluate the success and quality of your tear-and-share loaf.

3 Visit a local bakery, or use food magazines and recipe books on bread, to investigate various bread shapes. Sketch loaf shapes that could be pulled apart in sections to be eaten. Evaluate how easy it would be to pull apart each loaf instead of slicing it. Rate each shape as either easy or difficult to pull apart.

4 Develop a list of ingredients that would be suitable to use as garnishes on loaves of bread.

5 Undertake a taste test of a range of bread loaves and rolls of different shapes. Rank each bread based on how easy it was to tear apart into individual serves. Record your results in a table similar to the one below.

GENERATING

1 Use the Basic Bread recipe on Page 153 as a starting point for your design.

2 Complete the recipe map to develop two design ideas for your high-fibre tear-and-share loaf. Sketch and annotate your designs, including a garnish for each loaf.

3 Decide which is your preferred option. Justify your selection.

Type of loaf or roll	Appearance: sketch the shape and record the garnish	Texture	Flavour	Fibre content	Overall rating for tearing apart (1–5): 1 = poor; 3 = OK; 5 = excellent

PLANNING AND MANAGING

1 Using the information from your recipe map, prepare a food order and write up the method before making your product. Remember that the amount of moisture used in the recipe will need to be adjusted depending on which of the additional ingredients are used.

PRODUCING

1 Produce your preferred option.

EVALUATING

1 Answer your five evaluation criteria questions in detail.
2 Describe the sensory properties – appearance, aroma, flavour and texture – of your tear-and-share loaf of bread.
3 Did the bread pull apart easily into equal portions?
4 What, in your opinion, was the most difficult aspect of the production of the bread? Explain.
5 Discuss any improvements you would make to either the ingredients or the production process if you were to make the bread again.
6 What are some of the advantages and disadvantages of homemade bread?
7 Develop a class rating scale to evaluate the finish of students' loaves compared with those of a commercial bakery.

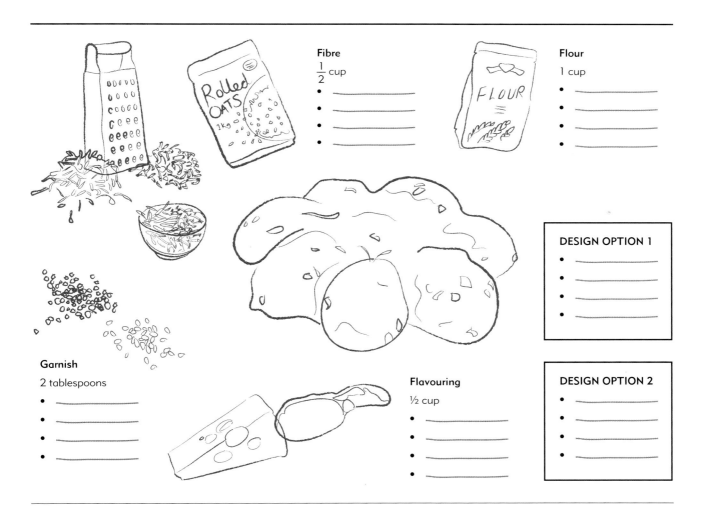

Recipe map for a loaf of tear-and-share bread

PLAIN RICE

Rapid boil

4 cups boiling water

⅔ cup long-grain rice

SERVES TWO

Absorption

1⅓ cups water

⅔ cup long-grain rice

SERVES TWO

*There are two simple methods for cooking plain rice, which is ideal for serving as an accompaniment. Try each method to see which you prefer.

METHOD

1. Bring the water to the boil.
2. Stir in the rice.
3. Rapidly boil uncovered for 12–15 minutes.
4. Test to see if the rice is tender by tasting a grain.
5. Drain in a strainer and serve.

METHOD

1. Bring the water to the boil.
2. Add the rice. Gently stir with a fork to separate the grains.
3. Place lid on the saucepan. Lower the heat and simmer for 12–15 minutes. Do not lift the lid during cooking.
4. Remove from heat. Keep covered with the lid and allow to stand for 5 minutes. Toss with a fork.

EVALUATION

1. Why is it necessary to stir the rice once or twice when you add it to the boiling water?
2. Explain why it is important to leave the lid on the rice when cooking by the absorption method?
3. Why is it necessary to allow the rice to stand for five minutes before serving when cooking by the absorption method?
4. Why does the rice increase in size once it has been cooked?
5. Describe how you can tell whether rice is cooked when using the rapid boil cooking method.

FOOD BY DESIGN

CREAMED RICE WITH ALMONDS AND APPLE PUREE

Creamed Rice with Almonds

¾ cup water

¼ cup short-grain rice

½ cup milk

¼ teaspoon vanilla essence

1 ½ tablespoons sugar

1 tablespoon flaked almonds

whipped cream to decorate

SERVES ONE

Apple Purée

1 Granny Smith apple

¼ cup water

small piece lemon zest

1 clove

1–2 teaspoons sugar (according to preference)

SERVES ONE

METHOD

1. In a small saucepan, bring water to boil and add rice. Reduce heat to simmer and cook until the water has been absorbed. Stir occasionally.
2. Add the milk, vanilla and sugar and simmer until the rice is tender. Stir occasionally and add a little more milk if necessary.
3. Spoon into serving bowl.
4. Lightly toast almonds under a grill.
5. Decorate the creamed rice with whipped cream and sprinkle with almonds. Serve with fresh or stewed fruit. (Layering the creamed rice and fruit in a tall glass looks impressive and tastes great!)

METHOD

1. Peel, core and slice apple.
2. Combine all ingredients in a small saucepan and cover with a lid.
3. Bring to boil. Reduce heat to simmer and cook until apple is tender. Check that water does not evaporate during cooking.
4. Puree the apple through a sieve or with a stick mixer.

EVALUATION

1. Explain what happens to the size and texture of the rice when it is boiled in water.
2. Why is it important to stir the rice mixture after adding the milk and sugar?
3. Why is it very important to watch the almonds carefully while toasting them under the griller?
4. Why are lemon zest and a clove added to the apple while cooking? Suggest some varieties of stewed fruit other than apple that could be used to accompany the creamed rice.
5. Evaluate the Creamed Rice with Almonds and Apple Puree by plotting its ingredients on a diagram of the Healthy Eating Pyramid or the Australian Guide to Healthy Eating. Write a brief paragraph to explain whether this dish would be suitable to serve as a healthy dessert.

SUSHI ROLL

Sushi Rice

2 cups Koshihikari or sushi rice

2 cups water

Sushi Su

½ cup su (rice vinegar)

2 tablespoons sugar

pinch of salt

prepared sushi rice

Assembling the Roll

ham

red capsicum

shredded egg

cucumber

avocado

pickled radish

prepared sushi rice

4 sheets nori

wasabi paste

egg mayonnaise

soy sauce, for serving

METHOD

Preparing the rice

1. Place the rice in a medium saucepan. Rinse several times with cold water until the water is clear. Drain thoroughly to remove excess starch.
2. Add the water to the rice. Stand for 10 minutes.
3. Bring to the boil. Boil for 1 minute. Turn the heat to low. Cover the saucepan tightly and simmer for 20 minutes.
4. Remove from the heat and allow to stand for 10 minutes. Do not uncover until ready. This will ensure the rice stays moist.

Preparing the sushi su

1. Mix all the ingredients well.
2. Transfer the freshly cooked rice to a large bowl and gently toss the rice with the sushi su. The rice is now ready to be used to fill the sushi rolls.

Assembling the roll

1. Cut filling ingredients into thin strips. Place half a sheet of nori onto a bamboo mat. Allow one edge to hang over the edge furthest away from you by about 1 centimetre.
2. Place a band of sushi rice in the centre of the nori. Spread evenly.
3. Place filling ingredients in layers across the centre of the rice to form a long band from left to right.
4. Use the mat to roll the up the sushi, rolling away from you. Press down firmly.
5. Remove the mat from the roll. Leave whole or cut into equal-sized serving portions. Serve with soy sauce.

EVALUATION

1. Why is a short-grain rice used to prepare sushi?
2. Explain why you need to rinse the rice several times in cold water before using it in this recipe.
3. Why is it important to make sure the lid is not removed from the rice while it stands for 10 minutes after cooking?
4. What other ingredients could be used to fill the sushi other than those suggested in the recipe?
5. Discuss why sushi is considered to be a healthy snack or lunch food.

BASIC BREAD

2½ cups bread flour

1½ teaspoons freeze-dried yeast

1¼ teaspoons bread improver

½ teaspoon salt

½ teaspoon sugar

½ teaspoon oil

200–250 millilitres warm to hot water

egg glaze

METHOD

1. Sift all dry ingredients together. Mix in the oil and approximately two-thirds of the hot water to make a moist dough. Gradually add the remaining water, if required.
2. Cover the surface with plastic wrap and leave in a warm place to prove and double in size.
3. Turn out onto floured board and lightly knead into a smooth dough.
4. Make into the desired shape and place on greased tray. Leave shaped loaf or rolls to double in size. Lightly cover with oiled plastic wrap.
5. Glaze with beaten egg.
6. Bake in a preheated oven at 230°C for 15–20 minutes or until golden brown.

EVALUATION

1. List the three environmental factors that are important to ensure that a yeast dough rises.
2. Why is it necessary to prove a yeast dough?
3. Explain why a yeast dough is kneaded before shaping.
4. Describe the sensory properties – appearance, aroma, flavour and texture – of the bread you have produced.
5. Discuss why bread is classified into the foundation layers of the Healthy Eating Pyramid.

MULTIGRAIN COB LOAF

1 cup white bread flour
1 cup multigrain flour
½ teaspoon salt
¼ teaspoon sugar
1 teaspoon dried yeast
1 cup warm water
canola oil spray for greasing tray

Topping

egg and milk glaze
poppy or sesame seeds

MAKES 1 LOAF

METHOD

1. Preheat oven to 210°C.
2. Sift flours, salt, sugar and yeast into a large bowl and make a well in the centre.
3. Add the warm water all at once and mix to a soft dough using a spatula.
4. Turn onto a lightly floured board and knead for 8–10 minutes, until dough is smooth and elastic.
5. Form into a cob-shaped loaf; slash the top with diagonal cuts and place on the greased baking tray.
6. Lightly cover with a clean tea towel or oiled plastic wrap and allow to prove in a warm place for 20 minutes.
7. Carefully brush with the egg and milk glaze and sprinkle with topping seeds.
8. Bake in the preheated oven for 30 minutes or until loaf is golden brown and sounds hollow when tapped.

EVALUATION

1. Explain why it is important to sift the dry ingredients before mixing in the liquid.
2. What would happen if cold water was used instead of warm water?
3. Why is it important to prove and knead the dough?
4. Describe the tests used to decide if the cob loaf was cooked.
5. Compare the nutrient value of the Multigrain Cob Loaf with that of a loaf of white bread.

FLATBREADS

Pitta Bread

1 cup unbleached flour

1 teaspoon freeze-dried yeast

½ teaspoon gluten flour

½ teaspoon salt

¼ teaspoon sugar

½ cup warm to hot water

 MAKES 4

METHOD

1. Sift dry ingredients, add water and mix into a soft dough.
2. Cover surface of dough with plastic food wrap and rest in a warm place for 10–15 minutes to prove the dough.
3. Turn onto a floured board and lightly knead. Divide into four portions.
4. Form each portion into a ball. Roll each ball into a circle approximately 14 centimetres in diameter. Allow to rest on the bench for 5 minutes so that a slight skin forms.
5. Preheat griller on high.
6. Cook pitta bread, one at a time, under hot griller until each puffs up and lightly browns, for approximately 1–2 minutes.
7. Turn and cook other side.
8. Allow to cool, then fill with accompaniments.

Naan

1 cup self-raising flour

1 tablespoon plain yoghurt (live if possible)

½ teaspoon salt

approximately 65 millilitres lukewarm water

 MAKES 4

METHOD

1. Put the flour, yoghurt and salt into a mixing bowl. Add the lukewarm water, a little at a time, working with your fingers into a soft, slightly sticky dough.
2. Cover with a damp tea towel and leave in a warm spot for the dough to ferment for 30–45 minutes.
3. Divide dough into four pieces and roll each into a ball. Roll out each ball on a floured surface into an oval 20–23 centimetres in length and 0.8 centimetres thick.
4. Preheat the griller on high. Grill the naans until they puff up and are speckled with brown spots, then turn over and cook the other side. (Naans cook very quickly, so keep watching them.)
5. Serve warm.

EVALUATION

1. Draw up a table similar to the one below to record the similarities and differences in the sensory properties of the homemade flatbreads and some commercial flatbreads.

TYPE OF FLATBREAD	APPEARANCE	AROMA	FLAVOUR	TEXTURE
Homemade pitta bread				
Commercial pitta bread				
Homemade naan bread				
Commercial naan bread				

2. Discuss the skills involved in the production of the flatbreads you made. What aspects of the production were difficult?
3. What are the advantages of the homemade flatbreads?
4. Of all the flatbreads you tested, which product did you like best? Why?
5. Suggest a range of fillings or toppings that could be used with any of the flatbreads that you evaluated.

PIZZA BASES

Wholemeal Pizza Dough

½ cup flour
½ cup wholemeal flour
1 teaspoon dried yeast
¼ teaspoon salt
¼ teaspoon sugar
1 teaspoon olive oil
100–125 millilitres warm water
1 tablespoon semolina

MAKES ONE MEDIUM PIZZA BASE

METHOD

1. Preheat oven to 210°C.
2. Sift the flours, yeast, salt and sugar into a medium bowl. Add the oil to the warm water, then mix into dry ingredients. Cover the surface of the dough with oiled plastic wrap and leave to prove for 10–15 minutes in a warm place.
3. Turn onto lightly floured board and knead lightly. Roll dough into the shape of the pizza tray.
4. Oil pizza tray, sprinkle with semolina then oil fingers and spread dough to fit the size of the tray.
5. Add toppings and bake in the preheated oven for 15–20 minutes.

Polenta (Cornmeal) Pizza Dough

1 cup flour
⅓ cup fine polenta (cornmeal)
2 teaspoons dried yeast
1 teaspoon sugar
1 teaspoon salt
⅓ cup milk, warm
½ cup warm water
1 tablespoon olive oil

MAKES 1 MEDIUM PIZZA BASE

METHOD

1. Preheat oven to 210°C.
2. Sift the flour, polenta, yeast, sugar and salt into a medium bowl. Mix in milk, water and oil. Cover the surface of the dough with plastic wrap and leave to prove for 10 minutes in a warm place.
3. Turn onto lightly floured board and knead lightly.
4. Oil pizza tray, then oil fingers and spread dough onto tray.
5. Add toppings and bake in the preheated oven.

EVALUATION

1. What are the three ideal environmental conditions yeast requires to grow?
2. Describe the physical changes that occur to the dough during the proving process.
3. Why is it important to knead a yeast dough after the proving process?
4. Which ingredients in these pizza base recipes are good sources of fibre?
5. Describe the sensory properties of a 'good' pizza base.

Pizza bases can be high and fluffy, like a bread crust, or thin and crisp. These recipes are for the thin, crispier style and have a higher fibre content than white, fluffier bases. Try one or the other with your favourite toppings!

NAPOLI SAUCE

2 teaspoons olive oil

½ small onion, finely diced

1 clove garlic, crushed

1 cup canned diced tomatoes or passata

½ teaspoon dried basil

¼ teaspoon salt

pepper

½ teaspoon sugar

🍴 MAKES SAUCE FOR 2 MEDIUM PIZZAS OR 1 SERVE OF PASTA

Napoli sauce is a concentrated tomato sauce from Italy. This rich, flavoursome sauce is used as a base for many pizzas. It is spread over the dough base before other toppings are added. Traditionally, fresh tomatoes are used; if in season, they have wonderful colour and flavour. However, canned tomatoes are a suitable alternative. This recipe is suitable to spread on a pizza base or to serve with freshly cooked pasta.

METHOD

1. In a small saucepan, heat the oil and sauté or gently fry onion and garlic until they are transparent.
2. Add the tomatoes, basil, salt and sugar and cook in an uncovered pan until the sauce thickens.
3. Season with the salt and pepper, to taste.
4. Cool before spreading on pizza base.

EVALUATION

1. Name the cooking process in which onion is fried until it is transparent.
2. What causes the sauce to thicken in Step 2 of the recipe?
3. Suggest some commercially available products that could be used on top of a pizza instead of making the sauce from scratch.
4. Why is the Napoli Sauce cooled before it is spread on the pizza base?
5. What are some other herbs that could be used in this recipe?
6. Discuss two safety rules to follow when cooking on the stovetop.

CALZONE

Dough

1 cup plain flour
1 teaspoon yeast
¼ teaspoon sugar
½ teaspoon salt
⅓ cup warm water
1 tablespoon olive oil

Filling

1 teaspoon olive oil
¼ medium onion, finely chopped
1 clove garlic, crushed
1 medium Roma or egg tomato, chopped, or ⅓ cup diced, canned tomatoes
1 tablespoon fresh basil, chopped
2 slices mild salami, sliced
2 small bocconcini, sliced, or 2 slices mozzarella cheese
2 teaspoons milk for glaze

MAKES 2 CALZONE

A calzone, or Italian pocket pizza, is a half-moon-shaped pizza with the filling folded inside a pocket of pizza crust. It is usually made as an individual serve and with a range of filling ingredients such as meats, vegetables and cheese. The cheese is either bocconcini or mozzarella, both of which are stretchy cheeses when heated. A calzone can be baked or deep-fried.

METHOD

1. Sift the flour, yeast, sugar and salt into a large bowl. Stir in oil and warm water and mix to a dough.
2. Knead dough on floured surface until smooth and elastic.
3. Place dough in an oiled bowl, cover with plastic wrap and stand in a warm place for 15–20 minutes.
4. Preheat the oven to 200°C.
5. Heat oil in a pan and fry onion and garlic until soft.
6. Add tomatoes and cook until the liquid has evaporated.
7. Remove from heat and add basil and salami. Allow mixture to cool.
8. Turn dough onto a floured surface, cut in half and roll each half into the shape of a dinner plate.
9. Top half of the round with the filling mixture, top the mixture with cheese and fold the top half over to enclose the filling. Dampen one edge with a little milk or water to help hold the dough closed.
10. Roll and twist the edges closed to seal the calzone.
11. Place each calzone on a greased tray and glaze with milk. Cut two slits in the top to vent the dough. Bake in the preheated oven for 20 minutes or until well browned.

EVALUATION

1. Why does the Calzone's design make it ideal as food to be eaten 'on the run'?
2. Explain why the filling is cooled before being wrapped in the dough.
3. What may happen if you forget to cut slits in the dough before baking?
4. Discuss any changes would you make to the filling ingredients if you made this recipe again.
5. Plot the ingredients for the Calzone on a diagram of the Healthy Eating Pyramid or the Australian Guide to Healthy Eating. Comment on the health rating you would give this recipe.

ROASTED VEGETABLE PIZZA

½ potato

150 grams sweet potato

150 grams pumpkin

¼ parsnip

¼ zucchini

¼ red capsicum

⅙ eggplant

2 mushrooms

1 clove garlic, peeled

1 tablespoon olive oil

salt and pepper

1 quantity wholemeal or polenta pizza base

½ quantity Napoli Sauce or ⅓–¼ cup commercial equivalent

60 grams cheddar cheese, grated

MAKES 1 MEDIUM PIZZA

Vegetables roasted in the oven develop a delicious brown crust, their flavour is intensified and their texture is softened, making them a tasty topping for a pizza. Root vegetables such as potato, sweet potato and parsnip are particularly good roasted, and pumpkin, mushrooms, eggplant, zucchini and garlic contribute a range of colour, textures and flavours to the end-product.

METHOD

1. Preheat oven to 210°C.
2. Wash all the vegetables except the mushrooms. Wipe the mushrooms with a clean, damp cloth.
3. Peel the potato, sweet potato, pumpkin and parsnip and cut into 1–2-centimetre cubes.
4. Cut the zucchini, capsicum and eggplant into similar-sized cubes. Cut the mushrooms in half.
5. Combine all the vegetables and the clove of garlic in a bowl and toss in the olive oil.
6. Cover a baking tray with a sheet of baking paper and lay out vegetables in a single layer. Season with salt and pepper.
7. Bake for 20–30 minutes or until vegetables are tender.
8. Roll out the pizza base thinly and spread with the tomato sauce.
9. Cover with a layer of roasted vegetables, then sprinkle on the cheese.
10. Bake in the preheated, very hot oven for 10–15 minutes or until the pizza crust is golden brown.

EVALUATION

1. Identify the vegetables in the recipe that could be described as root vegetables.
2. What are the benefits of roasting vegetables rather than boiling them?
3. Explain why the tomato sauce and grated cheese are essential in the making of the pizza.
4. Why are pizzas baked in a very hot oven?
5. Discuss why the Roasted Vegetable Pizza would receive a better health rating than an Aussie pizza.

FRAGRANT, FRUITY COUSCOUS SALAD

½ onion
¼ red capsicum
½ stick celery
1 spring onion
2 tablespoons dried apricots
½ apple
1 teaspoon oil
1 teaspoon ground cumin
1 teaspoon ground coriander
¾ cup vegetable stock
½ cup couscous
¼ cup peanuts
¼ cup sultanas

SERVES TWO

METHOD

1 Finely dice the onion and capsicum and slice the celery and spring onion.
2 Dice the dried apricot and apple and keep these separate.
3 Sauté the onion, capsicum and celery in the oil for 2 minutes, without browning.
4 Add the spices and cook for a further 30 seconds.
5 Add the vegetable stock and bring to the boil. Reduce heat to simmer and stir in the couscous.
6 Cover the couscous and turn off the heat. Allow to stand for 5 minutes with the lid on.
7 Fluff the couscous with two forks.
8 Stir through the peanuts, sultanas, dried apricots, apple and spring onion and serve immediately.

EVALUATION

1 Explain what it means to 'finely dice'.
2 Why is it important to dice the onion, apple and capsicum finely?
3 Define 'sauté'.
4 Why is it important to allow the couscous to stand, covered, for 5 minutes in Step 6?
5 Describe how you could garnish or decorate the finished dish.
6 Complete the below table of the key nutrients in your Fragrant, Fruity Couscous Salad.

INGREDIENT	NUTRIENTS	IMPORTANCE OF NUTRIENT TO GOOD HEALTH
Onion		
Capsicum		
Celery		
Dried apricots		
Apple		
Oil		
Couscous		
Peanuts		
Sultanas		

SWIRLING TWIRL DESSERT IN A GLASS

Poached Pears

1 cup water

¼ cup sugar

1 cinnamon stick

2 pears

¼ teaspoon vanilla

This recipe uses different methods of cooking to create the components of the Swirling Twirl Dessert in a Glass. Poaching is a moist method of cooking; frying the sweet, crunchy crumbs is a dry method. The preparation of the custard demonstrates the process of gelatinisation of starches.

METHOD

1. Combine the water and sugar in a saucepan and stir over low heat until the sugar dissolves.
2. Add the cinnamon stick and bring to boil. Reduce to simmer and cook uncovered for 5 minutes.
3. Peel, core and halve the pears.
4. Add the pear halves to the syrup and gently cook for 5 minutes or until tender.
5. Remove from heat, stir in vanilla and allow to cool.

Sweet, Crunchy Crumbs

1 cup fresh, dry breadcrumbs

1 tablespoon sugar

20 grams butter

METHOD

1. Combine the breadcrumbs and sugar.
2. In a frying pan, melt the butter over medium heat. Do not brown.
3. Add the breadcrumb mixture and fry until golden brown. Stir the crumbs so they brown evenly.
4. Remove from pan and allow to cool before use.

Custard Powder Sauce

1 tablespoon custard powder
1–2 teaspoons sugar
200 millilitres milk

Directions to make this product are on the powder packet. However, if the packet has been disposed of, use this recipe as a guide. Depending on the brand, the consistency may vary. Add more or less custard powder if a thicker or thinner custard is required.

MAKES 1 CUP

Mark Fergus Photography

METHOD

Custard powder sauce may be made on the stovetop or in the microwave.

On the stovetop

1. In a small bowl, blend the custard powder and sugar with 2 tablespoons of the milk.
2. In a small saucepan, heat the remaining milk until simmering. Remove from heat.
3. Pour hot milk into the blended custard powder and stir until combined.
4. Return custard mixture to the saucepan, then return to heat. Bring to boil, stirring constantly.
5. Cook for 1 minute and remove from heat.

In a microwave

1. In a microwave-safe bowl, blend the custard powder and sugar with 2 tablespoons of the milk.
2. Gradually add remaining milk. Whisk well.
3. Microwave on high for 2 minutes. Remove and whisk well.
4. Microwave on high for another 1–2 minutes or until the mixture thickens.
5. Remove from the microwave and whisk again.

Assembling the dessert

1. Select a special glass, or use a plastic take-away 'glass'.
2. Cut the pears into large cubes.
3. Layer the pears, custard and crumbs into the glass. Repeat until all the ingredients are used.

EVALUATION

1. Suggest a product you could use if you were too busy to poach the pears for yourself.
2. Describe what would happen if the sweet crumbs were not continuously stirred during frying.
3. Why is the custard powder blended with some milk before the heating process begins?
4. Explain why the custard is stirred during heating.
5. Describe the sensory properties – appearance, aroma, flavour and texture of your Swirling Twirl Dessert in a Glass.

8 MEATY IDEAS

KEY KNOWLEDGE

- What is meat?
  - Beef: from paddock to plate
- Sustainable farming of cattle and sheep
  - Meat for good health
  - How much meat should I eat?
  - The structure of meat
  - Selecting meat
  - Packaging meat for sale
- Poultry
  - Poultry for good health
  - Selection of poultry
- Fish
  - Fish for good health
- Top tips for working safely with meat, poultry and fish
  - Purchasing
  - Storing
  - Preparing and cooking
- Preparation for cooking meat, poultry and fish
  - Marinating
  - Crumbing
- Cooking meat, poultry and fish
  - Changes that occur during cooking
  - How to tell if it is cooked
  - Top tips for grilling
  - Top tips for stir-frying
  - Top tips for frying

KEY TERMS

connective tissue the tissue in meat that links and holds together muscles
marbling even distribution of deposits of fat cells in red muscle tissue
marinate to tenderise and/or enhance the flavour of meat or other food
modified atmosphere packaging (MAP) a method of packaging that causes change in the levels of gases inside a package in order to extend a product's shelf life of a product
muscle fibres cells that are bound into thin sheets of connective tissue; these bundles then form groups to create muscles

AUSTRALIAN CURRICULUM LINKS

DESIGN AND TECHNOLOGIES
- Knowledge and Understanding
- Processes and Production Skills

HEALTH AND PHYSICAL EDUCATION
- Personal, social and community health

GENERAL CAPABILITIES
- Critical and creative thinking

CROSS-CURRICULUM PRIORITIES
- Sustainability

WHAT IS MEAT?

Meat refers to the body tissues of animals, eaten as food. Early humans were hunter-gatherers, dependent on their immediate environment for food, and so meat became an important part of their diet.

Today, meat continues to be an important part of the diets of many Australians. The meat we eat generally comes from cattle (beef), sheep (lamb), pigs (pork) and poultry (chicken, turkey and duck). Beef, lamb and pork are often described as 'red meat' and poultry as 'white meat'.

Beef: from paddock to plate

PASTURE-FED CATTLE	FEEDLOT CATTLE
The cattle are kept on farms and graze on pasture in paddocks.	The cattle are kept in pens.
Cattle grazing in northern Queensland is undertaken on large cattle stations, where the cattle graze on native pasture. In the southern states, cattle graze on smaller farm holdings that are often sown with introduced pasture and fodder crops.	The cattle are fed in feedlots instead of grazing on pasture. Their diet is grain-based, high in energy and consists of wheat, barley, sorghum and canola seed.
The supply of pasture is dependent on rainfall and irrigation. Pasture is often not available during the winter months, therefore affecting meat supply.	A supply of high-quality beef is available all year, particularly during winter.
There is little control over how much the cattle eat. The weight and growth depends on the season and the amount of feed available. The meat produced is not as marbled.	The controlled diet means that the farmer can control the growth and weight of the cattle. Cattle remain in the feedlot for 180–360 days and produce meat that is marbled – that is, meat that contains fat that is evenly distributed throughout the muscle.
This method dominates the Australian industry.	This method ensures the continuous production of a consistent supply of high-quality beef for export – particularly to Japan – and for domestic markets.

Sustainable farming of cattle and sheep

Australian cattle and sheep farmers are committed to producing beef and lamb sustainably, and work hard to leave the land, waterways, vegetation and soils in better condition for future generations.

Target 100

Target 100 is an initiative by Australia's cattle and sheep industry to support farmers in the development of sustainable farming practices. The 100 strategies that form the basis of this campaign are aimed at improving the environmental sustainability of cattle and sheep producers, increasing their productivity and profitability and ensuring a sustainable food supply for generations to come.

The health of the soil is a key factor in ensuring environmental sustainability. Farmers are encouraged to improve soil health by maintaining good ground cover. Healthy ground cover plants reduce water run-off, reduce erosion and capture nutrients. Farmers need to ensure that paddocks are not overgrazed, since this can cause the soil to compact, reducing its ability to absorb water and grow pasture. Areas of the farm to be rested from grazing to improve soil health. Farmers are also encouraged to fence off dams to increase biodiversity on their properties.

Water is an essential resource for farmers to produce healthy pasture to maintain a sustainable food source for the animals. For hundreds of years, farmers have used windmills to pump water, and windmills continue to be an important source of renewable, cost-effective energy. Some farmers are investing in solar panels as a more efficient power source to pump water for irrigation and to power their homesteads.

Strategies to dispose of, or use, animal waste produced in both feedlots and abattoirs in a sustainable manner is another key focus of the Target 100 campaign. A range of initiatives have been developed to monitor and reduce methane production by animals to ensure that reduced levels of this greenhouse gas are emitted into the environment. Cattle and sheep farmers are also developing strategies to use animal manure on pasture as a natural fertiliser, instead of using artificial fertilisers.

Cattle and sheep producers are offered advice on strategies to improve animal welfare from a moral and ethical perspective. Some of these strategies include ensuring that the animals have food, water and shelter, and that they are protected from predators, pests and disease. They must also be provided with an environment in which they are able to express their normal behaviour. During transportation to market or processing, it is important that animals do not suffer from discomfort, injury, fear or distress.

Secondary processing of meat

After cattle are sold, they are transported to the abattoir for slaughter. Carcasses are cooled at the meatworks in large refrigerators or chillers. The carcass is cut in half or quarters and hung on large hooks. After the carcass is chilled overnight, it is ready to be sold to butchers, supermarkets, wholesalers or other processors, or to be shipped overseas for export. Live animals are also shipped to overseas countries for slaughter.

Meat Standards Australia

Australia has developed a beef grading and 'paddock to plate' computerised meat processing and tracking system. This system, Meat Standards Australia (MSA), introduced by Meat and Livestock Australia, ensures best practice standards in beef carcass and cut grading procedures to guarantee high-quality beef cuts. MSA grades each beef cut according to certain quality standards criteria. Information is provided to consumers on the best cooking method for each cut. When MSA beef is cooked as recommended, its eating quality is guaranteed.

Meat for good health

Lean meat is one of the important foods in the Australian Guide to Healthy Eating. It contains a high proportion of protein, which is essential for growth and to repair and maintain body tissues in active people. A 100-gram serving of lean meat – which is about the size of the palm of your hand – gives your body over half the protein it needs each day. The protein found in beef, lamb and poultry is 'complete'; that is, it has all the essential amino acids needed by the body in the correct proportions required for growth and repair of tissue. Plants do not contain all the essential amino acids, and so they must be combined to provide complete protein (for example, legumes with cereals).

Red meat, especially lean cuts, contain readily absorbed forms of iron and zinc. Iron helps oxygen to move around the body, and is important in producing energy. Zinc is vital to helping your body utilise carbohydrates and protein, and also to healing wounds. Meat is also a good source of niacin, riboflavin and some thiamine. These B group vitamins are essential for your body to utilise nutrients in food.

Iron is essential for red blood cells to carry out their function of transporting oxygen around the body to form energy for body functions.

The **protein** in red meat is high quality or complete, and therefore contains all the essential amino acids that are needed for growth and repair.

Vitamin B12 maintains a healthy nervous system and body cells and assists with healthy skin and good vision.

Zinc is vital for a healthy immune system and for growth and development. It helps the body use carbohydrates and protein for energy.

Nutrients in red meat

How much meat should I eat?

Red meat makes a major contribution to your protein and iron intake (specifically, haem iron), and also provides zinc and B group vitamins. The Australian Guide to Healthy Eating recommends that meat or meat substitutes should be eaten three to four times per week. Diets rich in lean red meat can still be low in saturated fat and not adversely affect plasma cholesterol levels. Try to choose lean cuts of meat that have little or no visible fat. A wide range of low-fat, nutrient-dense meals containing red meat can be prepared using the popular cooking methods of stir-frying, grilling, barbecuing (without charring) and roasting.

Meats that have a high fat content, such as bacon and some sausages, are in the 'Only sometimes and in small amounts' section of the Australian Guide to Healthy Eating.

The structure of meat

Meat is made up of 75 per cent water, 18 per cent protein and 3 per cent fat. The muscle fibres that form meat are cells bound by thin sheets of connective tissue. The bundles are organised in groups to form individual muscles, which are anchored to the bone by connective tissue. The longitudinal structure forms the grain of meat. Muscle fibres are small when the animal is young and the muscles have not been well developed. As the animal grows and exercises, the muscles enlarge, particularly those that are most frequently used, such as in the neck and legs.

Connective tissue links the muscles and holds them together. It is found between muscle fibres and between whole muscles. The more connective tissue the meat cut contains, the tougher the meat will be. Tougher cuts of meat usually come from the leg, shoulder and forequarter of the animal, because these are the parts of the animal that were the most

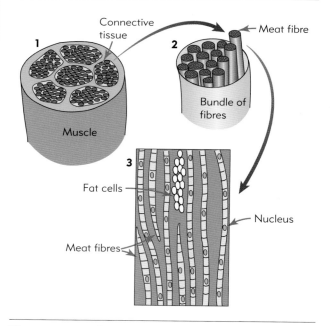

The structure of meat

exercised. When connective tissue is heated in a liquid, the insoluble collagen becomes gelatin, and becomes tender to eat. Cuts of meat with a lot of connective tissue are best if they are cooked slowly in a moist environment, because this softens the meat. Wet methods of cooking such as stewing, braising and casseroling tenderise tougher cuts of meat.

Fat tissue surrounds muscle tissue and is also incorporated in it. **Marbling** is the term used to describe the deposit of fat cells in the red muscle tissue. Fat has an important role in the sensory properties of flavour and texture when meat is cooked.

More tender cuts of meat are found around the ribs and back of the animal. These cuts have little connective tissue, and can therefore be cooked by dry radiant heat methods such as grilling, frying, roasting and barbecuing. Tender cuts of meat are therefore easiest to use for making quick and easy everyday meals. The meat is most tender if it is carved across the grain so that the muscle bundles are short and therefore easier to chew.

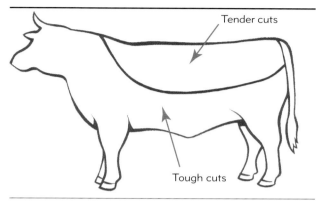

Tender and tough cuts of meat

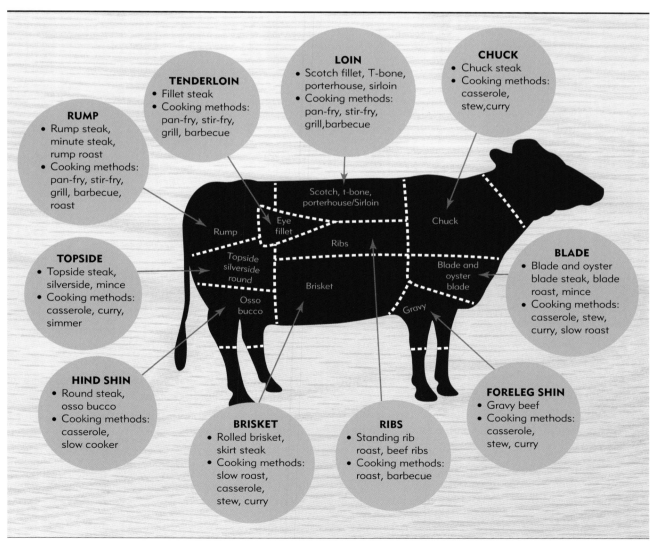

Basic cuts of beef

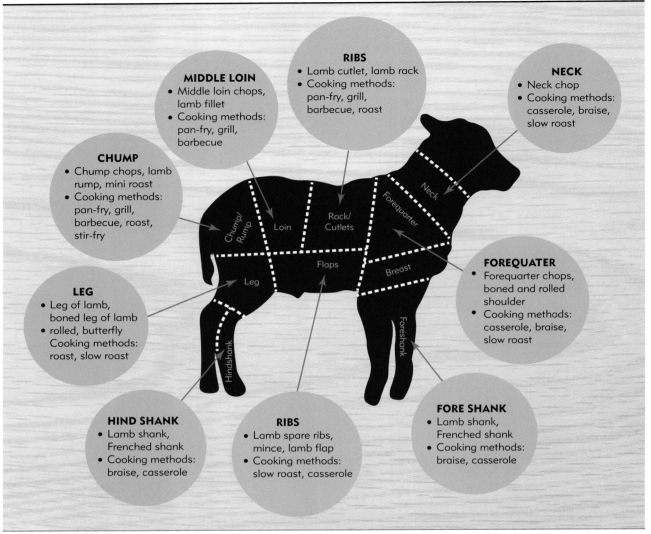

Basic cuts of lamb

ACTIVITY 8.1

Cuts of lamb and beef

Refer to the diagrams 'Basic cuts of beef' and 'Basic cuts of lamb' on Pages 167–8 and answer the following questions.

1. Identify three possible cuts of beef that are suitable to grill on a barbecue.
2. Explain why meat from the shin is unsuitable for grilling.
3. Name a cut of boneless beef suitable to roast for a family of four.
4. Identify the cut or portion of beef you would purchase to prepare a stir-fried beef and vegetable dish.
5. You have purchased a jar of curry simmer sauce to make a beef curry. Which beef cut would you purchase to prepare this dish? Explain why.
6. Discuss the advantages of roasting and serving an easy carve leg of lamb compared with a traditional leg of lamb containing the bone.
7. Why is the Frenched rack of lamb a popular cut on many restaurant menus?
8. Describe a suitable cooking method or recipe for using lamb shanks.
9. Outline the nutritional advantages of serving a trim lamb butterfly steak rather than a loin chop.
10. Which lamb cut would you choose to make a lamb casserole?

Selecting meat

Meat should be moist, have a fresh smell and be slightly springy to the touch. Good-quality beef will have bright-red flesh and creamy, yellow fat. The flesh of good-quality lamb is a pink–brown colour with a creamy-coloured fat. The flesh of good-quality pork is pale pink and the fat is a bright-white colour.

Packaging meat for sale

The term 'case-ready' refers to the preparation and packaging of fresh meat products for sale in retail outlets. Once this process is complete, meat products, such as beef mince and stir-fry strips, are ready to distribute. Modified atmosphere packaging (MAP) has been introduced as a method of packaging meat in the chilled cabinets of retail outlets such as supermarkets. This packaging method modifies the levels of oxygen, nitrogen, carbon dioxide and water vapour inside a package in order to extend the food's shelf life. MAP meats look more acceptable in the chilled fresh meat display of supermarkets, because they retain their bright-red colour, which consumers regard as an important indicator of freshness. Consumers often make choices between different packs of meat on this basis.

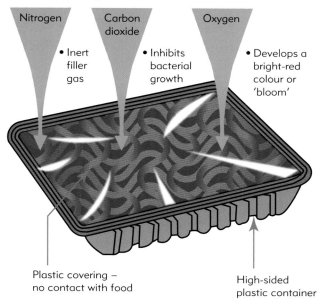

Modified atmosphere packaged minced beef

ACTIVITY 8·2

Marketing red meat

Visit the following websites to complete this activity:

- MEAT & LIVESTOCK AUSTRALIA (MLA)
- BEEFANDLAMB.COM.AU

1. Find information about the red meat industry. List five new things you have learnt about the production of lamb and beef.
2. Find information about the nutrients in meat. Identify four significant nutrients in meat and describe the health benefits of each of them for the human body.
3. Find information about the marketing of red meat around the world and answer the following questions.
    a. How many countries import Australian red meat?
    b. Explain why so many countries import Australian red meat.
4. Find information about the marketing campaigns for beef and lamb on television or in magazines in Australia. Select two such campaigns and complete the following table.

	Advertisement 1	Advertisement 2
Identify the title of the advertisement.		
Did the theme use fact, fiction or fear to attract consumers' attention?		
Outline the message about red meat in the advertisement.		
Identify the advertisement's target market or audience.		
Did the advertisement make you want to eat red meat? If so, why?		
Had you seen the advertisement before today?		

5. After completing your analysis of the two campaigns, identify the one that you think will most help to increase the sales of red meat. Discuss and justify your decision.

Testing knowledge

1. Explain the differences between pasture-fed and feedlot-fed cattle.
2. Which method of feeding cattle ensures that the meat is marbled?
3. Explain why Target 100 was developed by farmers, and describe two ways that farmers can ensure the sustainability of their land when grazing cattle and sheep.
4. Explain how the new beef grading system developed by Meat Standards Australia benefits the consumer.
5. What is complete protein and why is it valuable for our good health?
6. Identify the key nutrients in red meat and explain their functions in the body.
7. Outline the recommended amount of meat that should be consumed by people each week.
8. Discuss the differences between muscle fibres, connective tissue and fat tissue in meat. Sketch a diagram that indicates where the most tender and toughest cuts are found on an animal.
9. Describe the physical properties you would look for when selecting good-quality meat.
10. What is modified atmosphere packaging (MAP)? Explain why this packaging system is used for packaging fresh meat.

POULTRY

Poultry is the term used to describe any domesticated birds that are used as food. Chicken, turkey, duck, quail and pheasant are all part of the poultry group, of which chicken is the most popular. Today's chickens are descendants of wild fowl that roamed the dense jungles of primeval Asia. It was only after the Second World War that chicken became reasonably priced – prior to this, it was expensive and was served only as a roast meal on special occasions. Modern production methods have reduced the cost of this versatile food, and today, it is readily available in food stores. Poultry consumption in Australia has risen considerably over the last few decades due to its affordability, its availability, growing awareness of its nutrient value and its convenience, as it requires little preparation.

Poultry for good health

Chicken, like other meats, is a very good source of complete protein, and therefore helps to build and repair body tissue. However, it contains marginally less iron than red meats such as beef and lamb. Chicken also contains the B group vitamins thiamine and riboflavin, as well as some zinc. It has the same amount of protein as red and other white meat, but less

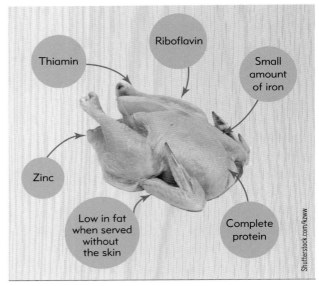

Nutrients in poultry

saturated fat and more polyunsaturated fat. Chicken is often thought to be a good substitute for red meats because, if it is prepared without the skin, it is very low in fat. Like other lean meats, chicken is an important food in the Australian Guide to Healthy Eating.

Selection of poultry

Chicken skin should be a light, creamy colour, and should be moist to the touch. The skin should not be broken and it should have no dark patches. The breast of the bird should be plump, and the tip of the breastbone should be flexible. Free-range chicken has a layer of fat over the breast and slightly yellow skin. The skin of a corn-fed chicken has a clear yellow tinge. The flesh of the poultry should smell fresh and be free from any film. The packaging of frozen poultry should be well sealed to ensure there has been no freezer burn or contamination.

Chickens are sold and categorised according to age and; for example, a number 16 chicken weighs no less than 1.6 kilograms. A poussin is a bird aged 3–4 weeks that weighs 400–500 grams.

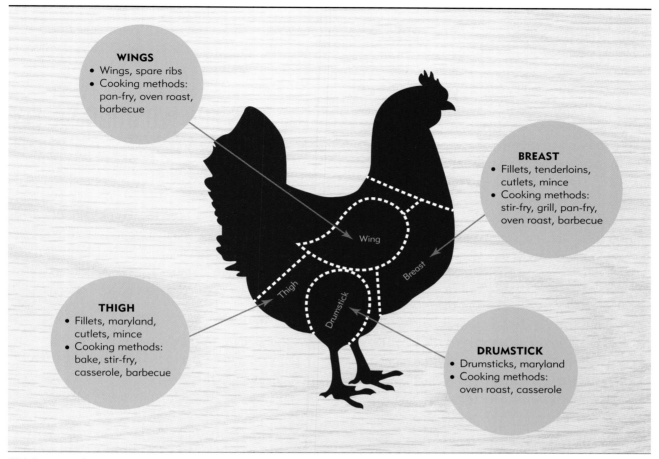

Chicken portions

A boiler is a hen that has stopped laying. Its flesh is very tough, so it must be cooked slowly by a moist method.

Chicken can be purchased whole for roasting or in portions such as breast fillets, thighs, drumsticks, wings, chicken chops, drumettes and winglets. Minced chicken is another option for preparing quick, easy everyday meals.

Today, there is a great variety of convenient products available for sale to add to chicken to make a quick, easy meal. Jars of sauces and packets of flavourings not only line supermarket shelves, but are also widely advertised on television and in magazines.

ACTIVITY 8·3

Using chicken in everyday meals

Undertake an internet search using the keywords 'how to cut and joint a chicken'. (The website Taste.com.au has a useful 'how to' video.) Review the information you find to answer the following questions.

1. Select two cuts of chicken. Describe their physical properties and recommended cooking methods.
2. Search for recipes that use the chicken cuts you have selected.
3. Draw a table like the one below and complete it using information from the website/s.

	Chicken cut 1	Chicken cut 2
Description of the chicken cut		
Physical properties of the chicken cut		
Name of a recipe that uses this chicken cut		
Main cooking method used in the recipe		
Four main pieces of equipment required to complete the recipe		
One health and safety issue to consider in the preparation or production of the recipe		

FISH

Fish makes a great ingredient for everyday meals because it contains only a very small amount of connective tissue and has only short muscle fibres, and so it cooks very quickly and is always tender. Because fish has such delicate flesh, it is best to cook fish just before you wish to serve it. Fish can be purchased in a variety of ways, such as whole, as fillets, as cutlets or canned. Fish is quick and easy to cook, and can be baked, grilled, pan-fried, deep-fried, poached, steamed, stewed, stir-fried, microwaved, barbecued or cooked in foil. Too many strong flavouring ingredients, however, can overpower the delicate flavour of fish.

Fish for good health

Health professionals encourage us to include up to two fish meals in our diets each week. Fish is a very healthy choice for meals, since it is an excellent source of complete protein and most varieties are naturally low in fat. Fish also provides the important omega-3 fatty acids, which are essential for brain and eye development. Eating fish on a regular basis is also thought to reduce the risk of childhood asthma and to help halve the likelihood of heart attack. However, fish does not contain any calcium, unless you eat the bones in canned fish such as salmon or sardines.

It is now recognised that, due to the pollution that is present in many of the world's rivers and oceans, larger fish may be contaminated with heavy metals such as mercury, so it is wise to minimise the consumption of fish such as shark (flake), swordfish and barramundi.

TOP TIPS FOR WORKING SAFELY WITH MEAT, POULTRY AND FISH

It is very important to take great care when preparing and cooking meat, poultry and fish. Because they are protein foods, they all need to be cooked in such a way that any harmful bacteria that are naturally present are destroyed by the heat. In particular, poultry must be handled with care, because it can be a major source of salmonella and so can cause severe food poisoning.

Purchasing

- After purchasing meat, poultry and fish, it is important to keep it cool while it is transported home to the refrigerator.
- Remove the chicken from the wrapping as soon as you bring it home from the supermarket or butcher. Place it on a clean plate and cover with plastic wrap before refrigerating.

Storing

- Meat, poultry and fish should be covered and stored in the coolest part of the refrigerator and placed well away from cooked foods or foods that will be eaten raw. If the item is purchased in prepacked trays, it will be date-marked and labelled to show the use-by date.
- Use fresh poultry portions, such as chicken breast or thighs, within two days of purchase.
- Frozen meat can be kept in the freezer for two to six months. It must be wrapped and sealed in plastic and stored as a flat package with the air squeezed out. Frozen meat must reach −18°C to freeze properly.

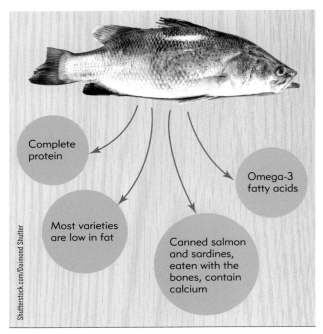

Nutrients in fish

PRODUCT	APPROXIMATE STORAGE LIFE (−18°C)
Beef roasts	4–6 months
Minced beef and lamb	2–3 months
Sausages	1–2 months
Steak	3–4 months
Lamb chops	2–3 months

- Meat, poultry and fish must be kept chilled until it is time to cook them. They must always be cooked properly to ensure that bacteria have been destroyed.
- When transporting meat, chicken or fish – for example, to a barbecue – it is important to carry it in an esky or insulated container with ice if the weather is hot.

Preparing and cooking

When preparing meat, poultry and fish, it is important to avoid cross-contamination or the transfer of bacteria between raw and cooked foods. Cross-contamination can be avoided by following these rules.

- Wash your hands in hot, soapy water before preparing food and handling raw meat, poultry or fish.
- Throughly clean knives and cutting boards in hot, soapy water after preparing the meat, poultry or fish.
- When preparing a variety of cooked and uncooked foods, use a different chopping board and utensils for each food type, or wash the boards thoroughly in hot, soapy water after each process.
- Do not place cooked foods on plates that have held raw meat, poultry or fish.
- Defrost frozen meat, poultry and fish in the refrigerator; only defrost in the microwave if it is to be cooked immediately.
- Never cook chicken that is partially frozen because, although the outside might look cooked, the centre could still be icy or undercooked, and the dangerous salmonella bacteria may not be killed during the cooking process.
- Do not refreeze meat, poultry or fish that has already been frozen without cooking it first.

ACTIVITY 8·4
Cooking safely with meat, poultry and fish

Work in small teams to prepare a digital notice board or fact sheet to inform consumers how to cook safely with protein foods such as meat, poultry and fish. Brochures or fact sheets will be available in supermarkets and most butchers. Your brochure or fact sheet should include:

- appropriate headings
- information about how to select good-quality meat, poultry or fish
- information about safely transporting meat, poultry or fish home
- information about strategies for storing meat, poultry or fish safely
- instructions for preparing and cooking meat, poultry or fish
- diagrams or photos, to help make the message clear.

PREPARATION FOR COOKING MEAT, POULTRY AND FISH

Marinating

A marinade is a flavoursome liquid in which meat, poultry or fish may be soaked to enhance its flavour. Marinating is a method used to tenderise and/or enhance the flavour of meats or other foods, and of tenderising these protein foods. The acid ingredient – either lemon juice, wine or vinegar – breaks down the structure of the muscle bundles so that the food becomes softer and easier to chew. If meat is not tenderised, its muscle fibres begin to shrink and toughen when they are subjected to dry heat. Red meat and poultry can be marinated for several hours or overnight, whereas fish only requires a short period of marinating, because it has very little connective tissue. Keep meat, poultry or fish in the refrigerator while it is marinating to prevent the growth of spoilage microorganisms. When selecting ingredients for a marinade, avoid those with a high sugar or thickener content, since they will burn during stir-frying. These ingredients can be added later in the production process.

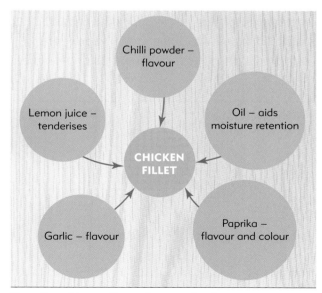

Functional ingredients in a marinade

ACTIVITY 8.5

Understanding marinades

1 Use recipe books, food magazines or websites to complete the research for this activity.

   a Select four recipes that involve the process of marinating.

   b Record the functional ingredients for each product in a table similar to the following.

Recipe and reference	Ingredient being marinated	Acid for tenderising and flavour	Oil for moisture	Flavourings
Beef Noodle Stir-Fry, *Food by Design*, Page 183	Beef	Rice wine vinegar	Oil	Oyster sauce; soy sauce

2 Go to the Food Standards Australia New Zealand (FSANZ) website and investigate the product MSG (monosodium glutamate). What is this ingredient, and what is its number under the Food Standards Code?

3 Explain why MSG is sometimes included in a marinade for meat.

4 Outline the effect that MSG can have on some people.

5 Some restaurants advertise that their food is MSG-free. Discuss why they might do this.

6 Analyse the label on a commercial meat-tenderising product. Predict which of its ingredients have a tenderising effect on the connective tissue in meat.

Crumbing

Crumbing is a process that can be used with meat, poultry and fish to create a crisp, golden-brown coating on the outside of the food. The process is best suited to small, tender pieces of food that can be cooked quickly. The food is first dipped in seasoned flour to create a dry surface, and is then dipped into beaten egg mixture, which makes a sticky surface for the breadcrumbs to adhere to. The crumbed food should be rested in the refrigerator for 30 minutes; this allows the moisture to be evenly distributed in the crumbing mixture and helps to prevent it from falling off the food during cooking. Finally, the crumbed food is fried in hot oil to create a crunchy, golden crust; the heat inside the crumb coating steams the food. Some examples of protein foods that can be crumbed and fried are lamb cutlets, beef schnitzel, chicken fillets and fillets of fish.

Marinades enhance flavour

Place flour on paper towel. Coat fish patty in flour.

Beat together egg and milk on a flat plate. Brush egg mixture over floured patties.

Place breadcrumbs on paper towel. Firmly press crumbs onto the patties with metal spatula.

Crumbing fish patties

COOKING MEAT, POULTRY AND FISH

Meat, poultry and fish are cooked in order to make them safe to eat and easier to chew and digest, and to destroy any harmful microorganisms. Methods of cooking can be divided into two groups – dry methods and moist methods.

The dry method of cooking is a fast way of cooking meat that is best suited to tender cuts. These cuts of meat, poultry or fish have enough water in their tissue to enable the conversion of the protein collagen to gelatin. Therefore, this cooking method uses little to no liquid, and the time taken to cook the meat depends on the size and thickness of the cut. Examples of this method are pan-frying, stir-frying, grilling, barbecuing and oven roasting.

The moist method of cooking is a slower way of cooking meat that is suitable for less tender cuts. These cuts do not have enough water in their tissues to convert collagen to gelatin. Therefore, the long, slow, moist cooking softens the connective tissue that makes meat tough. Examples of this method are casseroling, pot-roasting, braising and stewing.

Changes that occur during cooking

During the cooking process, a number of changes occur to the physical and sensory properties of meat, poultry and fish.

Texture
- When heat is applied, the protein in the meat, poultry or fish is denatured by coagulation. The protein food shrinks as some of the water is expelled and the texture becomes firm.
- Cooking improves the palatability of meat, poultry and fish by making it easier to chew.
- Connective tissue is tenderised and the collagen is converted to gelatin, if moisture is present. This method breaks down the protein in the muscle and connective tissue.
- Fat melts, giving meat a crisp, brown surface.

Colour
- The colour of meat changes from red to brown.
- Poultry changes from pale pink to white; the flesh of fish becomes white and opaque once it is cooked.

Flavour and aroma
- Flavour or extractives are squeezed out of meat and onto the surface, giving meat its flavour and aroma.
- Fat also adds flavour when meat and poultry are cooked.

Nutrient value

When meat, poultry and fish are cooked, a number of changes occur to its nutritive value:
- if overcooked, the protein becomes tough and indigestible
- fat-soluble vitamins A, D, E and K are not affected by cooking
- water-soluble vitamins (B group) are lost when meat is cooked.

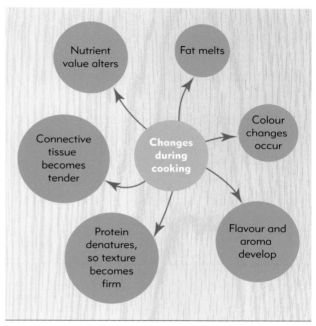

Changes that occur during the cooking of meat, poultry and fish

How to tell if it is cooked

Meat

The meat should be browned on the outside. Pierce the meat with a skewer – the juices should be pinkish to clear.

Poultry

Pierce the poultry with a skewer in the thickest part of the flesh – the juices should run clear. The chicken meat should no longer be pink. The skin should be golden brown and slightly crispy.

Fish

The fish should turn white and flake apart easily with a fork.

Top tips for grilling

Grilling is a quick method of cooking and is ideal for preparing family meals in a hurry.

1. For a grill separate to the oven, leave the grill door open during cooking to prevent heat building up inside the grill. For a grill inside the oven, grill with the door closed or follow the manufacturer's instructions.
2. Trim visible fat from the meat or poultry.
3. Use tongs for turning food during grilling.
4. If meat starts to curl while it is cooking, cut the curled-up edges with the point of a sharp knife.
5. Test to see if meat or poultry is cooked by gently pressing with blunt tongs. Do not cut the meat or poultry, as this allows the meat juices to escape. Fish will flake apart with a fork.
6. Soak bamboo skewers for kebabs in water before threading to prevent them from burning during grilling.
7. If meat or poultry is marinated with honey or sauces containing sugar, heat the grill to medium rather than high to prevent the marinade from burning.

Top tips for stir-frying

Stir-frying is a very quick method of cooking in a wok over intense heat. Tender cuts of meat and/or vegetables should be cut into equal bite-sized pieces prior to the cooking process. Meat and chicken can be marinated before stir-frying, but the marinating liquid should be drained off before cooking to prevent stewing.

1. Prepare meat strips by trimming off any fat and slicing across the grain. Strips should be five to eight centimetres in length. Cut chicken pieces in equal sizes so they cook evenly.
2. Add a small quantity of oil to the wok and swirl to coat its sides and base. Heat oil until smoke point is almost reached.
3. Drain off excess marinade, add the strips and stir-fry for two to three minutes. The meat strips should sizzle when added to the hot oil. Sear in small batches to prevent the meat from shedding its juices and stewing, and consequently toughening. Allow the wok to heat up again between batches.
4. Remove meat from the wok and stir-fry vegetables separately – firm vegetables first, then softer or leafy varieties. Return everything to the wok to warm before serving.

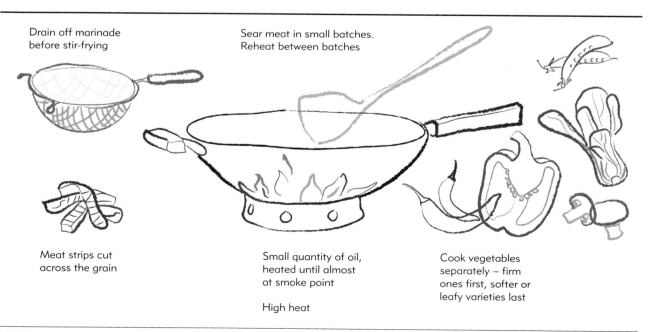

Drain off marinade before stir-frying

Sear meat in small batches. Reheat between batches

Meat strips cut across the grain

Small quantity of oil, heated until almost at smoke point

High heat

Cook vegetables separately – firm ones first, softer or leafy varieties last

Stir-fry cooking

Top tips for frying

1. Always have the oil hot enough so that when the food is added, it sizzles. Test with a cube of bread – it should turn golden in 30 seconds. Food soaks up cold oil like a sponge.
2. Make sure all portions of food are dry before placing them in hot oil. This prevents spitting.
3. Fry only small portions of food – they will cook quickly and more evenly.
4. Drain fried food well before serving.
5. Take care when frying because hot fat and oil reach very high temperatures. Use oven mitts to protect your hands.
6. Never leave a pan or deep-fryer unattended when cooking.
7. Keep the handles of the pan turned to the side of the stovetop to prevent it from being knocked over.
8. If the fat or oil catches alight, do not attempt to extinguish the flame. Instead, cover the cooking container with a lid or fire blanket to cut off the supply of oxygen and extinguish the flame.

Frying lamb chops

Testing knowledge

11. What is poultry, and why is it a valuable food to include in the diet?
12. Describe the physical properties you would look for when purchasing poultry.
13. Explain why fish can be cooked very quickly and is a good food to eat regularly.
14. Draw up a table like the one below and record at least two safe work practices that should be followed when using meat, poultry and fish in food preparation.

PURCHASING	STORING	PREPARING AND COOKING

15. What is a marinade, and why is it sometimes used in the preparation of meat, poultry and fish?
16. Discuss the functional role of an acid and an oil in a marinade.
17. Draw up a flow chart that demonstrates the process of crumbing food.
18. Describe some of the changes that occur to the texture, colour, flavour and nutrient value of meat when it is cooked.
19. Describe how you can test to determine whether or not meat, poultry and fish are cooked.
20. Sketch three concept maps that highlight best practice for grilling, stir-frying and frying food.

THINKING SKILLS 8·1

ANALYSE THE TWO METHODS OF BEEF PRODUCTION

1. In your analysis, consider the quality of the meat produced, animal welfare issues and the impact on the environment of the farming practice.
2. Summarise the similarities and differences between the two meat production systems.

DESIGN ACTIVITY 8·1

A gourmet hamburger

DESIGN BRIEF

Hamburgers are a popular food that are quick to prepare and can be eaten without cutlery. Preparing a hamburger from scratch allows you to incorporate interesting vegetables and flavouring ingredients to make a unique, gourmet product.

Design a gourmet hamburger that incorporates a range of vegetables and flavourings and would be suitable to pan fry or barbecue when friends come to visit.

1 Write a design brief that includes the following information.
- Who – who will be eating the gourmet hamburger?
- What – a gourmet hamburger featuring a variety of vegetables and flavourings
- When – when will the barbecue take place?
- Where – where will the barbecue be held?
- Why – why will you include a range of vegetables and flavourings in the hamburger?

2 After writing your design brief, develop four evaluation criteria questions that cover the specifications outlined your design brief to evaluate your gourmet hamburger.

INVESTIGATING

1 Undertake an internet search of ingredients that could be included in hamburger recipes. Complete the table below by identifying four ingredients suitable to include in each category.

2 Use the internet to find a recipe for barbecuing hamburgers (beef burgers).

List three important tips for successfully barbecuing hamburgers.

GENERATING

1 Using the information you have gained from your research into ingredients that could be included in hamburger recipes, select the ingredients for your new hamburger.

2 Using the basic recipe for Zucchini Barbecue Burgers (Page 184) as a guide to the proportion of ingredients required, write out the recipe for your new hamburger.

PLANNING AND MANAGING

1 Prepare a food order.

2 Write up a production plan for your new hamburger.

PRODUCING

1 Produce the hamburger you have designed.

EVALUATING

1 Using the words for describing food on Page 2, describe the sensory properties – appearance, aroma, flavour and texture – of your hamburger.

2 Evaluate the flavour of your hamburger. Did the flavours complement each other?

3 Identify two safety issues to consider when preparing and cooking the hamburger.

4 What changes or modifications would you make to your recipe if you were to make the hamburger again?

5 Use your knowledge of nutrition to check whether your hamburger contained all the key nutrients required for a healthy diet. Explain whether your hamburger was a healthy meal option.

Hamburger ingredient table

Meat	Vegetables	Flavourings	Herbs	Garnishes	Breads

DESIGN ACTIVITY 8·2

A stir-fried meal

DESIGN BRIEF

Stir-frying is a quick method of cooking a meal that usually contains both meat and vegetables. A stir-fry is very flexible – it can be made to feed either one person or a larger number of people. A wide variety of ingredients and flavours can be added either during the cooking process or once the stir-fry is plated.

1 Write a design brief for a stir-fry dish that uses marinated meat strips. The dish should include a cereal product that complements a stir-fry, and should serve one person. In your design brief, include information about:
- Who – who will eat the stir-fry?

- What – a stir-fry that includes four to six vegetables with a variety of colours and textures.
- Why – why is the stir-fry being served, rather than another style of meal?
- When – the occasion at which the stir-fry will be served.
- Where – the place you will serve the stir-fry.

2 After writing your design brief, develop five evaluation criteria questions that cover the specifications outlined your design brief to evaluate the success of your stir-fry.

INVESTIGATING
1 Complete Activity 8. 5, 'Understanding marinades' (see Page 174).
2 Undertake an internet search of ingredients that could be used to prepare the stir-fry.

GENERATING
1 Using the information you have gained from your research into ingredients, complete the stir-fry recipe map below.
2 Select the ingredients for your new stir-fry and write up your new recipe.
3 Prepare a food order.

PLANNING AND MANAGING
1 Write up a production plan.

PRODUCING
1 Prepare the product.
2 Note any modifications you make to the recipe during production.

EVALUATING
1 Evaluate your stir-fry according to your previously established evaluation criteria questions.
2 Comment on the overall success of your stir-fry meal.
3 Suggest ways in which you could modify your marinade to complement other cuts of meat.
4 Discuss the safe work practices you followed in preparing your stir-fry meal.
5 Plot the ingredients for your stir-fry on a diagram of the Healthy Eating Pyramid or the Australian Guide to Healthy Eating. Comment on the health rating you would give this meal.

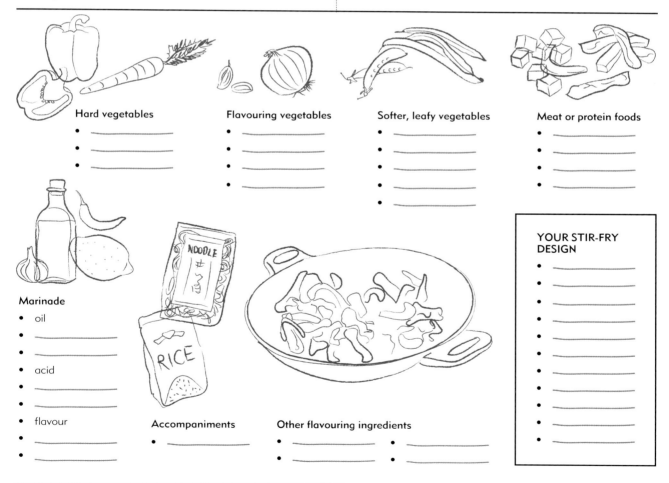

Stir-fry recipe map

DESIGN ACTIVITY 8·3

Meals using minced meats

Beef, lamb, pork and chicken can all be minced to create a versatile ingredient that can be used in sauces, as a filling in pies, pastries, crepes and pasta, and in balls, patties and loaves. This minced meat can be fried, baked, grilled, steamed or stewed to make a myriad of recipes suitable for serving as a meal.

1 Write your own design brief based on the five Ws – who, what, when, why and where. The theme is a main meal in which minced meat is the hero!
   - Who – who will be eating the main meal?
   - What – outline the occasion at which the main meal will be served, and the time available to prepare and serve it.
   - When – the time of day or season in which the main meal will be eaten.
   - Why – why the main meal will be eaten.
   - Where – where the main meal will be served.
2 Formulate the sentences or statements you developed in Step 1 into a paragraph that will become your design brief.
3 Use the specifications – that is, the constraints and considerations in your brief – to develop five criteria questions to evaluate the success of your main meal.

INVESTIGATING
1 Research and select two recipes that use minced meat, are suitable for a main meal and can be prepared in the time identified in your brief.

GENERATING
1 Sketch, in colour, the serving and presentation of each menu option. Annotate each option to include:
   a recipe title
   b list of main ingredients
   c major cooking methods
   d major processes
   e references.
2 Use a decision table similar to the one at the top right of the page to select the preferred option. Justify your choice.

PLANNING AND MANAGING
1 Calculate the ingredients required to prepare the recipe for the number of people you are serving.
2 Write up your new recipe.
3 Write up a production plan, including relevant health and safety considerations.

Decision table

Decision to be made:

Option 1	Option 2
Advantages:	Advantages:
Disadvantages:	Disadvantages:

Decision – preferred option:

Justify your decision based on the specifications in your design brief:

PRODUCING
1 Prepare the product.
2 Note any modifications or changes you made during the production of the recipe.

EVALUATING
1 Answer your evaluation criteria questions in detail.
2 What was the most challenging part of the production? Outline how you managed this challenge.
3 Comment on your overall management of time for this activity, including both your designing and planning and the production of the meal.
4 Plot all the ingredients of your meal on a diagram of the Healthy Eating Pyramid or the Australian Guide to Healthy Eating. Comment on the nutritional value of this meal.
5 If you were to produce this meal again, what changes would you make to improve it?

CURRIED MEAT AND VEGETABLE PIE

½ carrot, finely diced
½ potato, finely diced
2 teaspoons oil
½ medium onion, diced
125 grams lean minced steak
1 teaspoon curry powder
1 tablespoon flour
½ cup beef stock
1 teaspoon tomato paste
⅓ cup frozen peas
1 sheet frozen shortcrust pastry
½ sheet frozen puff pastry
egg glaze

 MAKES 2 PIES

METHOD

1. Preheat oven to 200°C.
2. Steam the diced carrot and potato until tender.
3. Heat oil in saucepan, add diced onions and sauté until tender – do not brown.
4. Add meat and cook until meat has changed colour. Use the potato masher or fork to break up lumps in meat.
5. Add curry powder and fry quickly with meat and onions.
6. Remove from heat, stir in flour and then slowly add stock and tomato paste and mix in well.
7. Return saucepan to the heat and bring to the boil, stirring constantly until the mixture boils and thickens. Remove from heat.
8. Add the steamed carrot and potato and peas and place the mixture on a plate, spreading it out so that it cools quickly.
9. Select 2 pie tins 8–10 cm in diameter. Cut the sheet of shortcrust pastry in half diagonally and press each piece into a pie tin, allowing a little to overhang. Dampen the upper edge with glaze.
10. Divide the filling in half and spoon into each pie tin.
11. Cut the sheet of puff pastry in half and press onto each pie as a lid.
12. Trim any excess pastry using the top edge of the tin as a guide.
13. Frill the edge to seal the layers. Cut a vent in the middle of each pie and decorate if desired.
14. Glaze the top of the pie, but avoid glazing the edge of the frill.
15. Bake for 20–25 minutes or until golden-brown.

EVALUATION

1. Explain why the meat and curry powder are fried before they are mixed with other ingredients.
2. What purpose does the flour have in the mixture?
3. Why is it important to boil the mixture after adding the flour?
4. Explain why you need to cool the filling before placing it into the pastry shell.
5. Why is shortcrust pastry used for the base of the pie instead of puff pastry?
6. Explain why pies should be an occasional food rather than an everyday food.

BAKED MEATBALLS AND SPAGHETTI

140 grams spaghetti

Meatballs

1 slice stale bread
2 tablespoons milk
200 grams minced beef or chicken
¼ onion, grated
½ egg, lightly beaten
1 tablespoon grated parmesan
pinch of cayenne pepper
pinch of salt and pepper

Tomato sauce

1 cup tomato passata or sugo sauce
1 cup canned diced tomatoes
¼ cup water
2 teaspoons sugar
2 teaspoons basil, finely chopped
2 teaspoons parsley, finely chopped
2 tablespoons parmesan cheese (extra as required)

SERVES TWO

METHOD

Spaghetti

1. In a large saucepan, boil the water, then cook spaghetti until al dente. Drain. While the spaghetti is cooking, prepare the meatballs.

Meatballs

2. Preheat oven to 200°C. Lightly oil baking tray.
3. Trim the crusts from the bread and tear into small pieces. Place in a small bowl, add the milk and allow the bread to soak for 5 minutes. Squeeze out excess milk.
4. Combine the minced meat, onion, egg, parmesan cheese, cayenne pepper, salt and pepper and the soaked bread. Mix well.
5. Using about 2 teaspoons of the meat mixture per meatball, shape into small balls and place on baking tray. Bake for approximately 15 minutes or until firm and lightly browned.

Tomato sauce

6. Combine the passata, canned tomatoes, water, sugar, basil and parsley in a saucepan and simmer for 10 minutes so the sauce can thicken a little.

Completing the dish

7. Add the cooked spaghetti, tomato sauce and meatballs into a small casserole dish. Gently stir to mix the sauce through the dish.
8. Sprinkle with extra parmesan and bake for 10–15 minutes until a light golden brown.

EVALUATION

1. What is the role of the soaked bread in the meatballs?
2. Discuss the preparation and nutritional benefits of baking the meatballs instead of frying them.
3. Describe a test you could use to determine whether or not the meatballs were cooked.
4. List some varieties of pasta other than spaghetti that could be used in this recipe.
5. If you were to make the Baked Meatballs and Spaghetti to serve at a later time, describe how you would store and reheat them safely.
6. Write a critique of the Baked Meatballs and Spaghetti recipe to justify its inclusion on the school cafeteria menu.

BEEF NOODLE STIR-FRY

200 grams rump steak, thinly sliced

Marinade
2 teaspoons oil

1 tablespoon oyster sauce

1 tablespoon soy sauce

1 tablespoon dry sherry or rice wine vinegar

Stir-fry
½ onion, cut into wedges

1 piece broccoli, cut into flowerets

⅓ carrot, julienned

4 snow peas

¼ red capsicum, julienned

½ zucchini, cut into batons

1 tablespoon bamboo shoots

1 tablespoon water chestnuts

200 grams fresh Hokkien noodles

2 tablespoons oil, for frying

3 centimetres fresh ginger, finely sliced

1 clove garlic, finely sliced

¼ cup water or stock

1–2 tablespoons satay sauce or sweet chilli sauce, to serve

SERVES TWO

METHOD

1. Combine the steak, oil, oyster sauce, soy sauce and dry sherry or vinegar and marinate in the refrigerator for 20–30 minutes.
2. Cut all the vegetables into the appropriate sizes.
3. Place the noodles in a large bowl, cover with boiling water and leave to stand for about 5 minutes.
4. Drain marinade from steak and heat one tablespoon of oil in wok on high.
5. Fry onion for 30 seconds then add half of the meat strips. Cook for 1 minute, then add the remainder of the meat. Stir-fry until almost cooked. Remove from heat and keep warm.
6. Wipe out the wok and heat the remaining oil on high.
7. Add the ginger and garlic, then the carrot, broccoli, capsicum, zucchini and snow peas, allowing 30 seconds between the addition of each ingredient.
8. Add the water or stock, bamboo shoots and water chestnuts. Cook until vegetables are just tender.
9. Drain noodles.
10. Return the meat and onion to the wok, then stir through the noodles and satay or sweet chilli sauce.

EVALUATION

1. Describe the best way to cut meat to ensure it is tender and easy to chew after being stir-fried.
2. Explain why the beef strips were marinated, and describe the functional role of each ingredient in the marinade.
3. Why is the meat cooked in two batches rather than all at once?
4. Identify two safety factors you took into consideration when preparing the noodle stir-fry.
5. Comment on the overall success of your Beef Noodle Stir-Fry and suggest some improvements you would make if you were to make the dish again.
6. Search a recipe website for stir-fry recipes. Discuss why there are so many stir-fry recipes. Consider preparation time, cuts of meat and the nutritional benefits of these recipes.

ZUCCHINI BARBECUE BURGERS

¼ medium zucchini, grated
100 grams lean minced steak
¼ onion, grated
½ small clove garlic, crushed
½ teaspoon soy sauce
2 teaspoons beaten egg
2 tablespoons fresh breadcrumbs
1 tablespoon flour
2 tablespoons oil, for frying
2 hamburger buns
lettuce, onion and tomato slices, to garnish

MAKES 2 BURGERS

METHOD

1. Grate the zucchini and allow to stand for 5–10 minutes. Squeeze the zucchini to remove as much moisture as possible.
2. Combine the zucchini, minced steak, onion, garlic, soy sauce, egg and breadcrumbs. If the mixture is too wet, add extra breadcrumbs.
3. Shape into two round patty shapes and dust with flour.
4. Barbecue on an oiled plate or fry in a frying pan until meat juices run clear.
5. Assemble on buns and garnish with lettuce and tomato slices.

EVALUATION

1. Why is it important to squeeze the moisture from the grated zucchini before adding it to the hamburger mixture?
2. Why is the onion grated and not diced in this recipe?
3. Explain the function of the egg in the hamburger mixture.
4. How could you test the hamburger to see if its juices run clear?
5. What other flavourings could be added to the mixture in place of the soy sauce?

LEMON CHICKEN SERVED WITH CHAT POTATOES, CARROT STRAWS AND SNOW PEAS

4 chat potatoes

spray oil

Chicken marinade

pinch of salt

¼ teaspoon ginger, grated

Rind of ½ lemon, grated

1 small egg (white only)

1 small chicken fillet

2 tablespoons flour

1 tablespoon oil

1 teaspoon butter

Garnish

1 spring onion, sliced

slice lemon

5 snow peas, steamed

½ carrot, cut into straws, steamed

SERVES ONE

METHOD

1. Preheat oven to 200°C. Set up steamer.
2. When the steamer is hot, steam the chat potatoes for 10 minutes.
3. Prepare the marinade. Using a fork, lightly whisk the salt, ginger and grated lemon rind into the egg white.
4. Add the chicken fillet and turn in to the mixture. Allow to stand for 10 minutes.
5. Remove potatoes and place on baking tray. Spray with oil, sprinkle with salt and bake for 20–30 minutes.
6. Drain the chicken fillet well, then place in a plastic bag with the flour and shake to coat.
7. Heat oil and butter in a pan. Fry fillet until golden brown and cooked through.
8. Remove fillet from the pan, drain on absorbent paper and serve on warm plate.
9. Garnish with spring onions and a lemon butterfly.
10. Serve with the steamed vegetables and baked chat potatoes.

EVALUATION

1. Explain why the fillet is dipped in egg white before it is dusted with flour.
2. Why is a combination of butter and oil used as a frying medium?
3. Outline three important safety considerations when shallow-frying.
4. Explain why draining the cooked fillet on absorbent paper is important.
5. Discuss the challenges you faced in preparing and serving this recipe.

SPICY 'SHAKE AND BAKE' CHICKEN WITH POTATO LATKES AND STEAMED BROCCOLI OR BEANS

Spicy 'shake and bake' chicken

3 tablespons rice flour

2 teaspoons Cajun spice

2 teaspoons olive oil

2 teaspoons soy sauce

8 chicken drumettes or 4 chicken drumsticks

olive oil spray

SERVES TWO

This recipe incorporates baking, frying and steaming in the preparation of this delicious meal.

METHOD

1. Preheat oven to 180°C. Cover a baking tray with baking paper.
2. Place the rice flour and Cajun spice in a plastic bag.
3. Mix together the oil and soy sauce and lightly brush over the chicken drumettes.
4. Place the drumettes in the plastic bag with the rice flour and Cajun spice. Shake well to coat.
5. Place the coated chicken drumettes onto the baking tray. Lightly spray with the olive oil spray.
6. Bake in the preheated oven for 20–25 minutes or until tender and juice runs clear when a skewer is placed in the thickest part.

Potato latkes

2 large potatoes

¼ onion

2 tablespoons plain flour

1 egg

salt and pepper

15 grams butter

2 tablespoons oil

SERVES TWO

METHOD

1. Peel and grate the potatoes. Firmly squeeze the potatoes to remove as much water as possible.
2. Peel and grate the onion and mix with the grated potato.
3. Add the flour and egg, then season with salt and pepper. Mix until smooth.
4. Heat the butter and oil in a frying pan over medium heat. Test the heat by dropping a small cube of bread in to the pan – if the butter and oil mix is hot, the bread will sizzle and brown immediately.
5. Drop ¼ cup of the potato mixture into the hot frying pan and pat down with a fork to form a small latke. Repeat with remaining potato mixture – it should make about four latkes.
6. Lightly fry for approximately 8 minutes until brown on one side and the bottom of the latke is crisp. Gently turn over and cook the other side for a further 8 minutes until brown and crisp and the potato is cooked through.
7. Remove from the frying pan and drain on absorbent paper.

Steamed broccoli or beans

4 small florets broccoli or
   10 beans

SERVES TWO

METHOD

1. Half-fill the base of a steamer with water and bring to the boil.
2. Top and tail the beans, or trim the base from the broccoli stem.
3. Place the broccoli or beans into the top of the steamer and steam for 4–5 minutes until just cooked. Test with a skewer.

EVALUATION

1. Why are oil and soy sauce brushed over the chicken drumettes before they are coated in the rice flour and Cajun spice?
2. List one important safety rule to observe when frying the latkes.
3. Describe another method of cooking the broccoli or beans that would retain a crisp texture.
4. Which part of this production was the most successful, and what did you find the most challenging about this production?
5. Plot the ingredients for the Spicy 'Shake and Bake' Chicken with Potato Latkes and Steamed Broccoli or Beans on a diagram of the Healthy Eating Pyramid or the Australian Guide to Healthy Eating. Comment on the nutritional value of your meal.

FISH PATTIES WITH COLESLAW

Fish patties

2 medium potatoes

220 grams canned tuna or salmon

1 tablespoon parsley, chopped

2 teaspoons lemon juice

salt and pepper

2 tablespoons plain flour

1 egg

1 tablespoon milk

⅓ cup dry breadcrumbs

2 tablespoons oil

Fish Patties with Coleslaw make a delicious light meal to serve in spring or summer, and are made using ingredients that are readily found in the pantry or refrigerator.

SERVES TWO

METHOD

1. Peel and quarter the potatoes.
2. Half-fill the base of a steamer with water and bring to the boil.
3. Place the potatoes into the top of the steamer and steam for 10–15 minutes until cooked.
4. Mash the potatoes with a fork or potato masher.
5. Drain the tuna or salmon and flake well with a fork. If using salmon, finely crush the bones and mix well through the salmon.
6. Add the flaked tuna or salmon, parsley, lemon juice and salt and pepper to the mashed potato. Mix well.
7. Form into four even-sized patties.
8. Beat together the egg and milk. Place on a flat plate.
9. Coat the fish patties in flour, then brush with the egg and milk mixture and coat in the breadcrumbs. Press the crumbs on firmly.
10. Heat the oil in a frying pan over a low heat. Place the patties in the hot oil and cook for approximately 3–4 minutes or until brown on the bottom.
11. Carefully turn over the patties and cook on the second side for a further 3–4 minutes or until brown.
12. Drain on paper towel.

Hint: Check the information on crumbing on Page 174. Refrigerate the patties for 15–30 minutes before frying; this helps the crumbs to stay attached to the patty.

Coleslaw

⅛ cabbage

½ carrot

¼ red capsicum

½ stick celery

1 spring onion

1 tablespoon coleslaw dressing

1 tablespoon French dressing

SERVES TWO

METHOD

1. Very finely shred the cabbage and place in a large bowl.
2. Grate the carrot and finely dice the capsicum and celery.
3. Mix the carrot, capsicum and celery with the shredded cabbage.
4. Toss the coleslaw and French dressings through the prepared vegetables.
5. Serve the coleslaw with the fish patties.

EVALUATION

1. Suggest another method you could use to cook the potatoes if a steamer was not available.
2. Why are the patties coated in flour before being brushed with the egg and milk?
3. List two safety rules to follow when frying the fish patties.
4. Discuss the part of this production that was the most successful and the part you found most challenging.
5. Plot the ingredients for the Fish Patties with Coleslaw on a diagram of the Healthy Eating Pyramid or the Australian Guide to Healthy Eating. Comment on the nutritional value of your meal.

9 EATING WELL FOR THE FUTURE

KEY KNOWLEDGE

- Influences on food choices
  - Adolescent food choices
- Food marketing
- Eating for good health
- The Australian Dietary Guidelines
- The Australian Guide to Healthy Eating
- Energy balance
- Sources of energy
  - Fat and sugar: what's the difference?
- Fat
  - Monounsaturated fats
  - Saturated fats
  - Trans fats
- Carbohydrates
  - The glycaemic index
- Using energy
  - Energy intensity
- Obesity
- Cardiovascular disease
- Diabetes
- Maintaining a healthy weight
  - Measure up!
  - LiveLighter campaign
  - Walking towards good health
  - Portion control
  - Making better snack food choices
- Osteoporosis
  - Healthy bones
  - Selecting foods high in calcium
- The importance of fibre in the diet
- Individual dietary needs
  - Food allergies
  - Food hypersensitivity
- Vegetarian diets
  - Top tips for vegetarian eating

KEY TERMS

cardiovascular disease a general term used to describe a range of diseases, including heart disease, stroke and blood vessel disease

coeliac disease a disease of the small intestine associated with permanent intolerance or hypersensitivity to gluten

diabetes a disease where the pancreas is unable to produce sufficient insulin to enable the glucose produced during digestion to be absorbed into the bloodstream

food allergy an abnormal immunological reaction to food

food hypersensitivity a reaction to food that is of a similar type to food allergies, but generally less severe

glycaemic index (GI) a ranking of carbohydrate foods based on the immediate effect they have on blood sugar levels

monounsaturated fats fats found in olives, olive oil, avocados and nuts that have been shown to reduce blood cholesterol levels

osteoporosis occurs when calcium is lost from the bones, making them very fragile and easily broken

saturated fats fats found mainly in foods of animal origin such as meat, cheese and butter that are linked to raised cholesterol levels; coconut oil and palm oil are also high in saturated fats

trans fats bad fats that can lead to serious health concerns and should be avoided; found mainly in hydrogenated vegetable oil used by food manufacturers in processed and fast foods

AUSTRALIAN CURRICULUM LINKS

DESIGN AND TECHNOLOGIES
- Knowledge and Understanding
- Processes and Production Skills

HEALTH AND PHYSICAL EDUCATION
- Personal, social and community health

GENERAL CAPABILITIES
- Critical and creative thinking

INFLUENCES ON FOOD CHOICES

The range of foods available to consumers today is remarkably wide, far greater than at any other time in human history. Fresh food markets, supermarkets, cafes, restaurants and fast-food outlets are full of exciting food options. Making a decision about which foods to choose from such a wide selection is often very difficult.

Your family, culture, personal likes and dislikes and mood can often be the deciding factors when you are selecting food. However, other influences, such as your concern for good health, your peers, advertising or accessibility to food venues, may also help to determine which foods you choose to eat.

Adolescent food choices

One of the most important aspects of adolescence is the development of independence. This striving for independence can be reflected in many ways, one of which is in the selection of food. While parents and family are still important in the lives of most young people, adolescence does provide an opportunity for young people to make many of their own decisions regarding the types of foods they will eat and when they will eat them.

Just as the influence of family begins to diminish during adolescence, the role of the peer group becomes far more important and can have a significant impact on the types of foods that young people consume. Some foods or food outlets are seen by adolescents to be more acceptable than others, and so can become an important part of their social lives. The increased mobility of young people, too, has had an impact on their ability to select and eat the types of foods that appeal to their age group. This appeal is often influenced by advertising and other marketing strategies employed by food producers. Many food advertisements are targeted at adolescents, highlighting issues such as lifestyle, self-image, the peer group and income.

The price of a food item is another important factor in the decision by young people to select particular foods. Most young people have limited incomes, derived mainly from allowances provided by their parents or part-time employment. Therefore, food needs to be relatively inexpensive and to represent good value for money.

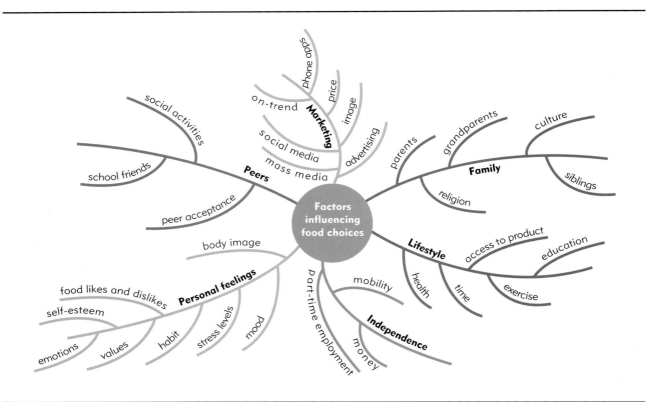

Influences on food choices

ACTIVITY 9·1
What influences your food choices?

1. Sketch your own 'spider map' like the one on the previous page to demonstrate the factors that influence your food choices. Your map should include the main branches of 'Family', 'Lifestyle', 'Independence, 'Peers', 'Personal feelings' and 'Marketing'.

2. Along each branch of the map, give specific examples of the main factors that influence your food choices. For example:
   - for 'Family', you might be influenced by your Asian, African or European culture, or by the favourite family dishes prepared by your parents or grandparents
   - for 'Marketing', you might include specific examples of food marketing that influence your food selection, such as a particular television advertisement or 'branding' at a sporting event
   - for 'Lifestyle', you could include your desire to make healthy food choices or the location of food outlets that you like near your school or on your way home.

3. Draw a table like the one below of 'Most important personal factors' and 'Most important community factors'. Based on the information in your map, list the four most important personal and community factors that affect the foods you select to eat.

Most important personal factors	Most important community factors
Foods I eat with friends need to be cheap, because I have a limited income.	I only go to those food outlets that are within walking distance of my home or accessible by public transport.

4. Write a paragraph to explain why these four factors influence your food selection more than other factors identified in your map.

FOOD MARKETING

While there are many factors that can affect your food choices, possibly one of the most influential is marketing. Food manufacturers use a wide variety of strategies such as advertising, development of cartoon characters, giveaways, two-for-one deals and inducements such as free toys or 'collectible' cards to encourage you to purchase their products. Today, advertising through various forms of social media such as Facebook, Instagram, mobile games (also known as 'advergames'), online activities, apps, email and SMS has become very prevalent. Food manufacturers also promote their products by sponsoring sporting events in which children participate. When combined with other, more traditional forms of advertising – on television and radio, in children's magazines, through direct mail, on food packaging or on outdoor media such as billboards – it is very hard to avoid.

Food manufacturers are highly strategic in their placement of advertising, aiming for the greatest impact on their target audience. Research shows that many children watch three hours or more of television daily. Evidence clearly shows that food manufacturers target children's viewing times to advertise their products, showing more than 70 advertisements for food products each week. Alarmingly, most of these advertisements are for foods such as highly processed breakfast cereals, confectionery, chocolate, soft drinks and snack foods, all of which provide little nutrient value. Constant repetition of these advertisements means that children are encouraged to pester their parents into purchasing the desired product. 'Pester power' has become one of the most influential marketing strategies.

Advertising can also highlight a particular property of a product that may be of interest to consumers. Many consumers today are keen to select foods that they think will provide them with a health benefit, such as those that are low in fat. However, while many products may have a lower fat content than similar traditional products, they may also be higher in sugar and, consequently, the product may not contain fewer kilojoules. Consumers therefore need to carefully read food labels so that they can analyse the nutritional information provided.

ACTIVITY 9·2
Junk food advertising in children's sport

Read the article 'Junk food advertising rife in children's sport' on Page 194, and then complete the activity that follows.

1. Working with a partner, complete a summary framework for the newspaper article based on the one on Page 195.

JUNK FOOD ADVERTISING RIFE IN CHILDREN'S SPORT

THE AGE

9 JUNE 2014

BY LUCY CARROLL

Children who participate in organised sport are being bombarded by junk-food advertising for up to four hours a week, a study shows.

Researchers found that children aged five to 14 who take part in rugby league, cricket and Little Athletics are most at risk of 'excessive' exposure to unhealthy messages through sponsorship by companies such as McDonald's and Coca-Cola Amatil.

Health experts say the saturation of fast-food and soft-drink promotion undermines the World Health Organization's (WHO) recommendation that the marketing of food high in saturated fat, sugar and salt be banned in children's settings.

'There is a fundamental conflict between sport and the promotion of sugary drinks,' says Jane Martin, executive manager of the Obesity Policy Coalition.

'This is banned in other countries because they take the WHO recommendations seriously. If this continues we will have a generation of children that die younger than their parents.'

The study, published in the Journal of Science and Medicine in Sport, combined a survey of almost 3500 parents about the amount of time their children spent playing popular sports with a survey of food marketing in NSW junior clubs. It found 75 per cent of junior athletics, track and field and rugby league clubs, and 42 per cent of cricket clubs in NSW are sponsored by food and drink makers.

At those clubs, children are exposed to unhealthy food messages for two to four hours each week. In comparison, fast-food promotion was virtually absent at swimming, tennis and martial arts clubs.

Lead author Bridget Kelly, a lecturer in public health at Wollongong University, said the huge promotional opportunities companies have through branding on signage, uniforms, certificates and vouchers was far greater than any funding given to the clubs.

'It certainly wouldn't spell the end of junior sport if this was banned,' Dr Kelly said. 'Multinationals like Subway and Domino's Pizza typically don't provide substantial amounts of money to the clubs, so they wouldn't fall apart if it was restricted.'

'Pervasive exposure' in environments that are supposedly healthy was sending contradictory messages to nearly two-thirds of Australian children who participate in sport regularly, she said.

'We know children are influenced by the sponsors.'

With about a quarter of Australian children overweight or obese, Ms Martin said parents were struggling to protect their children from the junk food industry.

'If these companies were really serious about not promoting to children, they wouldn't be doing this. To see little children targeted by McDonald's is exploitation.'

What is the issue outlined in the article?

What is the Obesity Policy Coalition? Why would this group be concerned about this issue?

Outline the two main arguments used in this article.

Argument 1:

Argument 2:

Plus (positives/supporting points):	Minus (negatives/concessions):	Interesting:

2 Individually, write a paragraph outlining what you think about the issue discussed in the newspaper article.

EATING FOR GOOD HEALTH

Good health is something we all aim to achieve so that we can enjoy life. Unfortunately, obesity, Type 2 diabetes and heart disease are some of the major health concerns that may confront many of us at some stage in our life. According to data from the National Health and Medical Research Council (NHMRC), 63 per cent of Australian adults and one in four children are overweight or obese. In addition, more than one million Australians have been diagnosed with Type 2 diabetes. Given that these health concerns have such a major impact on the lives of individuals in the community, health professionals have looked at ways of providing up-to-date advice on the amounts and types of food we should eat for good health. As a result, a range of tools to help us to wisely select food have been developed, including the Australian Dietary Guidelines and the Australian Guide to Healthy Eating.

THE AUSTRALIAN DIETARY GUIDELINES

In 2013 the Australian Dietary Guidelines were revised and updated to reflect the latest scientific evidence and expert opinion. In an effort to make the new guidelines easy to understand, the focus was changed to be on the foods, rather than the nutrients, that we should consume. The aim of this strategy is to help in preventing diet-related disease and to improve the health and well-being of the Australian community.

The new guidelines

1 To achieve and maintain a healthy weight, be physically active and choose amounts of nutritious food and drinks to meet your energy needs.
2 Enjoy a wide variety of nutritious foods from these five food groups every day: vegetables; fruit; wholegrains; lean meat, poultry, fish, nuts and seeds; milk, yoghurt, cheese. The guidelines also recommend drinking plenty of water.
3 Limit intake of foods containing saturated fat, added salt, added sugars and alcohol.
4 Encourage, support and promote breastfeeding.
5 Care for your food; prepare and store it safely.

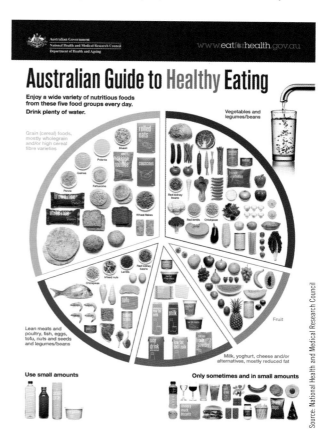

THE AUSTRALIAN GUIDE TO HEALTHY EATING

The Australian Guide to Healthy Eating has also been updated by the NHMRC. The Australian Guide to Healthy Eating is a visual representation of Guideline 2 of the Australian Dietary Guidelines: 'Enjoy a wide variety of nutritious foods from these five food groups every day.' It gives you a pictorial image of the foods in each of the five groups you should eat daily. The model is presented as a dinner plate containing the proportions of food from each group you should eat each day for good health.

The Australian Guide to Healthy Eating is based on eating a balanced diet that contains both plant and animal foods. However, it does contain alternatives for people who follow a vegetarian diet.

Source: National Health and Medical Research Council

ENERGY BALANCE

To maintain a healthy weight, it is necessary to achieve an energy balance, so that the energy you expend is equal to the energy you take in in the form of food. However, sometimes your body requires energy in addition to that supplied by the food you have just eaten. In these circumstances, your body needs to call on its reserves, which are stored in the fat tissue. If more energy is expended than is taken in, even over a fairly short period of time, weight loss may be the result. Alternatively, if you consume more food than your body is able to use in the form of energy, excess energy will be stored as fat tissue, leading to weight gain.

Testing knowledge

1. List five factors that may influence the foods we select to eat.
2. Explain why peers can influence the foods some adolescents choose to eat.
3. Why is the price of food seen to be an important factor in selecting food?
4. How do food manufacturers use social media to promote their products?
5. Explain why 'pester power' is considered to be an effective marketing tool.
6. Why has the National Health and Medical Research Council recently updated the Australian Dietary Guidelines?
7. Explain why the focus of the Australian Dietary Guidelines is now on the foods, rather than the nutrients, that we should consume rather.
8. Explain why the National Health and Medical Research Council developed the Australian Guide to Healthy Eating.
9. List three foods that are found in each of the following groups in the Australian Guide to Healthy Eating.
    - Cereals
    - Vegetables
    - Fruit
    - Milk
    - Lean meat and poultry
10. Make a list of the foods and drinks that the Australian Guide to Healthy Eating states we should:
    - use in small amounts
    - consume only sometimes and in small amounts.

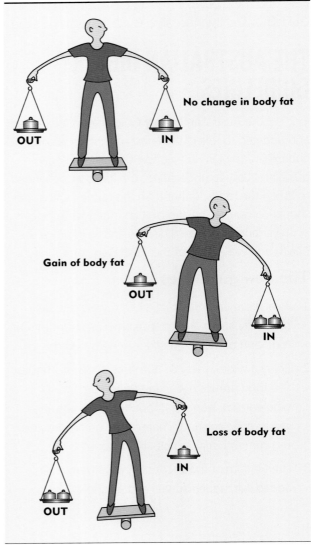

Energy balance

Although it would seem logical that two people who eat the same meals for breakfast, lunch and dinner would produce the same amounts of energy, this is in fact most unlikely. Each person's energy requirement is unique to them. There are, however, a number of factors that influence the amount of energy each person requires:

- The basal metabolic rate (BMR), or the amount of energy the body requires to function, varies according to a person's build. The greater the person's size, and the more muscle than fat they have, the higher their BMR will be.
- Generally, the younger a person is, the more energy they require.
- Gender is a factor in determining energy needs – males have a higher BMR than females.
- Children and adolescents who are rapidly growing have a higher BMR. Pregnant women, too, have an increased need for energy.
- People involved in heavy manual labour or athletes with intense training schedules also have an increased need for energy.

SOURCES OF ENERGY

To enable your body to function, you need energy. The body uses two main nutrients to generate energy – carbohydrate and fat. If these stores are depleted, the body uses its stores of protein to provide a secondary energy source. During digestion, carbohydrate is broken down – initially into glucose, and later into glycogen, which is stored mainly in the muscles of the body. Fat is broken down into fatty acids and stored as adipose tissue or fatty tissue if it is not required for energy production. Protein is broken down into amino acids and used for energy if all other sources have been exhausted.

The body can produce 16 kilojoules of energy from every gram of carbohydrate; 37 kilojoules of energy from every gram of fat; and 17 kilojoules of energy from every gram of protein. Nutritionists recommend that we obtain 55 to 60 per cent of our energy from foods that are high in carbohydrates, such as cereals, grains, breads, fruits and vegetables. Only 30 per cent of our energy should come from fats, and 10 per cent from protein.

Fat and sugar: what's the difference?

Many people are confused about whether there is a difference between fat and sugar and seem to think that they are the same thing! The truth is, they are completely different. Fat and sugar come from different food sources, have different functional properties and taste entirely different. Sugar is a form of carbohydrate, while fat is a lipid (which is another name for fat). Carbohydrates (including sugars) are more easily absorbed in to the body, but any that are not used to produce energy are stored in the body as fat.

Food manufacturers often combine fat and sugar in food products to make them more appealing to consumers. If you tried to eat a tablespoon of fat or sugar on its own, you would find it very unappetising. But when fat and sugar are combined, they become very palatable – and in some cases quite blissful. (Just think about eating chocolate!) Food manufacturers combine fat and sugar in many food items, such as chocolate, sweet biscuits, cakes or muffins, and even hamburgers, to make them appealing to consumers. This means that the sales of these items increase, and as a result, so do the manufacturers' profits. However, the problem for the consumer is that eating too many foods that are high fat and sugar can add many unwanted kilojoules to your daily intake.

FAT

As discussed earlier, fat, along with carbohydrate, is one of the two main nutrients that provide us with energy. However, the increasing incidence of obesity in society has led many consumers to view fat as 'the bad guy'. While it is true that we should limit the amount of fat we consume, not all fats are bad. We all need some fat in our diets, because it provides us with the essential fat-soluble vitamins A, D, E and K. These vitamins are vital for many aspects of good health, including encouraging healthy skin and bone growth and

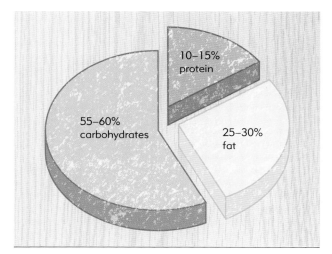

Main energy sources in the diet

preventing blood clotting. Rather than eliminating fat completely from our diets, it is more important to select foods that contain good fats.

Monounsaturated fats

Monounsaturated fats are the good fats. They can be found in olives, olive oil, avocados and nuts. It is also important to include the essential fatty acids omega-3 and omega-6, which are found in oily fish such as tuna.

Saturated fats

You should try to limit the amount of saturated fat, or bad fat, that you consume. Saturated fats are found in animal foods such as meat and dairy products, as well as in palm oil and coconut oil. Saturated fat can be a problem in the diet because it is high in cholesterol and can lead to heart disease. Many processed foods and fast foods are extremely high in saturated fat, and if eaten in excess, can lead to overweight and obesity.

Trans fats

Like saturated fats, trans fats are bad fats that can lead to serious health concerns and should be avoided. Trans fats increase the level of the bad low-density lipoprotein (LDL) cholesterol in your bloodstream, as do saturated fats. However, they also have the added problem of reducing the body's good high-density lipoprotein (HDL) cholesterol, which helps to protect you from heart disease. While trans fats are found naturally in small amounts in dairy products and meat, their main source in our diet is from the hydrogenation of vegetable oils. Food manufacturers use hydrogenated vegetable oils in the manufacture of processed foods and fast foods such as pastries, chicken nuggets, hamburgers and fried foods.

Characteristics, sources and health impacts of fats and oils

MONOUNSATURATED FATS	SATURATED FATS	TRANS FATS
Characteristics		
• Liquid at room temperature • Mainly vegetable sources • Fat-soluble vitamins A, D, E and K	• Solid at room temperature • Animal and vegetable sources • Concentrated energy source: 1 g = 37 kJ	• Semi-solid at room temperature • Unsaturated fat that behaves like saturated fat because of its chemical structure
Sources		
• Olive oil; peanut oil; canola oil	• Butter; cheese; cream; egg yolks; fat on meat • Coconut and palm oil; chocolate; hydrogenated fats	• Hydrogenated vegetable oils; processed foods
Health impacts		
• Lowers levels of blood cholesterol	• Cardiovascular disease • Raises levels of blood cholesterol	• Raises levels of blood cholesterol by increasing bad LDL cholesterol • Reduces good HDL cholesterol, increasing the risk of heart attack and heart disease

Testing knowledge

11 Explain the change in energy balance that can lead you to lose weight and to put on weight.

12 Define 'basal metabolic rate'.

13 List four factors that influence the amount of energy an individual requires.

14 Identify the two main nutrients used in energy production.

15 What is glycogen, and where is it stored in the body?

16 Explain what happens to fat during digestion and how fat is stored in the body for future use.

17 How many kilojoules of energy are provided by one gram of fat and by one gram of carbohydrate?

18 Outline the main differences between fat and sugar.

19 Discuss why it is important to include some fat in your diet. List four food sources of good fat.

20 What are trans fats, and why are they considered to be bad for our health?

CARBOHYDRATES

According to health professionals, we should get 50 to 60 per cent of our daily energy needs from carbohydrates. Carbohydrates are classified according to the number of molecules they contain. Monosaccharides are the simplest form of carbohydrate; disaccharides contain two monosaccharide molecules; and polysaccharides contain many monosaccharide molecules. Monosaccharides and disaccharides are both forms of sugar, whereas polysaccharides are found in the form of starch and cellulose in fruit, vegetables and wholegrain cereals. Health experts recommend that we select nutrient-dense carbohydrates such as pasta, wholegrain breads, fruit and vegetables as our sources of carbohydrate, rather than energy-dense foods containing sugars.

The glycaemic index

The **glycaemic index (GI)** ranks carbohydrate foods based on the immediate effects they have on blood sugar levels. Carbohydrate foods that release energy into the bloodstream over a prolonged period of time have a low GI rating. Carbohydrate foods that break down quickly during digestion, and therefore give an almost instant energy boost, have a high GI rating. All foods are ranked from 0 to 100:

- 55 or less = low GI
- 56–69 inclusive = moderate GI
- 70 or more = high GI.

The GI was originally developed to enable people with diabetes to better manage their blood sugar levels. Today, the GI is utilised by people who wish to lose weight or manage heart disease. Professional athletes also use the GI to enable them to develop their glycogen stores more effectively before competition, and also to recover quickly after an event.

However, one of the most important benefits of a low-GI diet is that it makes you feel full for longer. This means that a diet based on low-GI foods will reduce the likelihood that you will become hungry between meals, meaning you will be less likely to feel tempted to indulge in snack foods. Changing to a low-GI diet is not difficult – it simply involves swapping carbohydrate foods with a high GI for those with a low GI; for example, by selecting wholegrain or sourdough bread instead of white bread; by including at breakfast a cereal such as muesli that is made of oats and bran, and serving it with fruit; or by trying to cut down on potatoes and selecting pasta instead. It's easy!

GI rating of foods

FOOD	GI RATING
Prunes	29
Dried apricots	31
Fettuccini	32
Yoghurt (low-fat)	33
Chickpeas	33
Mixed-grain bread	34
Spaghetti	37
Apples	38
Sustagen (250 mL)	40
All-Bran breakfast cereal	42
Porridge	42
Oranges	44
Baked beans	48
Peas	48
Carrots	49
Bananas	55
Basmati rice	58
White bread	70
French fries	75
Corn Flakes	84
Potato (baked)	85

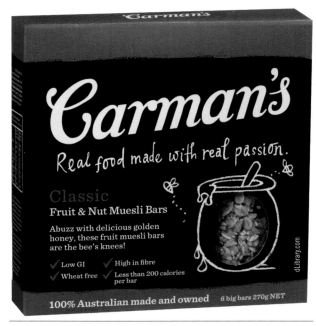

Low-GI snack bars

ACTIVITY 9.3

The glycaemic index

Access the Victorian Government's Better Health Channel website. Search the site using the keyword 'GI' and use the information you find to answer the following questions.

1. Explain why the GI is seen to be an important tool in selecting food.
2. List three factors that influence the GI ratings of food.
3. What is the meaning of the GI symbol that is found on some foods?
4. Using the information on the GI ratings of foods in the table on previous page or on the Diabetes Australia website, design a two-course meal that has a low GI rating. The taste.com.au website may also be helpful.

USING ENERGY

When you are sitting still or sleeping, your body requires energy to enable you to breathe, to pump blood around the body, to digest food and to keep the body temperature at 37°C. These processes all occur without your realising or having control over them. As a result, they are called involuntary life processes.

Energy is also used when you participate in physical activities such as walking, running, bushwalking or brushing your teeth. All these activities use different amounts of energy. They are called voluntary life activities, because you have control over the amount of work you ask your body to do.

Involuntary life processes

Voluntary life processes

ACTIVITY 9.4

Energy expenditure at different times of the day

Use the information in the graph below to answer the following questions.

1. At what times did Matthew get up and go to bed?
2. List the type of life processes that required energy during the sleep period.
3. Suggest why Sarah's energy expenditure rate rose between 8.30 am and 9 am.
4. When do you think Matthew had a Physical Education lesson?
5. What activities would Sarah most likely have been undertaking between 5.30 and 6.30 pm?

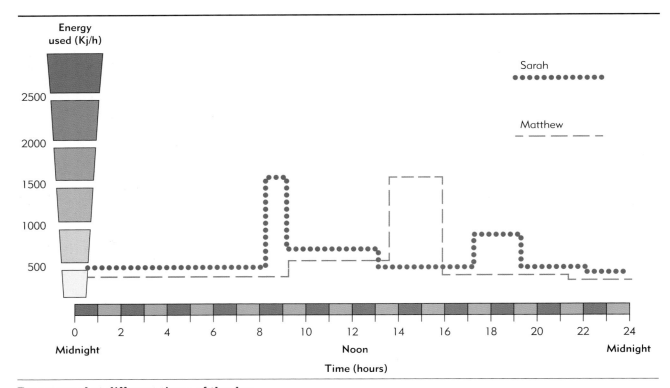

Energy used at different times of the day

Energy intensity

The amount of energy you use in undertaking an activity varies depending on your age, gender, height, weight and physical build. Other factors, such as the amount of time it takes you to complete the activity and the intensity of the activity, will also determine the amount of energy you use. Some activities, such as running and skipping, are very intense, and hence you use lots of energy when you participate in them. Other activities, such as sleeping, watching television or shopping, use hardly any energy, and so are considered very low-intensity activities. Your brain also uses energy to function, so when you are busy doing activities that require you to think, you also use more energy. For example, when you watch half an hour of television you use approximately 42 kilojoules, whereas playing a computer game for a similar amount of time uses about 86 kilojoules. While these two activities may seem similar, when you play a computer game, you are constantly thinking, and therefore using more energy.

Opposite is a list of the energy intensity levels of a range of activities.

Energy used in various activities

ACTIVITY	KILOJOULES EXPENDED PER 30 MINUTES, BASED ON A 16-YEAR-OLD, 60-KG PERSON
High-intensity activities	
Skipping	1560
Running	1443
Skiing (cross-country)	1130
Football (game)	1004
Tennis	1004
Cycling	941
Rollerblading	941
Basketball (game)	816
Moderate-intensity activities	
Dancing	628
Skateboarding	628
Weights training	628
Golf	602
Frisbee	377
Surfing	377
Walking	377
Low-intensity activities	
Driving	114
Vacuuming	90
Computer games	86
Eating	82
Ironing	70
Shopping	65
Personal care	55
Watching television	42
Sleeping	40

Cycling – high-intensity activity

Walking – moderate-intensity activity

Watching television – low-intensity activity

High-, moderate- and low-intensity activities

OBESITY

It is clear from the latest medical research that the spike in the number of Australian adults and children who are now overweight or obese is a direct result of the foods we consume.

The number of Australians who are overweight or obese has dramatically increased in recent years. The Australian Institute of Health and Welfare (AIHW) report, *Australia's Health 2014*, shows that 63 per cent of the Australian population is either overweight or obese. Overweight and obesity are major health concerns, because they are linked to the development of a range of serious health problems including type 2 diabetes, heart disease and some cancers.

Furthermore, the level of overweight and obesity in Australian children has risen sharply and is now among the highest in the world, with more than 25 per cent of young people being overweight or obese. It is estimated that by 2020, approximately 65 per cent of Australia's children will be overweight if current trends continue. As in other stages of life, obesity in childhood and adolescence can have a major impact on the health of young people. Type 2 diabetes and high blood pressure, once only seen in late adulthood, are now becoming more prevalent in young people who are overweight.

Much of the increase in the incidence of overweight and obesity in childhood is linked to the changing diets of Australian children. Although the amount of food young people eat has remained much the same as it was in the past, the energy density of the food being eaten has increased by approximately 11 to 15 per cent, resulting in a far higher kilojoule intake. This, combined with the fact that children today are far less physically active than children in past generations, has led to an increase in the numbers of children who are overweight or obese.

ACTIVITY 9.5

Australia struggles with obesity problem

Read the article 'Australia struggles with obesity problem: Heart Foundation', and then answer the following questions.

1. Identify four key points contained in the article.
2. Why would health officials be so concerned about the outcomes of this research on obesity rates?
3. List changes to the Australian diet and activity levels that have occurred in recent years and which could have led to the increasing rates of overweight and obesity.
4. Dietitian Karen Inge recommends making healthy food more affordable. Is this a realistic strategy, in your opinion? Justify your answer.
5. Compile a list of other recommendations that the community could adopt to reduce the levels of overweight and obesity among the Australian population.

AUSTRALIA STRUGGLES WITH OBESITY PROBLEM: HEART FOUNDATION

ALLAN RASKALL

SBS NEWS, 23 FEBRUARY 2014: 4.36 P.M.

Australians' waistlines are growing at a concerning rate, according to a new study from the Heart Foundation.

The study says men and women are much heavier than they were a quarter of a century ago, with two in three Australians considered to be overweight.

'That's a big problem because obesity leads to diabetes, some cancers and heart disease,' says Dr Rob Grenfell, National Director of Cardiovascular Health at the Heart Foundation.

More than four million Australians are considered to be obese while half a million are morbidly obese – they're over 40 kilos overweight.

The average Australian man now weighs 85.9 kilograms – up 6.5 kilos since 1989. The average woman is 71.1 kilos – 5.7kgs above the 1989 average.

Dietitian Karen Inge says people must take into account how small changes to a daily routine can affect their weight.

'Eating a little bit more, having that extra latte, not going for a walk – all these lifestyle changes and modifications make a difference to either gaining or losing weight.'

She says the key to a healthy nation is cheaper healthy food.

> 'This is what government needs to look at – making healthy foods more affordable. Because in the long run, prevention is better than cure.'
>
> But when it comes to healthy food, people often don't know what they're looking for, says Dr Grenfell.
>
> 'At the moment, we have a whole heap of gobbledygook on packets. There's organic, light, low-salt, low-fat. What does it all mean?'
>
> Experts say a lack of exercise and knowledge about healthy eating has led to the weight increase.

SBS Corporation

CARDIOVASCULAR DISEASE

Cardiovascular disease (CVD) is one of the major causes of death in Australia. CVD is a general term that is used to describe a range of diseases, including heart disease, stroke and blood vessel disease. During the year 2011–2012, approximately 3.7 million Australians had some form of CVD. In 2011, 45,600 Australians died of CVD – 31 per cent of all deaths during that year. It has been calculated that one Australian dies of CVD every 12 minutes.

Like other 'lifestyle' diseases, many deaths that result from CVD could be prevented. The main risk factors associated with the development of CVD are smoking, having high blood cholesterol, being physically inactive, being overweight or a diabetic, and having high blood pressure. It is therefore clear that, to minimise the risk of developing CVD, it is important to select a healthy diet, exercise on a regular basis and maintain a healthy weight.

DIABETES

Diabetes is the fastest-growing health condition affecting the Australian population. Diabetes occurs when the pancreas is unable to produce sufficient insulin to enable the glucose produced during digestion to be absorbed into the bloodstream.

There are two types of diabetes. Type 1 diabetes affects about 10–15 per cent of all people who suffer from diabetes. It occurs when the immune system damages the pancreas, meaning that the latter is unable to produce the hormone insulin. People who suffer from type 1 diabetes require a daily injection of insulin to break down the glucose in their bloodstream.

Type 2 diabetes is a far more common condition, accounting for 85–90 per cent of all diabetes cases. This type of diabetes is caused when the pancreas does not produce sufficient insulin to enable glucose to be absorbed into the bloodstream. The main risk factors for type 2 diabetes are being overweight or obese, being physically inactive and having another family member who has diabetes. Aboriginal and Torres Strait Islander people are much more likely to suffer from diabetes than other members of the Australian population.

According to health professionals, the number of Australians who are suffering from diabetes has doubled in the past 20 years. If the current trend continues, approximately three million people over the age of 25 will have developed type 2 diabetes by 2025. This increase in the incidence of diabetes is mainly a result of the increasing number of people in the community who are overweight or obese. Another major concern of the obesity epidemic is that type 2 diabetes is now beginning to be diagnosed in young people, rather than being confined to older adults, as was the case in the past.

Type 2 diabetes is often referred to as 'the silent killer', because some people who have it do not show any symptoms of the disease, leading to significant under-reporting of the condition. It has been suggested that the real number of people suffering from type 2 diabetes is likely to be double the number diagnosed with the condition. Diabetes can lead to CVD, severe kidney damage and eye disease, and can require the amputation of toes and limbs. The most effective way for people with type 2 diabetes to control their condition is through managing their food intake and exercise.

Testing knowledge

21. Identify the three main types of carbohydrate. Briefly explain the difference between them.
22. Why are some foods classified as having a low GI and others as having a high GI?
23. List three foods that have a low GI and three foods that have a high GI.
24. Explain the difference between voluntary and involuntary life processes.
25. Explain why one person may use more energy than another person when they both walk for 30 minutes.
26. What percentage of Australians are considered to be overweight or obese?
27. Outline two factors that have led to an increase in childhood diabetes. Explain how overweight and obesity can affect the health of young people.
28. List five of the main risk factors for developing cardiovascular disease.
29. Explain how diabetes occurs and how many people in the community are affected by this condition.
30. Outline the main health implications of diabetes.

MAINTAINING A HEALTHY WEIGHT

One of the keys to good health is to maintain a healthy weight range throughout life. But, given the ready availability of a wide variety of processed foods that are high in fat and sugar; the extra-large portion sizes commonly presented to us by food manufacturers and food retailers; and the impact of food marketing, this may be easier said than done! Two of the most important strategies in maintaining a healthy weight are to make sure you select food to minimise the intake of energy-dense rather than nutrient-dense foods such as snack foods, and to keep physically active. This will mean that your energy intake and energy output are more likely to be in balance, and weight gain will be minimised.

State governments and various health organisations have also developed a range of strategies to help you to maintain a healthy weight and to optimise your health.

Measure up!

The Heart Foundation has developed a simple guide to help adults determine whether they are overweight and at risk of developing a chronic disease. By measuring your waist circumference, you are able to check the amount of body fat you might have and where it is found. The more fat you have around your waist, the more likely it is that this fat will be deposited around your heart, kidneys, liver and pancreas. This greatly increases your risk of developing a serious health problem in later life. An adult woman should aim to have a waist measurement of less than 80 centimetres, while the waist measurement of an adult man should be no more than 94 centimetres.

The LiveLighter campaign

LiveLighter is a health campaign that was developed in Western Australia by the Heart Foundation and delivered in partnership with Cancer Council WA. In 2014 the campaign extended to Victoria where it is being delivered by Cancer Council Victoria and the Heart Foundation, with funding from the Victorian Government, and is viewed as a critical element of their Healthy Together Victoria program. The aim of the campaign is to encourage Australian adults to live healthier lives. A major focus of the campaign is to highlight the health problems associated with having excess body fat. The LiveLighter campaign describes excess body fat – especially fat that surrounds key organs in the body such as the heart and liver – as 'toxic' fat.

ACTIVITY 9·6
LiveLighter campaign

1. Access the LiveLighter campaign website.

   LIVELIGHTER

2. Review two television advertisements produced for the LiveLighter campaign.
3. Identify the techniques used in this campaign to highlight the issue of excess body fat.
4. Do you think this is an effective approach to addressing this important health issue? Why or why not. Justify your answer.
5. Find and read the fact sheet 'About sugary drinks'.
   a. Are you surprised by some of the facts presented about sugary drinks?
   b. Explain why this is seen as an important issue by the Livelighter campaign.
6. Complete the 'Sugary drinks calculator' to find out how much sugar you are drinking.
   a. Based on your results, do you think you need to reduce the amount of sugary drinks you consume each week?
   b. Discuss how the results you have calculated will impact on your life, both now and in the future.
7. The LiveLighter campaign's website provides consumers with a range of 'top tips' for improving their health and wellbeing.
   a. Examine two of these tips and discuss how easy they would be for members of the public to implement in their daily lives.
   b. Draw a conclusion about whether these tips could be effective in improving health.

ACTIVITY 9·7
Action on sugary drinks in schools

Worksheet

Read the article 'ACT Govt. Gets Gold Star for Action on Sugary Drinks in Schools', on Page 206 and then answer the following questions.

1. What new rules about sugary drinks has the ACT government introduced into schools?
2. Why does the Obesity Policy Coalition think that this is an important strategy for the long-term health of all Australians?
3. According to the article, cola drinks are the most popular form of sugary drinks consumed by Australians. Do you think this is true of you and your friends? Justify your answer.
4. What message do you think the ACT Government and the Obesity Policy Coalition is trying to send to children and parents through the introduction of this policy?
5. Do you think the strategy developed by the ACT Government will be valuable in addressing the issue of childhood obesity? Why or why not?
6. Would a similar strategy be successful if it was introduced into your school? What barriers might there be to its successful implementation? Explain.

Image courtesy of LiveLighter

ACT GOVT. GETS GOLD STAR FOR ACTION ON SUGARY DRINKS IN SCHOOLS

THE OBESITY POLICY COALITION (www.opc.org.au)
FRIDAY 21 FEBRUARY, 2014

The Obesity Policy Coalition (OPC) has today applauded the ACT Government for moving to phase out sugary drinks from government schools therefore removing one of the biggest contributors to extra sugar in children's diets.

Announced today by Chief Minister and Minister for Health Katy Gallagher, the new rules will see drinks with high sugar content, such as soft drinks, sports drinks and energy drinks, removed from vending machines by the end of the current term, and from canteens by the end of 2014.

The move was met with a strong endorsement by Jane Martin, Executive Manager of the Obesity Policy Coalition.

'The ACT Government gets a gold star for action and leadership with this initiative.

'We know that sugary drinks are a key contributor to excess sugar in children's diets and in turn if that sugar is not burnt off it can result in individuals becoming overweight or obese. With one in four Australian children overweight or obese, this step by the ACT Government shows commitment to the long-term health of its citizens.

'It takes real courage for a government to invest in preventative health measures, particularly in the face of pressure from industry. This announcement, which is part of the Towards Zero Growth-Healthy Weight Initiative shows the ACT Government is really leading the way in obesity prevention.

Sugary drinks are widely consumed by Australian adults and children. In the 12 months to October 2012, Australians bought 1.28 billion litres of carbonated/still drinks with sugar, with regular cola drinks being the most popular (447 million litres).

The Australian Dietary Guidelines recommend limiting sugary drinks and the World Health Organization also advises restricting consumption of sugary drinks.

'We know that a healthy diet is needed not only for good physical health, but also good mental health and optimal learning outcomes so it is pleasing to see the ACT Government creating learning environments where children have the best possible opportunities to grow into healthy, well-educated adults,' Ms Martin said.

It is also well established that obesity is a leading risk factor for type 2 diabetes, cardiovascular disease and some cancers.

'The removal of sugary drinks from schools will certainly contribute to reducing the growing burden of weight-related diseases like type 2 diabetes, heart disease and cancer on future generations.

'Many schools are investing time and energy into healthy lifestyle education, which is often undermined by the plethora of unhealthy food and drink available within the grounds, by removing sugary drinks children will receive a consistent message about healthy eating,' she said.

Walking towards good health

Health professionals constantly remind us that being involved in some physical activity each day is essential for good health because it helps us to maintain a healthy weight. It is recommended that, ideally, we should participate in 30 minutes of moderate-intensity activity on most days of the week. One strategy that has been developed is the 10 000 Steps program, which encourages us to try to increase our physical activity each day. This could involve participating in a specific exercise program, such as going for a walk each morning, as well the incidental exercise we undertake as a part of daily life. Incidental exercise is any type of physical activity or exercise you do in the natural course of your day, such as walking to and from school, walking between classrooms or walking around the local shopping centre.

When combined with a healthy diet, walking 10 000 steps each day will enable you to maintain a healthy weight and to minimise other lifestyle diseases such as CVD and diabetes. The problem is that we do not all consume a healthy diet all of the time. Snack bars, chips, chocolate and doughnuts are all delicious treats – but eating them comes at a price. These snacks are usually energy-dense, and consuming them on a regular basis can lead to weight gain. For example, a 60-kilogram person would take nearly 50 minutes to walk off the 1020 kilojoules of energy provided by a regular-sized (53-gram) Mars Bar.

Portion control

Another reason why people's waistlines have expanded in recent years is because we are now eating far more than we used to – and far more than we really need to! Food portions have dramatically increased in the past few decades, and, as the amount of food we eat has risen, so too has the amount of kilojoules we consume. In the 1980s, individual serves of soft drink were sold in 237-millilitre containers, whereas today, their usual size is between 500 and 600 millilitres. Equally, cupcakes and scones have doubled in average size over the same period from 40 grams to 80 grams.

King-sized, or 'Texas' muffins, have become the norm. Cakes, slices and biscuits sold in bakeries and cafes are now so big that they are in fact large enough for two or three people to share. Similarly, a giant serve of popcorn at the movies is enough for the whole family. Many other treats such as chocolate bars are now served in king-sized packets or twin packs. 'Super-sized' meals and 'meal deals' available in fast-food outlets also provide far more food (and, therefore, kilojoules) than we really need.

How has this happened? The food industry has been eager to increase serving sizes, which they dub 'upsizing'. Food manufacturers know that customers like to feel that they are getting value for money, and so will be happier to pay a little more for a larger serve than to pay what seems to be an expensive price for a smaller portion. Another trick manufacturers use is to bundle together food items to make a 'combo pack' or 'meal deal'. In such cases, the manufacturer wins, because the profit margin they make from the additional items, such as drinks or fries, is usually high, so the actual cost to them is almost negligible. Another problem is that in Australia, serving sizes listed on packages of food are often inconsistent, and the serving size is determined by the food manufacturer rather than by regulation.

Evidence clearly suggests that making even a small reduction in the amount of food you eat can make a big difference in managing your weight. So one of the best strategies is to resist king-sized treats and meal deals, and to only order small portions when eating out. Alternatively, share a sweet treat such as a muffin or cake with a friend. Eating more slowly will also enable your brain to register when you have had enough. Finally, only eat enough to satisfy your hunger – you can always leave some food on your plate!

Differing portion sizes of muffins

Making better snack food choices

One of the common characteristics of being a teenager is that you can constantly feel hungry. Adolescence is one of the periods of most rapid physical growth, during which your body demands food to satisfy its need for energy. (Sometimes, having three meals a day just isn't enough!)

One of the food habits that is common with many teenagers is the need to regularly snack. The problem is that snack foods may come to replace more nutritious foods in the diet. Snack foods and fast foods are generally low in important nutrients, but are usually high in fat, salt and/or sugar, and consequently provide a poor source of fuel for the body. These snacks are also high in kilojoules and can contribute to an increase in your weight. It is therefore important to remember that snack foods should be considered as 'extra foods' that should only be eaten sometimes.

While it may seem that eating fast food can be a health hazard, it is possible to choose foods that are lower in fat, sugar and salt such as a piece of fruit, a handful of nuts or a salad wrap.

Burritos – a delicious and healthy snack

ACTIVITY 9·8
Comparing snack foods

AIM
To compare the nutrient content of a range of snack foods

METHOD
1 Identify three different snack foods that you enjoy. You can select any of the items from the diagram below, or any other foods that you like to snack on.

2 Identify a commercial brand of each of the snack foods you have selected.

3 Research the company's website/s for information on each of the characteristics listed in the table. Alternately, use the Food Standards Australia New Zealand (FSANZ) nutrient database (NUTTAB 2010) for information.

4 Draw the following table and record your findings.

RESULTS

Comparison matrix for snack foods			
Characteristics	Items to be compared		
	Snack food 1:	Snack food 2:	Snack food 3:
Fat content			
Salt content			
Sugar content			
Kilojoule content			

ANALYSIS
1 When comparing the nutrient content of foods, is it more accurate to compare the portion size or the content per 100 grams of each food? Justify your answer.

2 Discuss the similarities and differences for each characteristic of the three snack food items you have compared.

3 Explain the implications for your long-term health of consuming each of these snack foods on a regular basis.

CONCLUSION
After comparing the fat, salt, sugar and kilojoule content of the three snack foods, what recommendations would you make to consumers about the link between long-term health and consuming these types of snack foods?

ACTIVITY 9·9
What's in snack food?

Many snack foods that students eat during study breaks are high in fat, sugar and salt. These ingredients can come in many forms and are included in many manufactured products. The appearance of the following terms on a label show that the food contains these ingredients, or equivalents.

Fat	Sugar	Salt
Beef fat	Aspartame	Baking power
Coconut	Brown sugar	Booster
Coconut oil	Corn syrup	Celery salt
Copha	Dextrose	Garlic salt
Cream	Disaccharides	Meat extract
Dripping	Fructose	Onion salt
Lard	Glucose	Monosodium glutamate (MSG)
Mayonnaise	Golden syrup	Rock salt
Sour cream	Honey	Sea salt
Nuts	Lactose	Sodium
Oil	Malt	Sodium bicarbonate
Palm oil	Maltose	Sodium metabisulfite
Vegetable oil	Mannitol	Sodium nitrate/nitrite
	Maple syrup	Stock (cubes)
	Molasses	Yeast extract
	Monosaccharides	
	Raw sugar	
	Sorbitol	
	Stevia	
	Sucrose	
	Xylitol	

Source: Government of South Australia: Department of Health (adapted)

1. Collect a range of snack foods such as single serves of potato crisps, savoury shapes, muesli bars, chocolate bars, noodles and low-fat yoghurt.
2. Analyse each product's label and record the following information in a table like the one below.

Name of snack food:			
Serving size:			
	Types of fat	Types of sugar	Types of salt
Total per serve			
Total per 100 grams			

3. Identify the types of fat, sugar and salt most commonly used in all the snack foods you investigated.
4. Discuss why food manufacturers use a wide range of fats, sugars and salts in the preparation of snack foods.
5. Explain why is it important that consumers look at the total amount of these ingredients, rather than rely on their identification in the ingredients list.
6. When comparing the fat, sugar and salt content of snack foods, explain why it is important to refer to the amount per 100 grams, rather than rely on the amount per serve.
7. Why is it important to look at the nutrient content of the food as a whole, rather than make your decision based on one nutrient alone?
8. Would you find this information on labelling useful when selecting snack foods in the future? Justify your answer.

OSTEOPOROSIS

Osteoporosis is a major health concern for the Australian population. Like many other health issues, it is a condition that only becomes evident in the later stages of life. Osteoporosis affects over one million Australians – approximately 50 per cent of women and 33 per cent of men over the age of 60.

Osteoporosis occurs when calcium is lost from the bones, making them very fragile and easily broken. A normal bone has a strong outer shell, but osteoporosis causes the outer shell of the bone to become thin. The internal structure of the bone is also affected, and instead of having a strong, mesh-like structure, the bone develops large holes, making it thin and very weak. The most common osteoporosis-related fractures occur in the hips, spine, pelvis and wrists. People who have osteoporosis often suffer from severe pain, especially in their backs. Height loss and a stooped appearance are other side effects of osteoporosis.

Healthy bones

While osteoporosis does not usually become evident until late adulthood, it is during adolescence and early adulthood that you can take steps to avoid developing this debilitating condition. To minimise your risk of developing osteoporosis, it is essential to develop peak bone mass during your teenage years. Peak bone mass can be achieved by combining a diet high in calcium with significant weight-bearing exercise during adolescence, when you are rapidly growing. Peak bone mass is usually achieved by the age of 18, but you can continue to gain bone mass until the age of around 30.

Protein, calcium, phosphorus and vitamin D are the nutrients that are essential for healthy bones. During adolescence, you need larger amounts of calcium to provide for the considerable increase in the bone structure of your body. Boys aged eight to 11 need approximately 800 milligrams of calcium daily, but between the ages of 12 and 15, their need for calcium increases by approximately 50 per cent, to 1200 milligrams daily. This is greater than the amount of calcium needed by girls, mainly because boys will generally grow to be taller than girls. Girls aged eight to 11 need slightly more calcium than boys of the same age (900 milligrams daily) because they start their growth spurt earlier; however, they need only 1000 milligrams daily between the ages of 12 and 15.

Foods high in calcium

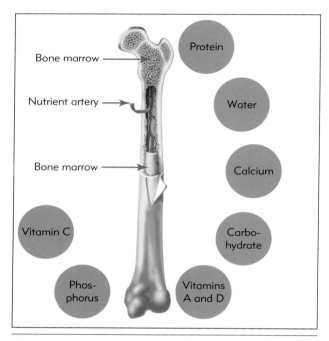

Nutrient needs for bone growth

Selecting foods high in calcium

Many adolescents and older Australians do not consume enough calcium to meet their daily needs. Calcium is found in a wide variety of foods, but some calcium sources are better than others. Remember that some foods and additives actually work against the absorption of calcium, preventing it from passing from the food during digestion. The foods most likely to work against calcium absorption are fats, the fibre in cereals, and some green, leafy vegetables. Salt, caffeine (in cola soft drinks and tea and coffee) and phosphates (which are added to processed foods and drinks) also increase the amount of calcium that is lost through urine.

The best food sources of calcium

FOOD SOURCE	AMOUNT REQUIRED	MILLIGRAMS OF CALCIUM
Whole milk	1 cup (250 mL)	310
Fortified milk	1 cup (250 mL)	438
Evaporated milk	1 cup (250 mL)	658
Skim milk	1 cup (250 mL)	310
Chocolate-flavoured milk	1 small carton (300 mL)	348
Fortified soy drink	1 cup (250 mL)	295
Yoghurt	1 small carton (200 g)	255
Cheddar cheese	1 slice (30 g)	240
Edam cheese	1 slice (30 g)	288
Swiss cheese	1 slice (30 g)	320
Parmesan cheese	40 grams	460
Chocolate (milk)	6 squares (30 g)	73
Canned salmon (eaten with the bones)	½ cup (125 mL)	325
Sardines	5 small	285
Dried figs	5	150
Baked beans	½ cup (125 mL)	47
Carrot	½ cup (125 mL)	23
Spinach or silverbeet	½ cup (125 mL)	40
Parsley	1 tablespoon (15 mL)	20
Honeydew melon	1 cup (250 mL)	64
Almonds	¼ cup (60 mL)	95

ACTIVITY 9·10
Taste testing cheese

Cheese is one product that is a very good source of calcium. For example, as shown in the list on Page 210, 30 grams of cheddar cheese contains approximately 240 milligrams of calcium. Your teacher will arrange to have a variety of different brands of cheddar cheese for your class to test.

Cheddar cheese

AIM
To compare the sensory properties and calcium content of a variety of cheddar, or tasty, cheeses

METHOD
1. Place samples of each cheese on a small plate and label each of them with a number.
2. Try some of each cheese and fill in a table similar to the one below. Leave the 'Brand of cheese' row blank for the moment.
3. When you have finished your tasting, your teacher will tell you the brand of each of the samples.
4. Read the label on each of the cheese packets to find out how much calcium is contained in each type of cheese.
5. Complete the table.

RESULTS

Sample number	1	2	3	4
Brand of cheese				
Colour (for example, dark, light, bright, dull)				
Flavour (for example, salty, strong, weak)				
Mouth feel or texture (for example, smooth, creamy, granular)				
Comment (for example, what you thought of the flavour)				
Milligrams of calcium per 100 grams				
Your rating (for example $7/10$)				

ANALYSIS
1. Which cheese had the best flavour?
2. In your opinion, which cheese had the best mouth feel?
3. Which cheese contained the highest amount of calcium?
4. Which cheese did you prefer overall? Why did you choose this one?

CONCLUSION
Which of the cheese samples would you recommend to your friends? Why? If they do not like cheese, suggest two other foods they could eat that are high in calcium.

Testing knowledge

31. Describe four strategies or tools that individuals can use to help them maintain a healthy weight.
32. Why do health professionals state that having excess fat around your waist is a danger to health?
33. Explain what is meant by the term 'incidental exercise', and why it is important as a part of a healthy lifestyle.
34. Explain how the portion sizes of food sold to consumers have changed in recent decades, and why this change has occurred.
35. Outline two strategies manufacturers have used to increase the serving size of food products.
36. Describe the effect of osteoporosis on the bones.
37. List the main nutrients that are needed for bones to grow.
38. What is 'peak bone mass', and how is it achieved?
39. Why do boys who are 12–15 years old need more calcium than girls of the same age?
40. Make a list of the factors that stop the body from absorbing calcium properly. Identify three foods that are the best sources of calcium.

THE IMPORTANCE OF FIBRE IN THE DIET

Dietary fibre is essential to good health. Foods that are high in dietary fibre have been shown to help lower blood cholesterol, reduce glucose absorption and prevent diseases of the bowel such as constipation and diverticulitis. High-fibre foods can also help you to control your weight because they provide greater satiety (that is, make you feel full for longer). Health professionals recommend that adults should consume approximately 25–30 grams of dietary fibre each day. Unfortunately, most Australians do not include sufficient dietary fibre in their diets, consuming only 18–25 grams of fibre daily.

Dietary fibre is found in the cell walls of all plant foods, such as fruits and vegetables, peas, beans and cereals. Foods that have a high level of dietary fibre are usually low in fat, salt and sugar.

There are two main types of dietary fibre: insoluble dietary fibre and soluble dietary fibre. The body is unable to digest or absorb insoluble dietary fibre, but it is nevertheless very important to include it in your daily meals because it adds bulk to the diet and helps to eliminate the waste material from your body. By helping food pass through the digestive system more quickly, insoluble dietary fibre helps to maintain your bowel health and prevents you from becoming constipated. A diet high in insoluble dietary fibre is also known to increase the number of good bacteria, or intestinal flora, that live in your gut. These good bacteria help to prevent serious diseases of the bowel such as bowel cancer.

Soluble dietary fibre is found in oatmeal, legumes and the bran from rice and barley.

One of the most important functions of soluble dietary fibre is to help to reduce blood cholesterol levels. It has also been found to be helpful in managing diabetes by stabilising blood glucose levels.

Many vegetables are high in insoluble dietary fibre

Insoluble dietary fibre is found in a wide range of vegetables.

The table below lists the amount of insoluble dietary fibre found in some popular vegetables.

VEGETABLE	INSOLUBLE FIBRE (RAW) PER 100 GRAMS (G)
Beans	2.7
Broccoli	4.1
Capsicum	0.9
Carrot	3.3
Cauliflower	1.8
Celery	1.8
Cucumber	1.1
Eggplant	2.3
Lettuce	1.7
Mushroom	2.5
Onion	1.5
Parsnip	2.5
Peas	2.3
Potato	1.7
Pumpkin	1.5
Spinach	2.7
Sweet corn	4.5
Sweet potato	2.0
Tomato	1.2
Zucchini	1.6

Oatmeal, served here with milk and caramelised apple, is high in soluble fibre.

ACTIVITY 9.11
Comparing a commercially processed fast food and an equivalent homemade product

Worksheet

AIM
To compare the physical and sensory properties of a commercially processed fast food and an equivalent homemade product

EQUIPMENT
- 1 quantity of homemade sausage rolls
- Commercial sausage rolls
- 2 oven trays

METHOD
1. Prepare the Traditional Sausage Rolls from the recipe on Page 224. Note: the sausage rolls should be approximately the same size as the commercial sausage rolls.
2. Place the commercial sausage rolls on an oven tray and heat according to the manufacturer's instructions.
3. Complete a comparison of the fibre content and the physical and sensory properties of both types of sausage rolls. Use a table similar to the one below. Refer to the sensory wheel on Page 2 for additional words to assist you.

RESULTS

Physical and sensory properties	Traditional sausage rolls	Commercial sausage rolls
Quantitative measures		
Weight (in grams) of one sausage roll		
Height and length (in centimetres) of one sausage roll		
Colour of cooked product		
Preparation time		
List of ingredients high in dietary fibre (refer to product label for information)		
Qualitative measures		
Appearance		
Aroma		
Flavour		
Texture or mouth feel		
Overall appeal		

ANALYSIS
1. Describe the similarities and differences in the heights, weights and colours of the homemade and the commercial sausage rolls.
2. Did the time involved in the preparation of the Traditional Sausage Rolls detract from their overall appeal?
3. Which product is more likely to be higher in dietary fibre? Why?
4. Which variety of sausage roll had the most appealing appearance?
5. Which variety of sausage roll had the most appealing aroma?
6. Which variety of sausage roll had the most appealing flavour?
7. Which variety of sausage roll had the most appealing texture or mouth feel?
8. Which variety of sausage roll was most appealing overall?
9. Which product would be preferable to include in a healthy diet? Why?

CONCLUSION
After analysing your results, decide which product you would prefer to use again. Explain your reasoning.

Homemade sausage rolls

INDIVIDUAL DIETARY NEEDS

There are many people who need to manage their diets to avoid particular foods to which they may be allergic such as peanuts or eggs. Some people may have a hypersensitivity to a food – for example, wheat or certain fruits – and so they must also be careful about the foods they select to eat. Some people choose to follow a vegetarian diet; they too have individual dietary needs that must be planned for.

Food allergies

A food allergy is an abnormal immunological reaction to food. A foreign substance, usually a protein, enters the bloodstream, and an antibody is produced to fight it. Each time the foreign substance enters the body again, more antibodies are produced.

The reaction caused by a food allergy is usually physical and occurs within an hour of exposure to the food. Symptoms such as hives, rashes, hayfever, asthma, stomach pain or diarrhoea, headache or swelling of the face or eyelids may occur. In some cases, the physical symptoms of a food allergy can become more severe with each exposure, and can even be life-threatening. There is no cure for a food allergy, so the treatment is simply to avoid the problem food. It is recognised that the risk of developing an allergy is much higher if another member of your family also has an allergy. Foods that can cause allergies in some people include milk, fish, shellfish, peanuts, eggs and legumes. Many children grow out of food allergies. However, peanuts are an exception, and this allergy is often severe and lifelong.

People with food allergies must read and understand food labels to ensure they do not eat foods that are toxic to their bodies. FSANZ aims to assist people with a food allergy by requiring food manufacturers to include information on food labels if the food contains an ingredient that may cause a severe allergic reaction such as anaphylaxis, regardless of how small the amount added. People involved in food preparation, either in the home or in the hospitality industry, should also take particular care when preparing and serving food to people with allergies to ensure their food is not contaminated.

Food hypersensitivity

Some people are born with a condition that makes it impossible for them to metabolise a particular food or nutrient. They may lack an enzyme or be unable to produce an enzyme in sufficient amounts to digest certain foods. Food hypersensitivity reactions are similar to food allergy symptoms, but are generally less severe. The reactions are delayed for 24 to 48 hours after exposure to the food, and the severity of the symptoms usually decreases the more the food is avoided. Foods that may cause hypersensitivities include chocolate, wheat, cola, eggs, garlic, cucumbers and certain fruits; for example, oranges, strawberries, pineapples and tomatoes.

Food hypersensitivities that are relatively common in our society include lactose intolerance and gluten intolerance. Lactose is the sugar (carbohydrate) found in milk and milk products. People who are lactose intolerant lack the enzyme lactase in their system, or have it in insufficient amounts. They suffer from bloating of the stomach, abdominal cramps and diarrhoea when they eat dairy products.

Coeliac disease

Coeliac disease is a disease of the small intestine associated with permanent intolerance or hypersensitivity to gluten, the protein in wheat. According to Coeliac Australia, approximately one in 100 Australians are affected by coeliac disease. However, it seems that almost 75 per cent of people who have the disease remain undiagnosed. This

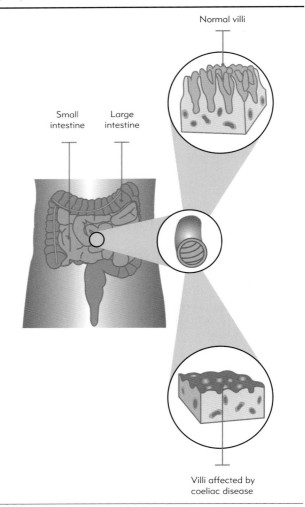

The effect of coeliac disease on the digestive system

means that approximately 160,000 Australians who have coeliac disease do not yet realise that they have it.

When someone who has a predisposition to coeliac disease eats foods containing gluten, damage occurs to the lining mucosa of the small intestine. Normally, food passes from the stomach through the duodenum and into the small intestine, where it is gradually digested and absorbed into the system. Leftover material passes into the large intestine (colon) and is eventually passed out of the body as faeces. The small intestine is a long tube lined by folds called villi. These 'villi' project into the intestine, like fingers, and increase the surface area of the intestine, so much so that it is around the size of a tennis court. In untreated coeliac disease, these villi are damaged by gluten, and the mucosa becomes flat and inflamed so that the area available for absorption is reduced (to the size of a card table). Because of this, unabsorbed food passes down to the large intestine and out in the bowel motions. Diarrhoea is often quite severe, and the abdomen may be bloated or distended by gas and undigested food. Poor growth and weight loss may result from the malnutrition caused by loss of food material.

The removal of gluten from the diet allows the lining of the intestine to return to normal. The diet must be gluten-free, rather than just low in gluten, and must be maintained for life. Gluten occurs in wheat, rye, oats, triticale and barley. It is therefore present in a range of commonly eaten foods such as breads, pizza, pasta, cereals, cakes, biscuits and pies. It is also used as a thickener in processed food.

Cereal foods suitable for people with coeliac disease

The following foods are made from cereals that do not contain gluten:
- rice and rice products, including rice pasta, rice cakes, rice crackers, puffed rice, baby rice cereal, rice noodles and rice bran
- buckwheat and sorghum
- maize (corn) and related products, including polenta, maize cornflour, pure corn chips, taco shells and popcorn
- sago, tapioca and arrowroot
- lentils and chickpea flour
- soy and soy products, including soy bran and potato flour
- gluten-free mixes for bread, pastry, pizza, cakes and biscuits.

ACTIVITY 9·12

Gluten-free patty cakes

AIM
To determine the acceptability of small patty cakes made from gluten-free flours

METHOD

Work in a group to prepare a batch of patty cakes using each of the following flours.
- Buckwheat flour
- Gluten-free flour mix
- Rice flour
- Soy flour

See the recipe for Gluten-Free Patty Cakes on Page 300 in Chapter 12. Share the patty cakes among the group and evaluate the sensory properties – appearance, aroma, flavour and texture – of each batch. Remember to have a drink of water in between tasting each variety of patty cake.

RESULTS

Type of flour used	Appearance	Aroma	Flavour	Texture
Gluten-free				
Rice				
Soy				
Buckwheat				

ANALYSIS
1. The batch of patty cakes made with gluten-free flour could be described as a control recipe. What does this term mean, and why is a control recipe important when testing recipes?
2. Which flour produced the best shape and colour for a patty cake?
3. Which flour produced the patty cake with the best flavour and texture?
4. Which flour produced the patty cake with the most inviting aroma?
5. Explain why accurate measurement is important when testing recipes.
6. Why should you drink water before testing the flavour and texture of each batch of patty cakes?

CONCLUSION
1. What are your recommendations for the most acceptable gluten-free flour for patty cakes? Explain your answer.
2. Identify the group of people in the population who would be most interested in the findings of this recipe test.

VEGETARIAN DIETS

There are some people who, because of their religious beliefs, are opposed to the killing of animals for food or who, because of their views on health, follow a vegetarian diet. There are two main types of vegetarians.

1. Lacto-ovo-vegetarians do not eat meat, but do eat dairy products and eggs.
2. Vegans do not eat meat or any other product that comes from animals.

It is essential that vegetarian diets contain an adequate supply of complete protein. Soybeans and soybean products such as soy milk, tofu and textured vegetable protein are good sources of complete protein.

While some plant foods contain protein, they are generally lower in one or more of the essential amino acids that make up complete protein. To overcome this problem, it is important to complement proteins from plant sources to make up a complete protein. This is done by combining foods from cereals such as wheat, rice and pasta with pulses such as dried beans (soy, haricot or cannellini beans), lentils, chickpeas or nuts in the one meal.

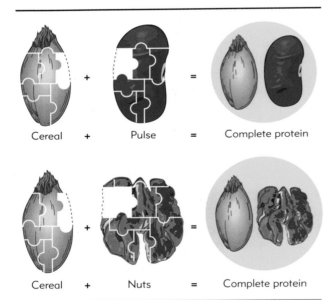

Complementing proteins

Because vegetarians do not eat meat, they may also have trouble gaining adequate amounts of iron in their diets. They need to make sure they eat lots of green, leafy vegetables, wholegrain cereals, dried fruit and legumes that contain non-haem iron. It is also important to eat foods high in vitamin C at the same time, because this vitamin helps the body to absorb the iron from these plant sources.

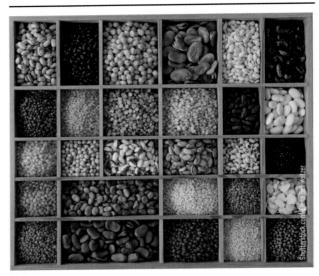

Pulses

Top tips for vegetarian eating
Meal planning

1. Try to maintain your energy levels by frequently eating small meals and snacks during the day.
2. Eat a wide variety of foods to obtain all the nutrients needed for good health.
3. Remember to base everyday meals on cereals such as rice, pasta or bread, and to include pulses such as lentils, soybeans, tofu and nuts so that you have a complete source of protein.
4. Have a glass of orange juice for breakfast or include capsicum or tomato in a salad for lunch or dinner; the vitamin C in these foods can help to increase the amount of iron the body absorbs from each meal.
5. Make sure you still include lots of dairy products such as milk, cheese and yoghurt in your daily meals to provide sufficient supplies of calcium.

Food preparation

1. Dried legumes such as chickpeas need to be soaked overnight and then cooked for several hours to make them soft enough to eat.
2. Canned chickpeas and beans are a great alternative to using dried beans in vegetarian recipes.

3 Overcooked lentils become mushy. Red lentils take only about 20 minutes to become soft. Green or brown lentils need to be cooked for approximately one hour.

4 Tofu is a delicious source of protein and iron, but it has a delicate texture, so it is important to prepare and cook it with care.

ACTIVITY 9·13
Selecting vegetarian meals

1 Access the Sanitarium website and search for information about vegetarian diets.
2 Explain three health benefits of following a vegetarian diet.
3 Write down three food and nutrition tips for people who follow a vegetarian diet. Ensure these tips are different from those listed in this chapter.
4 Examine two recipes on the Sanitarium website and identify the ingredients in each of them that would provide a source of complete protein.
5 Using the information from the website, list four ways you could increase your consumption of wholegrain cereals.
6 Write out a one-day eating plan for a vegetarian that includes breakfast, lunch, dinner and snacks based on the information you have discovered on the website.

Testing knowledge

41 Explain why it is important to include dietary fibre in your diet every day.
42 What is soluble dietary fibre? Explain two health benefits of including soluble dietary fibre in your diet.
43 Define a food allergy. List the foods most likely to cause food allergies.
44 Identify some of the physical symptoms that may occur if a person has a food allergy.
45 What is food hypersensitivity? List the foods that may cause food hypersensitivity.
46 What is coeliac disease, and how does it affect the human body?
47 List some of the cereal foods that are suitable for people who suffer from coeliac disease.
48 Explain the difference between a lacto-ovo-vegetarian and a vegan.
49 Define the term 'complementing protein'.
50 Discuss how vegetarians can make sure they obtain an adequate supply of iron in their diet.

THINKING SKILLS 9·1

'The youth of today need to take greater responsibility for their health and future wellbeing.'

Work in small teams to develop arguments for the affirmative and negative in response to this statement.

DESIGN ACTIVITY 9·1
Risotto

Worksheet

DESIGN BRIEF
Coeliac Australia is about to update the recipe section of its website. The organisation is looking for new and exciting recipes for risotto that are suitable for people who have coeliac disease.

1 Write your own design brief for a risotto dish based on a classic risotto recipe. Develop your own specifications based on the five Ws – who, what, when, why, where.
   - Who – for whom will the risotto be suitable?
   - What – the risotto must reflect the latest trends in ingredients and flavours
   - When – in what season will you serve the risotto?
   - Why – why will the risotto be useful to include on the Coeliac Australia website?
   - Where – will the risotto be served as a home-style meal or a as a cafe meal?
2 Format sentences or statements based on each of the above specifications into a paragraph that will become your own design brief.
3 Based on the specifications in your design brief, develop four to five evaluation criteria questions by which you can judge the success of your finished product.

INVESTIGATING
1 Trial the Basic Risotto recipe on Page 222 before beginning your design to develop an understanding of the ingredients and processes involved in preparing a risotto.
2 Research a variety of food magazines and websites to identify current trends in ingredients and flavours. Make a list of the most popular flavouring ingredients.
3 Look at a recipe website or the Sunrice website to answer the following questions.
   a What information is provided about the different rice varieties?
   b What are the main nutritional benefits of rice?
   c List the main uses of arborio rice.

CHAPTER 9 EATING WELL FOR THE FUTURE

d List four important tips for cooking rice.
e List the important tips for cooking with arborio rice.
f Risotto is a traditional Italian dish. List three other rice dishes that are traditional in other cultures' cuisines.
g Examine recipes for risotto on the website/s and note their combination of flavour and textural ingredients. Identify the order in which the ingredients are added to the risotto.

GENERATING

1 Develop two design options based on the specifications in your design brief. Use the recipe map below as a guide. Select the flavouring and complementary ingredients to develop your own design.
2 Use a decision table (see example on Page 9) to help you select your preferred option.

PLANNING AND MANAGING

1 Prepare a food order.
2 Write up a production plan, including relevant health and safety issues.

PRODUCING

1 Prepare the product.
2 Note any modifications or changes you made during the production of the recipe.

EVALUATING

1 Evaluate the success of your risotto using the previously established criteria.
2 Was the flavour of the product appetising? In your opinion, does the recipe require any further modification to enhance the product's flavour? Would you add or omit some ingredients if you were to make this product again?
3 What aspect of the production did you find most challenging? Outline how you managed this challenge.
4 Comment on your overall management of time for this task – discuss your designing and planning as well as the production of the meal.
5 Plot all the ingredients of your meal on a diagram of the Australian Guide to Healthy Eating. Comment on the nutrient value of the risotto you produced.

Recipe map for risotto

DESIGN ACTIVITY 9·2

Vegetable lasagne

DESIGN BRIEF

A relative has asked you to produce a vegetable lasagne for them that is suitable to serve as an evening meal. The lasagne should include some dairy products and feature a range of different-coloured vegetables to produce a nutritious meal. It should use a pasta that cooks quickly so that the vegetable lasagne can be prepared in 80–100 minutes.

Worksheet

1. Write a design brief based on the five Ws.
   - Who – who is the lasagne to be made for?
   - What – what are the key ingredients to be included in the lasagne?
   - When – at what time of day or in which season will the lasagne be serve?
   - Where – where will the lasagne be served?
   - Why – why has your relative requested the meal?
2. Develop four or five questions to evaluate the success of your new lasagne product.

INVESTIGATING

1. List four examples of prepared sauces available in supermarkets that would be suitable for use as a component of a lasagne recipe.
2. List two vegetables for each of the colours below that could be used in a vegetable lasagne.
   - Orange
   - Yellow
   - White
   - Red
   - Brown
   - Green
   - Purple
3. Instant-dried and vacuum-packed lasagne sheets absorb a lot of moisture during cooking. What modifications would be required to a traditional lasagne recipe to ensure that the pasta sheets were tender?

GENERATING

1. Use the recipe for Vegetable Lasagne on Page 221 and your research to complete the following recipe map.

Recipe map for a vegetable lasagne

Vegetable sauce	White sauce	Lasagne
•	•	•
•	•	•
•	•	•
•	•	•
•	•	•

2. Using the information in your recipe map, develop two design options for your vegetable lasagne.
3. Construct a decision table. Select the option you would prefer and explain your choice.

PLANNING AND MANAGING

1. Complete a food order.
2. Before producing your vegetable lasagne, write up a production plan, noting any safe work practices to be followed, and identify the major processes to be used.

PRODUCING

1. Produce your preferred option.

EVALUATING

1. Use the questions you developed to evaluate the success of your product.
2. Discuss the most challenging part of the production. Outline how you managed this challenge.
3. Comment on your overall management of time for this production – discuss your designing and planning as well as the production of the meal.
4. Plot all the ingredients of your lasagne on a diagram of the Australian Guide to Healthy Eating. Comment on the nutritional value of the meal.

Ingredients for a vegetable lasagne

DESIGN ACTIVITY 9.3

A two-course family meal

DESIGN BRIEF

Plan a two-course meal for your family based on Guideline 2 of the Australian Dietary Guidelines: 'Enjoy a wide variety of nutritious foods from these five food groups every day: vegetables; fruit; wholegrains; lean meat, poultry, fish, nuts and seeds; milk, yoghurt, cheese.'

1. Write a sentence or statement that answers each of the five Ws.
   - Who – which family members will share the meal?
   - What – an everyday, two-course meal or a meal for a special family occasion based on Guideline 2 of the Australian Dietary Guidelines?
   - When – at what time of day or in which season of the year will the meal be prepared?
   - Where – will the meal be served indoors, or as a more casual meal outdoors?
   - Why – an explanation of the occasion at which the meal will be served.
2. Format the sentences or statements into a paragraph that will become your own design brief.
3. Use the specifications – that is, the constraints and considerations in the brief – to develop four to five evaluation criteria questions that will allow you to judge the success of the product.

INVESTIGATING

1. Using the recipe index, write down the name and a brief description of two recipes that you think would be suitable for your meal. Choose two courses that your meal will consist of from:
   - starters
   - main courses
   - desserts.

GENERATING

1. Complete a decision table similar to the one below.

Decision table

Decision to be made: a menu for the family meal	
Meal 1	Meal 2
Course 1:	Course 1:
Course 2:	Course 2:
Advantages:	Advantages:
Disadvantages:	Disadvantages:

- Decision made (preferred meal):
- Identify the menu selected.
- Justify your decision – use the information in your advantages and disadvantages in your discussion.

PLANNING AND MANAGING

1. Write up a food order for your two-course family meal.
2. Prepare a production plan so that you can prepare the meal in the time available.

PRODUCING

1. Prepare your family meal and record any changes you make to the ingredients or method of the recipe during production.

EVALUATING

1. Answer the evaluation criteria questions you developed during the design stage.
2. Describe the sensory properties – appearance, aroma, flavour and texture – of your two-course family meal.
3. What was the most challenging part of the production? Outline how you managed this challenge.
4. Comment on your overall management of time for all of the tasks in this design activity – discuss your designing and planning as well as the production of the meal.
5. If you were to make this meal again, what changes would you make to improve it?
6. Plot the ingredients of your meal on a diagram of the Australian Guide to Healthy Eating. Discuss how well your meal meets Guideline 2 of the Australian Dietary Guidelines.

An everyday meal

VEGETABLE LASAGNE

Vegetable sauce

- 2 teaspoons olive oil
- ½ onion, finely diced
- 1 clove garlic, crushed
- 100 grams mushrooms, sliced
- ½ zucchini, grated
- 200 grams canned tomatoes, diced
- 1 tablespoon tomato paste
- ¼ carrot, grated
- ¼ cup celery, finely diced
- ¼ red capsicum, finely diced
- ¼ Granny Smith apple, peeled and grated
- ¼ teaspoon dried oregano, basil and rosemary

White sauce

- 1½ cups milk
- ⅛ onion, finely diced
- 1 clove
- pinch of nutmeg
- 30 grams butter
- 2 ½ tablespoons flour
- salt and pepper

Assembling the lasagne

- 3 vacuum-packed sheets of fresh lasagne pasta
- 2 slices mozzarella cheese
- 1 tablespoon parmesan cheese, grated

MAKES 1 LARGE OR 2 SMALL PORTIONS

METHOD

Vegetable sauce

1. Heat oil over medium heat and sauté the onion and garlic. Cook for 1 to 2 minutes but do not brown. Add the mushrooms and cook until soft.
2. Add the remaining ingredients to the saucepan, cook for 3 minutes and remove from heat. The sauce will be sloppy.

White sauce

1. Place the milk, onion, clove and nutmeg in a small saucepan and heat until simmering. Remove from heat and allow to cool.
2. Drain and retain the cooled, flavoured milk.
3. Melt butter in small saucepan and blend in flour. Cook for 30 seconds, taking care not to brown. Remove from heat.
4. Gradually stir in flavoured milk. When blended and lump-free, return to heat, stirring constantly. Cook until boiling, then remove from heat.
5. Season with salt and pepper.

Assembling the lasagne

1. Preheat oven to 180°C.
2. Line the baking tray with one pasta sheet, then alternately layer the vegetable sauce, white sauce, mozzarella and more pasta. Finish with the white sauce and sprinkle with parmesan cheese.
3. Bake for 20 minutes or until golden brown.

EVALUATION

1. Describe the benefits of using the vacuum-packed fresh lasagne sheets over traditional dried pasta sheets.
2. Why are canned tomatoes often described as a 'convenience product'?
3. List two reasons why the vegetables in the sauce are finely cut or grated.
4. Define the cookery term 'to infuse'. In which step in the recipe was this process used?
5. Identify the steps in the recipe that help to ensure the white sauce is lump-free.
6. Write a paragraph that outlines why the Vegetable Lasagne would be considered a healthy meal.

BASIC RISOTTO

15 grams butter
2 spring onions, diced
¼ red capsicum, diced
⅓ cup arborio rice
1 ½ cups chicken stock
4 mushrooms, diced
⅓ cup peas
black pepper
2 tablespoons parmesan cheese

SERVES ONE

METHOD

1. Sauté the spring onion and capsicum in butter until soft and lightly coloured.
2. Add the rice and cook for a further 1 minute or until the rice becomes opaque.
3. Bring the chicken stock to the boil, then add ½ cup to the rice.
4. Cover with a tight-fitting lid and simmer very gently until the rice has absorbed the stock.
5. Add a further ½ cup of chicken stock and continue to simmer very gently until the rice has absorbed the stock.
6. Add the remaining stock and stir gently to loosen the rice from the bottom of the saucepan. Add the diced mushrooms and peas. Season with a little black pepper. Cover with the lid.
7. Continue to very gently simmer until all of the stock has been absorbed and the rice is plump and creamy. The risotto should take approximately 20–25 minutes to cook.
8. Add parmesan cheese and serve immediately.

EVALUATION

1. Describe the sensory properties – appearance, aroma, flavour and texture – of your risotto.
2. Why are the onion and capsicum sautéed before the rice is added?
3. Explain why it is important to barely simmer the risotto during the cooking process, rather than boil it quickly.
4. List the important health and safety steps to follow when preparing a risotto.
5. Which aspect of this production did you find most challenging? Why?

CHEESY PASTA BAKE

150 grams penne pasta
½ onion, finely diced
2 rashers bacon, finely diced
250 grams ricotta cheese
1 egg
125 millilitres evaporated milk
¼ cup parmesan cheese
1 tablespoon parsley, chopped
salt and pepper
40 grams cheddar cheese, grated

SERVES TWO

METHOD

1. Preheat oven to 180°C.
2. Lightly grease two small ovenproof dishes or foil containers.
3. Bring a large saucepan of water to the boil.
4. Add penne to boiling water. Stir once or twice. Cook the penne for approximately 12 minutes or until al dente. (To test for al dente, bite a piece of pasta. It should be firm, but not hard. You should not be able to feel or see the hard centre.) Drain and keep warm.
5. Cook the finely diced onion and bacon in a non-stick frying pan for 4–5 minutes or until soft and just beginning to brown.
6. Mix the ricotta, egg, evaporated milk, parmesan cheese, parsley and salt and pepper in a bowl until well combined.
7. Place a quarter of the pasta in the base of each of the two ovenproof dishes. Spread with half the onion and bacon.
8. Top with a quarter of the ricotta and egg mixture. Add the remaining penne. Pour the remaining ricotta and egg mixture over the top.
9. Sprinkle with the cheddar cheese.
10. Bake in the preheated oven for 20–25 minutes or until golden on the top.

EVALUATION

1. Why is it important to add the penne to boiling water?
2. Describe the sensory properties – appearance, aroma, flavour and texture – of the Cheesy Pasta Bake.
3. Describe two safety rules to observe when using the oven.
4. Refer to the table on Page 210 that shows sources of calcium. Calculate the amount of calcium this recipe would provide for one person.
5. Evaluate the Cheesy Pasta Bake by plotting its ingredients onto a diagram of the Australian Guide to Healthy Eating.

TRADITIONAL SAUSAGE ROLLS

1 slice bread, crumbled
2 tablespoons milk
200 grams sausage mince
½ onion, finely diced
½ medium carrot, grated
½ zucchini, grated
1 sheet puff pastry
2 tablespoons flour
1 tablespoon milk, for glazing

🍴 MAKES 6 LARGE OR 12 SMALL SAUSAGE ROLLS

METHOD

1. Preheat oven to 200°C.
2. Soak the bread in the milk for 10 minutes, then squeeze out the excess liquid. Combine the sausage mince, diced onion, grated carrot, grated zucchini and soaked bread. Thoroughly mix.
3. Cut the sheet of pastry in half lengthwise.
4. Sprinkle a board with flour. Roll the sausage mixture into two rolls the same length as the pastry.
5. Place one of the rolls of sausage mince along the edge of one half of the pastry. Moisten the edges of the pastry with water.
6. Roll the pastry over the meat with the fold side under. Repeat with the second pastry sheet.
7. Cut the rolls into even-sized pieces. Place on the baking tray and mark with a knife.
8. Glaze with milk and bake at 220°C for 10 minutes, then at 190°C for a further 10–15 minutes.

EVALUATION

1. Describe the sensory properties – appearance, aroma, flavour and texture – of the sausage rolls.
2. Discuss the benefits of adding carrot and zucchini to the sausage rolls.
3. Explain why the sausage rolls are initially cooked at 220°C and then with the temperature reduced for the remainder of the cooking time.
4. List the important health and safety steps to follow when preparing the Traditional Sausage Rolls.
5. Which aspect of the production did you find most challenging? Why?

Mark Fergus Photography

SPICY POTATOES WITH CHICKPEAS

1 medium waxy potato (150 grams)
150 grams sweet potato
1 tablespoon oil
½ onion, finely diced
1 clove garlic, crushed
1 teaspoon ground cumin
½ teaspoon garam masala
½ teaspoon ground coriander
½ teaspoon ground fennel
¼ teaspoon ground turmeric
pinch of cayenne pepper
220 grams canned tomatoes, chopped
½ cup vegetable stock
⅓ cup frozen peas
½ cup (100 grams) chickpeas, rinsed and drained

SERVES TWO

This recipe served with rice is high in complete protein and makes a delicious meal for vegetarians.

METHOD

1. Peel the potato and sweet potato and cut into 2-centimetre cubes.
2. Heat the oil in a medium saucepan and sauté the onion and garlic until soft but not brown.
3. Add the spices and cook for 30 seconds.
4. Add the diced potato and sweet potato and lightly toss to coat in the spices.
5. Stir in the undrained tomatoes and vegetable stock.
6. Bring to the boil. Reduce the heat and simmer for approximately 30 minutes or until the potato is tender.
7. Stir in the peas and chickpeas and heat through.
8. Serve the Spicy Potatoes with Chickpeas with Plain Rice (Page 150) accompanied by Pappadams (Page 28) and Minted Garlic Dip (Page 267).

EVALUATION

1. What is the purpose of frying the spices for 30 seconds in step 3?
2. Identify the ingredients in the Spicy Potatoes with Chickpeas and the rice that are good sources of protein for vegetarians.
3. Why are the frozen peas and canned chickpeas only heated through in Step 7 and not added with the potatoes in Step 4?
4. Which part of the production was the most successful, and which part did you find the most challenging? Why?
5. Plot the ingredients for the Spicy Potatoes with Chickpeas, Plain Rice, Pappadams and Minted Garlic Dip on a diagram of the Australian Guide to Healthy Eating. Comment on the nutritional value of your meal.

CRUSTLESS QUICHE

2 slices wholemeal bread (does not need to be fresh)

2 rashers bacon, diced

½ onion, finely diced

1 small clove garlic, crushed

1 ½ cups vegetables; for example, sweet potato, carrot, pumpkin, zucchini, capsicum, celery or sweet corn

2 eggs

⅓ cup milk

¼ cup self-raising flour

½ cup cheese, grated

2 tablespoons oil

pepper

SERVES TWO

METHOD

1. Preheat oven to 180°C.
2. Grease two small ovenproof dishes or foil takeaway containers.
3. Use a food processor to process the bread into breadcrumbs.
4. Place the breadcrumbs in the bottom of the ovenproof dishes.
5. Heat the oil in a small frying pan.
6. Place the diced bacon, onion and garlic in the frying pan and sauté gently until softened and lightly browned. Cool.
7. Grate or finely dice the vegetables into a medium bowl.
8. Add the sautéed bacon, onion, garlic, eggs, milk, flour, grated cheese, oil and pepper. Stir until well combined.
9. Pour the mixture over the breadcrumbs.
10. Place dishes on a baking tray.
11. Bake in the preheated oven for approximately 20 minutes or until the quiche is set.

EVALUATION

1. Refer to the words for describing food on Page 2. Use some of these words to help you describe the sensory properties – appearance, aroma, flavour and texture – of your Crustless Quiche.
2. If you were to make this recipe again, would you change it in any way to alter the flavour?
3. Do you think your recipe is high in dietary fibre? What ingredients are included in your recipe that increase the dietary fibre content of this dish?
4. Make a list of the health and safety issues you needed to consider in the production of your Crustless Quiche.
5. Ask a friend or family member to taste your Crustless Quiche and to comment on its flavour and texture. Record their observations.

MARGHERITA PIZZA WITH A CAULIFLOWER CRUST

½ small cauliflower (approximately 300 grams), cut into florets

⅔ cup (85 grams) gluten-free pizza dough mix

⅓ cup (30 grams) parmesan, finely grated

1 egg, lightly beaten

salt and pepper

2 tablespoons Napoli sauce (see Page 157)

½ cup mozzarella, grated

6 basil leaves

MAKES 2 × 16-CENTIMETRE PIZZAS

This recipe for pizza is a great alternative to traditional pizza for someone who has coeliac disease. Instead of using wheat flour to make the pizza base it uses finely chopped cauliflower and gluten-free flour or ground almonds to give structure to the pizza base.

METHOD

1. Preheat the oven to 200°C. Draw a 16-centimetre circle onto two sheets of baking paper. Turn the paper upside down and place each sheet on a baking tray.
2. Whiz the cauliflower in a food processor until it resembles couscous. You will need approximately two cups of 'cauliflower couscous'.
3. Combine the cauliflower in a bowl with the gluten-free pizza dough mix, parmesan cheese and egg. Season to taste with salt and pepper.
4. Place half the cauliflower mixture on to each sheet of paper and shape each into a 5-millimetre-thick pizza base.
5. Bake the pizza bases for 20 minutes or until dark, golden and firm.
6. Spread two tablespoons of Napoli sauce over each pizza base. Top with the mozzarella cheese.
7. Bake for a further 5 minutes or until the cheese is browned and melted.
8. Take care when transferring the pizza to a serving plate; the base will be more delicate than a traditional pizza's.

Note: you can substitute the gluten-free pizza dough mix with ⅔ cup of ground almonds to give a slightly nutty flavour to the dough.

EVALUATION

1. Explain why this recipe would be ideal to serve to someone who suffers from coeliac disease.
2. Describe the sensory properties – appearance, aroma, flavour and texture – of your Margherita Pizza with a Cauliflower Crust.
3. Describe some other ingredients that would make delicious toppings for the cauliflower base.
4. What aspect of this production did you find most challenging? Explain why.
5. Discuss the management skills you used in preparing, cooking and serving the pizza.
6. Plot the ingredients for the Margherita Pizza With a Cauliflower Crust on a diagram of the Australian Guide to Healthy Eating. Comment on the health rating you would give this recipe.

10 INDIGENOUS FOOD AND FLAVOURS

KEY KNOWLEDGE

- Indigenous food customs
  - Collection of food
  - Bush foods from animal sources
- The traditional diet of Aboriginal and Torres Strait Islander people
  - Traditional cooking techniques
- European settlement
  - Impact of European settlement on Aboriginal and Torres Strait Islander health
- The Aboriginal and Torres Strait Islander Guide to Healthy Eating
- Native foods in today's menus
  - Kangaroo meat
- Cooking with bush flavours
  - Aniseed myrtle
  - Bush tomato (or desert raisin of akudjura)
  - Lemon myrtle
  - Native pepper (or mountain pepper)
  - Wattleseed
  - Macadamia nuts

KEY TERMS

bush tucker the huge variety of edible native Australian herbs, spices, mushrooms, fruits, flowers, vegetables, animals, birds, reptiles and insects

ground oven an oven made by lining a pit in the ground with hot stones, placing wrapped food on the stones and covering the hole over with grasses, leaves and sand

risk factors habits that can increase the chance of developing a disease or health condition

AUSTRALIAN CURRICULUM LINKS

DESIGN AND TECHNOLOGIES
- Knowledge and Understanding
- Processing and Production Skills

HEALTH AND PHYSICAL EDUCATION
- Personal, social and community health

GENERAL CAPABILITIES
- Critical and creative thinking
- Intercultural understanding

CROSS-CURRICULUM PRIORITIES
- Aboriginal and Torres Strait Islander histories and culture

INDIGENOUS FOOD CUSTOMS

Before European settlement, the Aboriginal and Torres Strait Islander people of Australia successfully lived off the land. These first Australians enjoyed a varied and nutritious diet. A large part of each day was spent hunting and gathering food from the forests, mountains, rivers, bush and deserts. In very dry areas, this often meant travelling long distances in search of food and water. Aboriginal and Torres Strait Islander people would collect the morning dew from leaves or squeeze the water from a frog's pouch to drink. Women spent many hours of the day gathering fruits, berries, grubs and small game while the men hunted and fished. Following the cycle of the seasons, each tribe moved through various parts of their territory, hunting and gathering food.

Each Aboriginal and Torres Strait Islander person was vitally involved in the business of food and required the ability to plan, an eye for detail and well developed bush skills. Women learnt the locations of every yam patch and fruit tree in their territory and the times at which the trees bore fruit. Bees were followed to their hives and small animals were stalked. Men learnt from their elders to become clever hinterland hunters and to trap birds; they learnt the art of camouflage and were able to swim under water.

Collection of food

Indigenous food is collected by hunting and gathering, rather than by the European means of crop cultivation and the domestication of animals. Groups of Aboriginal and Torres Strait Islander people moved around the land from season to season in search of food. Detailed knowledge of the environment and careful planning were essential to ensure their survival as well as that of the environment. Different species of animals and plants were found in different climate zones and rainfall areas. Native flora and fauna were used as food in every area, and food collection, preparation, cooking and distribution were major daily activities. Generally, Aboriginal and Torres Strait Islander people hunted and gathered sufficient foods for their needs at the time. They were concerned always concerned to ensure that the animals and plants in an area could regenerate and replenish themselves naturally, and so continue to provide a wide range of foods into the future.

Aboriginal and Torres Strait Islander people divide their food into categories according to the way it is obtained:

Indigenous foods

METHOD OF OBTAINING FOOD	FOOD TYPE	FOOD FUNCTION
• Spearing	• Animal foods – kangaroo; fish	• Protein for growth and repair
• Digging	• Plant foods – grains; roots; edible grubs	• Carbohydrates; fats and oils for energy
• Collecting in a dilly bag	• Plant foods – fruit; vegetables; seeds; kernels • Sweet foods – nectar; honey ants • Eggs	• Vitamins and minerals to protect and regulate body systems

Bush foods from plant sources

FRUITS	VEGETABLES	SEEDS	ROOTS	NUTS	FLOWERS
• Bush tomato • Lilly pilly • Native passionfruit • Bush banana • Illawarra plum • Quandong • Nonda plums	• Bulrushes • Pigweed • Waterlily • Mangrove	• Wattle (acacia) • Wild rice • Millet and grass seeds	• Yam • Bush potato • Waterlily root	• Bunya • Macadamia • Kurrajong • Moreton Bay chestnut	• Native fuchsia • Honey grevillea • Flowering gum

Quandong

Moreton Bay chestnut

Desert yam

Bush tomato

Wattleseeds

Lilly pilly

Bush banana

Bunya nuts

Bush foods from animal sources

MAMMALS	BIRDS	INSECTS	REPTILES	SEAFOOD	
Kangaroo	Duck	Ant	Crocodile	Barramundi	Clam
Wallaby	Emu	Locust	Turtle	Crab	Oyster
Koala	Pigeon	Witchetty grub	Snake	Frog	Pippie
Wombat	Cockatoo	Honey bee	Goanna	Eel	Yabby
Bandicoot	Parrot	Moth		Freshwater bream	
Water rat	Mallee fowl				

Witchetty grub

Water rat

Bandicoot

Wombat

Barramundi

Goanna

Mallee fowl

Crocodile

THE TRADITIONAL DIET OF ABORIGINAL AND TORRES STRAIT ISLANDER PEOPLE

The Australian environment supplies a wide range of foods, providing a nutritious diet for Aboriginal and Torres Strait Islander people. These foods are known as **bush tucker**, a term that refers to the huge variety of edible native Australian herbs, spices, mushrooms, fruits, flowers, vegetables, animals, birds, reptiles and insects.

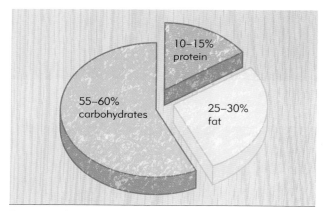

Sources of energy in the Aboriginal and Torres Strait Islander diet

A traditional Indigenous Australian diet consists mainly of plant foods that have had little or no processing, and which therefore supply high levels of fibre. These plants and fruits are very rich in nutrients, particularly vitamins and minerals. The seeds of many native grasses, in particular, often contain much higher levels of protein and fat than cultivated cereal crops. In a traditional Indigenous Australian diet, animal food contributed approximately 50 per cent of the total energy intake. The staples were kangaroo, freshwater bream and yams.

Traditional cooking techniques

Many bush foods, such as fruits and berries, are eaten raw, as they are collected. When a large animal such as a kangaroo is killed, it is cooked in its skin over an open fire. The skin protects the meat during cooking and helps to retain the meat juices and nutrients. Small birds and fish are wrapped in paperbark to protect the delicate flesh from the naked flame of the open fire, and to retain and create moisture when they are placed under the ashes and allowed to steam. Foods are often surrounded with clay and then cooked in the hot ashes. Seed damper is placed in the hot ashes to cook.

On other occasions, a **ground oven** is made by lining a pit in the ground with hot stones, placing wrapped food on the stones and covering the hole over with grasses, leaves and sand. This method keeps in the heat and allows the food to steam for some time.

1 A fire is lit under a layer of wood and with stones covering the pit.

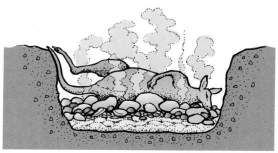

2 The heated stones and hot coals fall into the pit. The meat is placed over the hot coals.

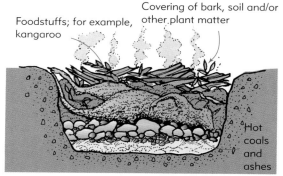

3 The steam is sealed in with a layer of paperbark, and then earth.

Traditional ground oven

ACTIVITY 10·1

Traditional cooking methods

1. Investigate cooking methods used around the world today and find some techniques that are comparable to those used by traditional Aboriginal and Torres Strait Islander people.

Worksheet

2 List the advantages and disadvantages of traditional Indigenous Australian food preparation and cooking methods and techniques. Consider:
   - fuel economy
   - nutrient retention
   - flavour
   - environmental impact.
3 Paperbark is often used as a wrapping for cooking tender pieces of food such as fish. List some types and cuts of fish or other meats that would be suitable for cooking using this technique.
4 Identify some bush spices that would be suitable to use when cooking with paperbark. Refer to Page 236 for information on cooking with bush flavours.

EUROPEAN SETTLEMENT

To the first European settlers, it seemed that the Australian country was harsh and barren and that the Aboriginal and Torres Strait Islander people lived a 'primitive' lifestyle. The new inhabitants quickly began to mould the landscape to the needs of European agriculture by clearing the land, introducing new plants and seeds for cultivation and new and exotic species of domestic animals. Little attempt was made to understand or appreciate native plant or animal foods or the hunter-gatherer means of existence. Australian cuisine quickly embraced the 'new' foods that migrants brought with them, but, strangely, the uniqueness of bush foods, which represent the culture and cuisine of native Australia, was not appreciated.

Impact of European settlement on Aboriginal and Torres Strait Islander health

European settlement gradually had a great influence on the diet and health of Aboriginal and Torres Strait Islander people, and disrupted their food supply. Their nomadic, hunter-gatherer lifestyle was replaced by life in settlements. Many Indigenous Australian people went to work on farms or to live in missions where the food supplied was highly refined and processed. Over the years, this new diet, based on white flour, white sugar, salt and tea, slowly replaced traditional foods. As a result, many Aboriginal and Torres Strait Islander people suffered serious health problems; malnutrition was common. The diet of Aboriginal and Torres Strait Islander people underwent a rapid change, from one containing high-protein, low saturated fat and fibre-rich foods to one dominated by refined carbohydrates and saturated fats. The nomadic lifestyle required high levels of physical activity to hunt and gather food, but this was largely replaced by a sedentary lifestyle often featuring little exercise. These changes in diet and lifestyle have led to high levels of obesity in Indigenous communities, as well as the resulting risk factors or habits that can increase the chance of developing a disease or health condition such as cardiovascular disease (hypertension and stroke), type 2 diabetes and kidney disease.

THE ABORIGINAL AND TORRES STRAIT ISLANDER GUIDE TO HEALTHY EATING

One of the objectives of the *Food and Nutrition Policy (1992)* of the Australian Government and the Department of Health is to help Australians develop the skills and knowledge necessary to choose a healthy diet. The Australian Guide to Healthy Eating was developed as a national food selection guide for use as a nutrition education and information tool. It reflects the multicultural nature of the population.

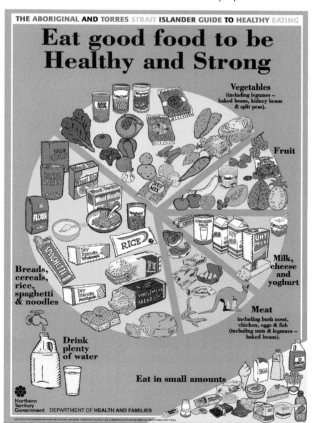

The Aboriginal and Torres Strait Islander Guide to Healthy Eating is reproduced with permission of the Northern Territory Government, Department of Health

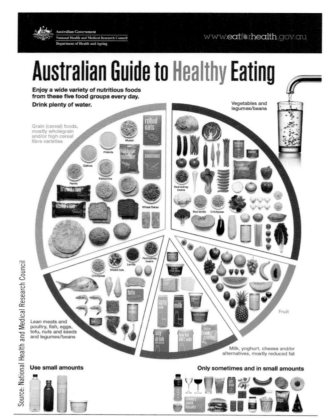

The Australian Guide to Healthy Eating

The Australian Government is also committed to the delivery of effective and efficient health services for Aboriginal and Torres Strait Islander people. The Aboriginal and Torres Strait Islander Guide to Healthy Eating was developed by the Northern Territory Government specifically to educate and help Aboriginal and Torres Strait Islander people to understand the importance of choosing a healthy diet that is balanced and varied. The foods in the guide are readily available at stores, and also include some local bush food and seafood.

Both guides are scientifically-based nutrition education resources that explain how healthy eating throughout life can help to reduce the risk of chronic diseases such as type 2 diabetes, heart disease, cancer and obesity.

ACTIVITY 10-2

Comparing food guides

1 Describe the similarities between the Australian Guide to Healthy Eating and the Aboriginal and Torres Strait Islander Guide to Healthy Eating; consider shape, food groupings and the proportion of each group.

2 Explain the main differences between the foods contained in the vegetables section of both guides. Give reasons for the differences.
3 Describe the differences between the meat sections of each guide.
4 Use a Venn diagram to compare the fruit sections of both guides.
5 Why are tinned, dried and long-life foods included in most sections of the Aboriginal and Torres Strait Islander Guide to Healthy Eating?
6 Explain why it was necessary for the Northern Territory and the National Government to develop a separate guide to healthy eating for Aboriginal and Torres Strait Islander people.

Testing knowledge

1 Make a list of the foods traditionally supplied by Aboriginal and Torres Strait Islander women and those supplied by Aboriginal and Torres Strait Islander men.
2 Discuss the impact that weather and the seasons had on the traditional diet of Aboriginal and Torres Strait Islander people. Explain why a knowledge of the environment was essential for Aboriginal and Torres Strait Islander people in their search for food.
3 List the nutritional advantages of the traditional Indigenous diet.
4 Based on the information on pages 230–1, list three dietary guidelines that people following a traditional Indigenous Australian diet would meet.
5 Explain the technique of cooking in a ground oven.
6 Why would paperbark be used as a material for wrapping around fish or small birds in traditional Indigenous cooking?
7 Describe how European settlement had an influence on the diet and health of Aboriginal and Torres Strait Islander people.
8 Explain the impact that a westernised diet has had on the health of Aboriginal and Torres Strait Islander people.
9 Describe the factors that have led to the increasing incidence of type 2 diabetes among Aboriginal and Torres Strait Islander people.
10 Explain how the Aboriginal and Torres Strait Islander Guide to Healthy Eating differs from the Australian Guide to Healthy Eating.

NATIVE FOODS IN TODAY'S MENUS

Australian native foods are being rediscovered by both Aboriginal and Torres Strait Islander people and Australians of other heritages. There is a growing pride in Indigenous Australian culture, and the increasing number of restaurants in Australian cities has established a potential market for exciting, different foods. In the late 1980s, due to the popularity of the ABC television series *Bush Tucker Man* and the pioneering work of its host Les Hiddens, consumers were educated about the benefits and tastes of Australian bush foods. To begin with, the bush foods supplied to restaurants and manufacturers were all gathered in the wild, but it soon became obvious that if the industry was to be more than a novelty, wild harvesting was unsustainable and the crops would have to be farmed. Some native foods are now being cultivated as environmentally friendly crops. The Outback Pride Project is promoting the Australian native food industry by developing a network of production sites within traditional Aboriginal and Torres Strait Islander communities. The company has created systems of propagation and cultivation for up to 64 bush food species to be grown commercially for supply to restaurants, supermarkets and markets.

Bush food products

Kangaroo meat

The kangaroo is a marsupial that is native to Australia and was one of the principal animals hunted and eaten by Aboriginal and Torres Strait Islander people and the early European settlers. As more land became available for grazing, beef and lamb replaced kangaroo meat. Wild meats such as kangaroo are lower in fat than the meat of domesticated animals. Wild animals must scavenge for food and move around the countryside in search of food. Their energy intake and expenditure are closely balanced, meaning fewer fat stores are accumulated.

Kangaroo meat has been on menus in South Australia and Tasmania since the 1980s, but with recent changes in legislation, it is now widely available throughout Australia and exported overseas. Kangaroo meat is cooked like any other game meat – that is, seared on a high heat to seal in meat juices, cooked lightly, and then rested to tenderise the meat. Overcooking should be avoided, since the meat is lean (low in fat) and quickly dries out and becomes tough. Kangaroo meat can be purchased as a boneless leg, rump, tender loin fillet, strip loin or loin fillets, or it may be diced or minced. The prime cut is the long fillet.

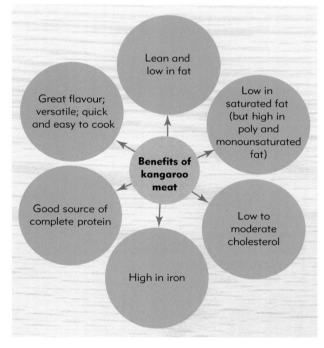

Kangaroo fillet

ACTIVITY 10.3

Analysing kangaroo meat

KANGAROO MEAT COMPARED WITH OTHER MEATS

Examine the table below and visit the Kangaroo Industry Association of Australia website. Find information about different cuts of kangaroo meat and kangaroo recipes, and then answer the following questions.

1. Why does kangaroo meat have one of the lowest fat contents of any meat?
2. Compared with beef, what are the nutritional advantages of kangaroo meat?
3. Compared with chicken, what are the nutritional advantages of kangaroo meat?
4. With other members of your class, brainstorm answers to the following questions:
   a. Do you believe that kangaroo meat is a viable alternative to other meats? Remember to think about cost, flavour, time of cooking, nutrient value, ethics and environment.
   b. Why do you think kangaroo meat is not commonly seen on a large number of menus?
   c. Why are many consumers reluctant to eat kangaroo meat?
   d. What strategies can you suggest to increase consumer awareness of the benefits of eating kangaroo meat?
5. Identify a cut of kangaroo meat that would be suitable to use in a stir-fry recipe.
6. What is the recommended cooking time for stir-fried kangaroo?

	Protein (%)	Fat (%)	Iron (mg/100 g)	Cholesterol (mg/100 g)	Kilojoules (per 100 g)
Kangaroo	24	1–3	2.6	56	500
Lean lamb	22	2–7	1.8	66	530
Lean beef	22	2–5	3.5	67	500
Lean pork	23	1–3	1.0	50	440
Lean chicken breast	23	2	0.6	50	470
Rabbit	22	2–4	1.0	70	520

COOKING WITH BUSH FLAVOURS

Aniseed myrtle

Ground aniseed myrtle has an aniseed/pernod flavour. It can be used to flavour cakes, muffins, biscuits and desserts, particularly ice-cream.

Bush tomato (or dessert raisin or akudjura)

One popular flavour is the fruit (about the size of a blueberry) of the small shrub known as the bush tomato. The fruit must be allowed to ripen on the bush; otherwise, it will be toxic. This fruit is a good source of carbohydrates and vitamin C. It is best used after being ground to a powder in a food processor. The ground bush tomato is called akudjura, and has an intense flavour.

The bush tomato can be:
- used to enhance the flavour of savoury dishes and give them a spicy, piquant taste
- added to soups, casseroles, pasta sauces, chutneys, relishes, pizzas, risotto or sauces – in fact, any dish that traditionally uses cultivated tomatoes.

Bush tomatoes

Lemon myrtle

Lemon myrtle is derived from the aromatic leaf of a large tree found in Queensland. The leaf contains essential oils, which give it a wonderful perfume and spicy lemon flavours.

Lemon myrtle

Lemon myrtle can be:
- infused in hot liquids (this releases the oils) and used in recipes in which it is inappropriate to use dried herbs
- used to flavour tea, vinegar, oil dressing or a dessert
- used as a fresh herb – for example, draped over a fish or chicken fillets before baking, or as a substitute for kaffir lime leaves
- dried, crumbled and sprinkled over meats or fish before baking
- added to breads, muffins, biscuits, sauces and mayonnaise
- used as an oil to flavour cream or yoghurt.

Native pepper (or mountain pepper)

The large, aromatic leaves of the native pepper tree have a hot flavour between a pepper and a chilli.

Native pepper

The native pepper can be:
- dried and then crumbled or ground in to curries, dishes containing chilli, breads, pastries, or chutneys for use on meat or fish
- used as an infusion to make a refreshing herbal tea, vinegar, oil or dressing.

Wattleseed

Wattleseed is one of the most popular and best known of all Indigenous Australian foods because it is Australia's national floral emblem. The small, black seeds of the wattle are ground after dry roasting to produce grounds that contain coffee, chocolate and hazelnut flavours and can be used in a variety of ways. A small quantity of wattleseed is brought to the boil to soften the grounds. The liquid is then drained off and used as flavouring. The remaining solids (grounds) can also be used.

Wattleseed

Wattleseed can be used:
- as a substitute for coffee to make 'wattlechinos' (wattle cappuccino); wattleseed is caffeine-free – use one teaspoon of ground seeds per cup
- to flavour syrup to be poured over pancakes and puddings
- to flavour cream, ice-cream, mousses and meringues
- in biscuits and cakes (by using the grounds)
- in bread, muffins, pastry, pancakes and pasta; always add the wattleseed towards the end of the mixing, because it affects the gluten and toughens the flour.

ACTIVITY 10.4
Understanding bush flavours: bush chips

AIM
To understand the sensory properties of bush flavourings

EQUIPMENT
- 4 different ground bush flavourings such as aniseed myrtle, akudjura (ground bush tomato), lemon myrtle or native pepper
- 2 small pitta breads, cut open
- macadamia oil or spray
- oven tray

METHOD
1. Preheat the oven to 180°C.
2. Lightly spray one side of the bread with macadamia oil, or brush on oil with a pastry brush.
3. Lightly sprinkle with ground flavouring – use one of the selected flavours per piece of pitta bread.
4. Place on an oven tray and bake for approximately five to 10 minutes or until crisp and lightly browned.
5. Complete a taste test comparison using a table similar to the following. Refer to the sensory wheel on Page 2 to assist you.

Bush flavouring	Appearance	Aroma	Flavour
1			
2			
3			
4			

ANALYSIS
1. Which flavour was the strongest?
2. Which flavour would, in your opinion, have the greatest consumer appeal?
3. Which flavour would be best to serve with each of the following foods: pasta dish, chicken, meat, vegetables, and cheese?

CONCLUSION
Which bush flavour did you enjoy the most? Which bush flavour would be the easiest to identify? Note: The pitta wedges can be used to produce tasty alternatives to bread or crackers to serve with pâté or dips, or to accompany soup.

Macadamia nuts

The macadamia nut is an ingredient that is considered to be typically Australian. It is native to south-east Queensland and northern New South Wales, where it is grown in the rainforests, close to streams. In 1881, Americans took the seeds of this nut to Hawaii, where it was originally used as an ornamental tree, later being grown in plantations.

After harvesting in late spring, the nuts are dehusked and spread to dry, protected from the sun, for two to three weeks. After drying, the shells are removed and the nuts can be left raw or roasted.

The nut has a high oil content, and is consequently high in fat and energy. The nuts can be purchased raw or oven-roasted. The raw nuts are best used in cakes, biscuits and puddings; the oven-roasted nuts are best in ice-cream and mousses. Macadamia nuts can be used in recipes such as tarts, cakes and biscuits to replace walnuts, almonds and hazelnuts, or be combined with breadcrumbs to cover fish or chicken. Macadamia nut oil can be extracted from the nuts; it can be used in place of other oils in salad dressings, for frying, or in cakes to replace butter. If you would like to find more information about macadamia nuts, visit a website such as that of Australian Macadamias.

Macadamia nuts

THINKING SKILLS 10.1

1. Develop a promotional video clip or television advertisement to promote kangaroo meat as a viable ingredient in everyday meals. The promotional material should include the following information:
   - the nutrient value of kangaroo meat
   - why kangaroo meat is a versatile ingredient to use
   - recipe ideas
   - other relevant information.
2. Write a paragraph for a food website that discusses:
   - ethical issues that would support kangaroo meat production as a food source
   - arguments against the use of kangaroo meat as an ingredient in meals

   Suggest an appropriate platform to post your arguments.

Testing knowledge

11 Explain why kangaroo meat is considered to be a healthy food choice.

12 List three bush tucker flavours that complement savoury foods such as soups and fish.

13 Identify the traditional name given to the bush tomato. List three main uses of this Indigenous Australian food.

14 Explain why bush tomatoes would have been a good nutritional supplement in a traditional Indigenous Australian diet.

15 Describe the main flavour of wattleseed. List the steps you would need to take to prepare wattleseed for use in a recipe.

16 Why is it important to add wattleseed towards the end of mixing when adding it to bread or muffins?

17 Which bush tucker food can be used as an oil?

18 When are macadamia nuts harvested, and how are they prepared for sale?

19 Identify the main nutrient content of macadamia nuts.

20 List five ways in which macadamia nuts can be used in cooking.

DESIGN ACTIVITY 10.1

An Australian biscuit or sweet treat

DESIGN BRIEF

A chain of gourmet food stores called Share a Taste of Australia operates at all major airports in Australia. These stores sell a range of local and native Australian food items for international visitors to buy as souvenirs as they leave the country. Biscuits are a popular purchase with tourists.

Share a Taste of Australia wishes to introduce a new gourmet biscuit that reflects authentic Australian ingredients and flavours. The biscuits must be suitable to package and transport without damage and reflect the Australian theme.

1 Write a design brief based on the five Ws.
   - Who – who will the biscuits be made for?
   - What – a new gourmet biscuit that utilises and reflects native Australian ingredients and flavours
   - When – at what time of year will the biscuits be sold?
   - Where – where will the biscuits be sold?
   - Why – why does Share a Taste of Australia want the new range of biscuits?

2 Develop five evaluation criteria questions to judge the success of your finished biscuits.

INVESTIGATING

1 Research suitable basic recipes for the biscuits and possible native ingredients.

GENERATING

1 Write up the new recipe for your gift biscuit.

2 Design a label for the biscuits. Your label should include:
   - the name of the biscuits
   - an illustration or sketch – of the biscuits
   - a list of ingredients, in descending order of quantities
   - the name and address of the biscuits' manufacturer.

PLANNING AND MANAGING

1 Prepare a food order.

2 Write up a production plan.

3 Make a list of the aspects of the production task that will rely on you and your bench partner sharing and working collaboratively.

PRODUCING

1 Prepare the product.

2 Note any modifications made to the recipe during production.

EVALUATING

1 Answer in detail your five evaluation criteria questions to consider the success of your biscuits.

2 Describe the sensory properties – appearance, aroma, flavour and texture – of your Australian biscuits.

3 What, in your opinion, was the most difficult aspect of the production? Explain.

4 Discuss any improvements you would make if you were to make the biscuits again.

5 Evaluate the suitability and appeal of the label you have designed.

DAMPER WITH EUCALYPTUS OR GUMLEAF OIL BUTTER

Damper

1 ½ cups milk

1 tablespoon wattleseed

1 tablespoon bush honey

2 cups self-raising flour, or 1 cup white self-raising flour plus 1 cup wholemeal self-raising flour

1 teaspoon turmeric

Eucalyptus or gumleaf oil butter

½ cup softened butter

3–4 drops eucalyptus or gum leaf oil (ensure that you use the variety suitable for cooking, not medicinal purposes)

few drops of green food colouring

SERVES FOUR

METHOD

Damper

1. Preheat the oven to 220°C.
2. Place the – milk, wattleseed and honey in a saucepan and bring to the boil. Allow to cool.
3. Sift flour and turmeric into a basin.
4. Add cooled milk mixture and mix to soft dough with a spatula.
5. Lightly knead the mixture and form into six individual mini dampers or one large, round damper.
6. Place on a greased oven tray, or place the large round in a greased cake tin.
7. Bake at 220°C for 20–30 minutes or until the damper sounds hollow when tapped. Mini dampers will only take 15–20 minutes to cook.
8. Serve the damper with eucalyptus or gumleaf oil butter.

Eucalyptus or gumleaf oil butter

Blend together all ingredients to form a spreadable butter.

EVALUATION

1. Describe the sensory properties – appearance, aroma, flavour and texture – of your damper.
2. Why is it important to warm the wattleseed with the honey and milk?
3. Explain the effect kneading has on the damper.
4. List three tests, other than tapping the base, that could be used to determine whether or not the damper is cooked.
5. What modifications would you make to the Damper with Eucalyptus or Gum Leaf Oil Butter if you were to make it again?

KANGAROO MEATBALLS

1 small onion, grated

1–2 tablespoons oil

¼ cup fresh breadcrumbs

1 tablespoon commercial bush tomato sauce or tomato relish

1 small egg, beaten

4 shakes of pepper

250 grams kangaroo meat, finely minced

1 tablespoon cream

¼ cup cornflour

 SERVES TWO

Meatballs can be served as an entrée or be eaten as finger food. They could be served with a suitable accompaniment using Indigenous Australian plants, such as bush chutney, as flavouring.

METHOD

1. Gently fry the onion in 2 teaspoons of oil until soft. Cool.
2. Mix breadcrumbs with tomato sauce or relish.
3. Add beaten egg, pepper and cooked onion.
4. Combine the meat, breadcrumb mixture and cream and mix well.
5. Roll mixture into balls about 3 centimetres in diameter and roll in the cornflour.
6. Heat remaining oil in a frying pan and cook meatballs one layer at a time, shaking the pan until all sides are brown.
7. Drain on absorbent paper and pierce each meatball with a toothpick.

EVALUATION

1. Why does this recipe use fresh breadcrumbs instead of dry bread crumbs?
2. What is the purpose of the cream in this recipe?
3. Suggest some other bush flavours that could be used in the Kangaroo Meatballs.
4. What other low-fat meats could be used in this recipe if kangaroo meat was unavailable?
5. What was the most challenging aspect of this production? Why?

WATTLESEED PAVLOVAS WITH MACADAMIA CREAM AND SUGAR BARK

Wattleseed pavlovas

1 teaspoon wattleseed

2 egg whites

½ cup caster sugar

½ teaspoon vanilla essence

½ teaspoon vinegar

Sugar bark

⅔ cup caster sugar

⅓ cup water

Macadamia cream

150 millilitres cream

40 grams macadamia nuts

1 tablespoon Nutella

MAKES 6

METHOD

1. Preheat oven to 160°C. Cover baking tray with baking paper and draw 6 circles approximately 6 centimetres in diameter.
2. Pour 1 tablespoon of boiling water over the wattleseed and soak for 10 minutes. Strain through a fine strainer. Retain the grounds.
3. In a medium bowl, beat the egg whites into a stiff foam. Gradually add the caster sugar, one tablespoon at a time. Beat until mixture is glossy and stiff.
4. Fold in the vanilla essence, vinegar and wattleseed grounds. Divide the mixture into 6 equal portions and shape into nests within each circle.
5. Place in oven at 160°C, then immediately reduce heat to 130°C and cook for 20–30 minutes or until the crust has dried. Cool.

Sugar bark

1. Combine sugar and water in a saucepan and stir over low heat until the sugar has dissolved. Wash down any sugar crystals from the sides of the saucepan with cold water and a pastry brush.
2. Bring syrup to boil and reduce heat until the colour is pale gold.
3. Remove from heat and allow to rest until bubbles have subsided. Pour onto a foil-lined tray. Cool.
4. Break the toffee into strips that resemble bark.

Macadamia cream

Beat the cream to soft peaks. Finely chop the macadamia nuts and fold through the cream with the Nutella.

To assemble the pavlovas

Divide the cream filling between the pavlova nests. Break the sugar bark into shards and place vertically in the filling to decorate.

EVALUATION

1. When making the pavlova, what steps should be taken during preparation to ensure the egg whites beat into a stiff foam?
2. Why is the caster sugar added into the beaten egg whites a little at a time instead of all at once?
3. Why are the vanilla, vinegar and wattleseed grounds folded through the pavlova mixture instead of being beaten in to it?
4. Why is it important to dissolve the sugar before bringing the syrup to the boil?
5. Identify two safe work practices that should be followed when using electric handheld beaters and making the sugar bark toffee.

CHOCOLATE AND MACADAMIA BISCUITS

- ⅓ cup macadamia nuts, finely chopped
- 1 ¼ cups plain flour
- ½ teaspoon baking powder
- 125 grams butter
- ¼ cup soft brown sugar
- ⅓ cup caster sugar
- 1 teaspoon vanilla
- 1 egg, lightly beaten
- ½ cup white chocolate buttons, roughly chopped

MAKES ABOUT 24 BISCUITS

METHOD

1. Preheat oven to 180°C.
2. Line two biscuit trays with baking paper.
3. Spread the chopped nuts onto one of the baking trays and cook for 6–8 minutes or until a light gold. Allow to cool.
4. Sift the flour and the baking powder.
5. Cream the butter and sugars until light and fluffy. Beat in the vanilla.
6. Gradually beat in the egg.
7. Stir in the sifted flour and baking powder.
8. Add the cooled toasted nuts and chocolate and mix well.
9. Place dessert-spoonfuls of mixture onto the two biscuit trays covered with baking paper, leaving enough room for spreading.
10. Bake at 180°C for 12–15 minutes or until golden and firm to touch.
11. Remove from the oven and allow to stand on the tray for 5 minutes.
12. Carefully remove with a spatula and transfer to a wire rack. Allow to cool.

EVALUATION

1. What physical property of macadamia nuts causes them to brown quickly?
2. How did you know when the butter and sugar were sufficiently creamed?
3. Why is it important to sift the flour and the baking powder together?
4. Why are the biscuits allowed to stand on the tray for five minutes before they are transferred to the wire rack?
5. Comment on the overall success of your Chocolate and Macadamia Biscuits.

11 GLOBAL GOODIES

KEY KNOWLEDGE

- Influences on eating in Australia
- Influences from Italy
  - Pasta
  - Top tips for cooking pasta
  - Gnocchi
- Influences from Asia
  - Hawker-style foods
  - Flavours of Asia
  - Ingredients and flavourings of Asia
  - Tools and equipment
  - Asian cooking processes
- Vietnamese cuisine
  - Influences on Vietnamese cuisine
  - Characteristics of Vietnamese cuisine
- Thai cuisine
  - Influences on Thai cuisine
  - Characteristics of Thai cuisine
- Mexican cuisine
  - Influences on Mexican cuisine
  - Characteristics of Mexican cuisine

KEY TERMS

hawker-style foods small portions of food traditionally prepared, cooked and served in front of the customer on the street

Pacific Rim cuisine foods originating from countries around the edges of the Pacific Ocean

Thai flavour wheel a diagram that shows the five flavours – salty, sweet, sour, spicy and bitter – that are blended in Thai foods

AUSTRALIAN CURRICULUM LINKS

DESIGN AND TECHNOLOGIES

- Knowledge and Understanding
- Processes and Production Skills

GENERAL CAPABILITIES

- Critical and creative thinking
- Intercultural understanding

CROSS-CURRICULUM PRIORITIES

- Asia and Australia's engagement with Asia

INFLUENCES ON EATING IN AUSTRALIA

Before you can fully understand the evolution of Australian cuisine, you need to reflect on the past and trace the influences on food consumption and choices since the earliest days of the colony.

- **Pre-1800s**: Before the arrival of Captain Cook, foods native to Australia were eaten by Aboriginal and Torres Strait Islander people. (For more information on this subject, see Chapter 10.)
- **1800s**: An Anglo-Saxon diet came to Australia with the English colonial settlers. Early European settlers ate a diet that largely consisted of meat supplemented with rations of flour, sugar, salt and tea, but contained little fruit and vegetables. This diet was nutritionally poor, containing large quantities of saturated animal fats and sugar and little fibre, and was hardly appropriate for the warmer Southern Hemisphere climate.
- **1850s**: The gold rush saw an influx of immigrants, including many from China, who brought with them new food and cooking skills. The Chinese quickly became market gardeners and greengrocers, and they adapted their recipes to suit the mainly Anglo-Saxon miners by adding more meat.
- **1850–1900**: The dairy and wine industries began to develop due to the influence of immigrants from Denmark, Sweden and Germany. There was expansion in the food processing, industries, with the production of jams, cordials, cheeses, tomato sauce and preserved meats. Households preserved their own fruit and vegetables and made their own jams, pickles and sauces, although much of Australia's food was also imported.
- **1900s**: The early part of the 20th century saw further developments in technology, which greatly influenced food trends. Gas and electric stoves became available and the domestic refrigerator slowly began to replace the ice chest. Methods of canning and freezing were introduced, and these had a significant impact on food types, storage and shopping patterns. The government approved the establishment of operations of foreign food companies such as Peters, Kraft, Nestlé, Cadbury and Kellogg's. Nutritional advice became more readily available, and advertising of food began. During the Great Depression, many people lived in poor economic circumstances, and many meals were based on cheaper foods such as bread and dripping and food from the land, especially rabbits.
- **1940s**: The Second World War led to significant food shortages. Rationing was introduced to ensure the fair distribution of food and proper nutrition of the population. American-style foods were introduced as American military personnel on leave in Australia demanded products such as hamburgers, canned sweet corn and Coca-Cola. Post-war immigration introduced new skills, products and knowledge to the restaurant trade. Takeaway food outlets also began to appear. After the War the government adopted a policy of increasing Australia's population, and subsidised a scheme of migration. During this period, many people migrated to Australia from Britain, Italy and Greece.

Early Italian food in Australia

- **1950s**: Coloured packaging arrived in stores, and self-service began to replace counter service in stores.
- **1956**: The Olympic Games in Melbourne brought many overseas chefs to the Olympic village. Many of these chefs stayed on in Australia, introducing new skills and expertise. The introduction of television added a new dimension to food advertising.

- **1960s**: More women entered the paid workforce. This meant more disposable income for families, less time for food preparation and more money for leisure activities. These factors greatly influenced the sales of convenience foods. People also began to buy a larger proportion of their food items from supermarkets.
- **1970s**: More people moved to the cities from country areas; their incomes increased, and as a consequence, their consumption of animal proteins, sugar, fats and takeaway foods also increased. The growth of foreign travel resulted in familiarity with new ethnic cuisines, particularly Italian, Greek and Chinese. Pre-prepared, or convenience foods flooded supermarket shelves, accompanied by marketing of instant, no-fuss, frozen and ready-to-eat foods. New styles and skills in food preparation were used, and outdoor cooking became popular. Boys joined girls in school Home Economics and Food and Technology classes. During this period, migration from Asian countries such as Vietnam and Cambodia increased, and the food from these countries had a significant influence on Australian cuisine.
- **1980s**: Nutrition became a very important issue and people generally turned to a healthier lifestyle and diet, with brown rice, yoghurt, low-salt and low-fat products increasing in popularity. Food was generally simpler, lighter and had a greater emphasis on vegetables and fish.
- **1990s**: Many people became increasingly concerned about their health and focused on diet and exercise. More pre-prepared foods began appearing in supermarkets as the demand for 'shortcut', quick meal solutions increased. The 1990s also saw a growth in the number of fast-food outlets and a greater consumption of food outside the home. During this period, **Pacific Rim cuisine** — that is, foods from countries around the edges of the Pacific Ocean — gained in popularity and led to the availability of a wide range of ingredients used in Asian dishes.
- **2000s**: Busy lifestyles created time-poor consumers, many of whom had few skills in food preparation. As a result, many consumers became reliant on convenience food for family meals instead of preparing food from scratch. Ready-to-go meals that required only heating and meal solutions such as pasta and simmer sauces that required only three or four extra ingredients, occupied significant shelf space in supermarkets. The range of foods with additional health benefits and those that catered for specific allergies or hypersensitivities increased, as did the popularity of organic food. Many refugees and migrants arrived from countries in Africa and South America, and many new restaurants appeared. There was increased interest in Indigenous Australian foods as television shows about bush ingredients educated the population and small product ranges of native ingredients began to appear in supermarkets.
- **2010s**: Farmers' markets have become increasingly popular in cities and regional areas with shoppers who are keen to purchase foods directly from growers. Food manufacturers have developed a greater range of functional foods with an emphasis on reduced-fat and low fat products and foods with added vitamins and minerals. The range of foods designed to meet specific dietary needs, such as gluten-free and lactose-free foods, has expanded. Organic foods and processed products are widely available in supermarkets and food shops.

Some consumers are becoming more conscious of their health, and 'superfoods' such as goji berries, quinoa, acai berries and kale have become prominent because they are natural foods rich in particular nutrients that are considered to have health benefits.

Many restaurants are now serving small-portion dishes, such as tapas, and meals designed to share with fellow diners at the table rather than individual courses. Pop-up food trucks are another new source of takeaway foods, and there has been an extraordinary growth of cafes serving coffee. Television programs such as *Master Chef* and *My Kitchen Rules* have captured viewers' attention and heightened interest in food preparation.

A ready-to-eat meal

ACTIVITY 11.1

Australia today

Australians' eating habits and trends in eating are undergoing significant changes. There is an enormous variety of foods available in Australia today, many of which are from various ethnic groups from around the globe.

Worksheet

1 List the different nationalities represented in your school.
2 List the country or countries of birth of people from three generations of your family.
3 Make a list of the foods your family eats that are typical of your cultural background.
4 Make a list of any foods your family eats that are from other cultures.

ACTIVITY 11.2

Looking back over three generations

The following activity will help you to understand the foods and traditions of family meals in past generations.

1 From your experience of your own family and the families of your friends, write a definition of the term 'generation'.
2 Interview someone from your parents' generation and someone from your grandparents' generation to determine what they ate when they were your age. Interviewees could come from your own family or be friends or neighbours.
3 Interview someone from your own generation about the foods they eat.
4 Use the following points as the basis of a questionnaire that could be used to examine the eating patterns of the past three generations.
    - Name
    - Age
    - Who does/did the shopping in the household
    - Who does/did the cooking for the household
    - Where meat, vegetables and groceries are/ were purchased
    - How these food items are/were purchased
    - What food is/was produced at home; for example, vegetables, homemade cakes and sauces
    - What is/was eaten for the main meal of the day
    - The time of the day at which this main meal is/was served.
5 You could record both of your interviews using a tablet or other electronic device.

ANALYSIS

1 What similarities do you notice between the eating patterns and the ways in which food is/ was purchased and prepared in households over the three generations?
2 What are the main differences between the eating patterns and the way in which food is/was purchased and prepared in households over the three generations?
3 At what time was the main meal served in the past two generations compared with today?

CONCLUSION

What are the main changes that have occurred over the past three generations? Why do you think these changes have occurred?

INFLUENCES FROM ITALY

After the Second World War, many migrants came from Italy to settle in Australia. These Italians introduced many food items from their native country to Australia, such as olive oil, Roma tomatoes, eggplant, garlic, pasta, pizza and gnocchi.

Eating, shopping and preparing food are very important parts of Italian life. Cafes and bars are a focal point in town squares all over Italy.

In the north of Italy, polenta and rice are important staple foods. Dairy foods – particularly cheese – and seafood are key ingredients in many dishes.

The centre of Italy is a major gastronomic region, with food such as fresh pasta, hams, salamis and cheeses being widely produced. Balsamic vinegar, used to flavour salads and vegetables, is also produced in this region.

In the south of Italy, sun-ripened vegetables such as tomatoes, olives and fruit are popular, as are pasta and cheeses. Some of the most popular dishes in this region are pizzas topped with fresh mozzarella or goat's cheese, spaghetti with tomatoes and meat sauce and granita made from lemons.

Pasta

Although pasta is considered to be a staple food of Italy, today, it is widely eaten in many countries throughout the world. In Australia, pasta has become a very popular food because it is quick and easy to prepare and can be served in a variety of ways, with a wide range of sauces to suit different tastes.

Pasta can be purchased either fresh or dried in supermarkets and other stores, or can be easily made at home. Today, pasta dough is often flavoured with other ingredients, including spinach or tomato paste. It can also be made with wholemeal flour to increase the proportion of dietary fibre it contains.

Pasta for good health

Like other foods made from cereal grains, pasta is an important food to include in a well-balanced diet because it is high in carbohydrates and is also a good source of B group vitamins. Wholemeal pasta is also a good source of dietary fibre. Because it contains such a high proportion of carbohydrates, pasta is thought to be a good energy food. Some people think pasta is fattening, but this is only the case if a rich, creamy sauce or other foods that are high in fat are added to it.

Medical experts believe that increasing the amount of pasta and other cereal-based foods in the diet may help to reduce the risk of some diet-related diseases such as coronary heart disease, obesity, bowel cancer and diabetes, provided they are served with low-fat sauces.

Pasta is considered to be an ideal food for athletes, since a 100-gram serve provides approximately 500 kilojoules of energy. The energy obtained through pasta has a low GI rating, and is mainly in the form of complex carbohydrates. Pasta (served without a sauce) is very low in fat, containing only 0.3 grams of fat per 100-gram serve. Pasta also contains 12 per cent protein.

Types and shapes of pasta

In Italian, the word 'pasta' means pastry or dough. Pasta dough, once prepared, can be cut or moulded into an endless variety of sizes, shapes and designs. There are hundreds of different shapes of pasta available, some of which may have several different names depending on the region in which they originated.

Pasta is, however, usually divided into four main groups according to the way in which it is used:
- small shapes used in soups: anelini, creste di gallo, ditalini, farfallini, ruoti
- longer lengths of pasta used for boiling and coating in a sauce: spaghetti, lasagnette, capellini, vermicelli, bucatini, fettuccine, tagliatelle, penne, linguine, farfalle
- sheets or shaped pasta used for baking: lasagne, cannelloni, conchiglie, rigate
- shaped pasta that can be filled: agnolotti, cappelletti, tortellini, ravioli.

ACTIVITY 11.3

Identifying pasta shapes

Types of pasta

Worksheet

Pasta	Number	Use
Spirale		
Tagliatelle		
Farfalle		
Lasagne		
Spaghetti		
Creste de galli		
Gnocchi		
Tortellini		
Penne		
Ruoti		
Fettucine		
Conchiglie		

1 Use a range of recipe books or food magazines to help you match each pasta shape below with its correct name from the table.

2 Copy the table into your workbook. Match the number of the pasta to its name in your table.

3 Find out the main use of each pasta type; for example, in soups, in salads, filled, coated with sauce or baked.

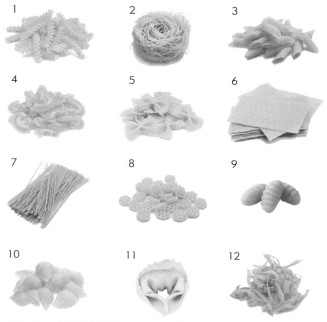

Shutterstock.com/Onigiri studio, Shutterstock.com/Noraluca013, Shutterstock.com/Ilizia, Shutterstock.com/Levent Konuk, Shutterstock.com/Viktor1

Pasta-making

All pasta, regardless of its shape or size or whether it is filled or unfilled, is made with the same basic ingredients: water and hard wheat flour (preferably durum wheat) or semolina. Homemade pasta is made from wheat flour or semolina and eggs that are combined into dough. Sometimes oil is added to the pasta to improve the dough's elasticity. Commercially prepared pasta is made from semolina and water, and generally does not contain egg.

Semolina is the coarsely milled outer layer of the wheat grain (or endosperm) and is composed almost entirely of protein, with very little free carbohydrate. Pasta made with semolina does not require as much water to mix it to a firm dough, and is preferred by commercial manufacturers because it dries far more easily than pasta made with plain wheat flour. Machine-made doughs using semolina are also easier to roll because they form a very strong dough. The semolina dough is also ideal for machine-made pasta because it is not as brittle, and therefore does not break as easily as doughs made with flour.

A recipe for Basic Pasta Dough is included on Page 261.

Making basic pasta

ACTIVITY 11.4

Investigating pasta

Visit the website of a major pasta manufacturer (such as Latina, Barilla, Vetta or San Remo) and a supermarket's website to answer the following questions.

1. Make a list of the main brands of fresh and dry pasta that are available in your local supermarket.
2. What nutritional information is provided on the packaging by the pasta manufacturers? What difference is there in the amount of carbohydrates present in commercial wholemeal pasta compared with white pasta?
3. What additional information about the nutritional value of pasta is provided by one of the commercial pasta manufacturers?
4. List four important tips to follow when cooking with pasta.
5. List the main methods used to package the pasta sauces available in your local supermarket. What would be the advantage to consumers of purchasing a pasta sauce in clear plastic packs compared with either glass or cans?
6. What are the four most common flavours of pasta sauce available in each type of packaging?
7. Are any low-fat sauces available in the supermarket? What marketing strategies have been used by the manufacturers of these products?
8. Search one of the pasta websites for ideas for a flavoured pasta meal that is low in fat and would be suitable for someone wishing to reduce their kilojoule intake.

Top tips for cooking pasta

1. Use plenty of water – specifically, 1 litre of water for every 100 grams of dried pasta – so that the pasta can swirl and roll around to cook evenly.
2. Add the pasta to the water only after it has reached a rolling boil.
3. Quickly stir the pasta when it first hits the water to prevent it from sticking together.
4. Add a teaspoon of oil to the water – this will help to prevent the water from boiling over, and will also stop the pasta from sticking together.

5. Add salt to the cooking water to prevent the pasta from tasting bland.
6. Do not cram too much pasta into one saucepan; this will mean the water takes too long to reboil and the pasta will stick together.

Gnocchi

In Italy, gnocchi is a traditional favourite Sunday dish. Each region of Italy has a different way of preparing it. In the north, gnocchi is made with flour and water; in Rome, with semolina; and in the south, with flour, potatoes and eggs.

Gnocchi are small dumplings that can be served with either a tomato or pesto sauce, or a heavier meat sauce or ragu. It is important to select potatoes of a starchy variety that mash well to ensure that the dumplings are light.

1. Boil potatoes until soft and tender. Drain, then push through the ricer.

2. Mix with other ingredients and form into a soft dough.

3. Divide the dough into portions and roll out into sausages, each about two centimetres thick.

4. Cut the gnocchi into two-centimetre-long pieces and roll over lightly with thumb to create a small indentation for the sauce.

5. Place eight to 10 pieces of gnocchi into boiling water. When they float on the surface, the gnocchi is cooked. Remove from saucepan and cook the next batch.

Steps in making gnocchi

Testing knowledge

1. Read the section 'Influences on eating in Australia' on pages 246–7, and then create a similar timeline that identifies major events and influences in your family that have shaped your eating patterns.
2. Outline four examples of how migration has influenced the cuisine of Australia.
3. Explain how concern for good health has influenced the foods available to consumers.
4. Discuss how the lifestyles of consumers have had an impact on the types of foods that are available in supermarkets today.
5. List 10 foods that Italian migrants brought to Australia after the Second World War that form an important part of our diet today.
6. Discuss why pasta is considered to be good for your health.
7. What is semolina, and why is it preferred for commercial pasta-making?
8. Why is it important to use plenty of water when cooking pasta? Outline two strategies to make sure that pasta does not stick together when cooking.
9. Explain why starchy potatoes are used for making gnocchi.
10. Describe how you can tell when gnocchi are cooked.

INFLUENCES FROM ASIA

In Australia, there is now an exciting East-meets-West cuisine derived from our neighbours in South-East Asia and East Asia, including Thailand, Malaysia, Indonesia, Vietnam, China and Japan.

Chinese people first came to Australia in large numbers during the gold rush in the 1850s and 1860s. As the gold ran out, many of the Chinese miners started businesses and stayed permanently in Australia. Many opened up Chinese restaurants and grocery stores in cities and in many country towns.

Other Asian foods and flavours began to appear on Australian menus in the late 1970s, in the wake of the Vietnam War. With continuing waves of multicultural immigration, mainly drawn from the Asian region, there has been an explosion of restaurants producing different cuisines, such as Thai, Vietnamese, Malaysian and Japanese. Traditional European restaurants also began to add Asian

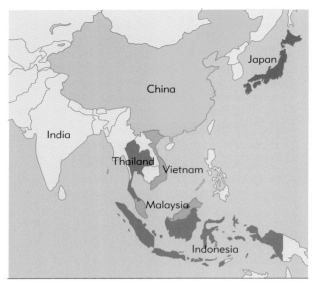
Asian countries neighbouring Australia

ingredients to their dishes and menus and create dishes that fused eastern and western flavours.

During the 1970s and 1980s, the availability of cheap travel enabled many young Australians to travel to Asian countries, and therefore to taste Asian flavours, which they often continued to eat when they returned home. An influx of Asian students studying in Australia has also increased the demand for traditional Asian foods and restaurants.

Australians are becoming more conscious of their health and demanding lighter, smaller, low-fat dishes that include a greater quantity and range of vegetables and complex carbohydrates such as rice- and noodle-based dishes. Asian dishes are quick to prepare and require minimal equipment – for example, a complete main course for a family can be prepared in one wok. Food manufacturers have responded to this trend by producing marinated meats, prepared sauces and frozen vegetables ready for stir-frying. Many Asian ingredients are grown locally; and those that are not are imported.

Hawker-style foods

Hawker-style foods are small portions of food that are traditionally prepared, cooked and served in front of the customer on the street. A 'hawker' is someone who tends a mobile food stall in the street. Hawker-style foods are traditionally eaten on the run.

In Malaysia, hawkers sell buns filled with sweet barbecued pork for breakfast, soup, laksas or won ton noodles for lunch, satays for snacks and a variety of noodle and rice dishes for the evening meal. In Thailand, charcoal-grilled chicken and sticky rice, soups and noodle dishes are readily available on street corners. In the past, hawkers often delivered their foods to people's homes. Today, hawkers still sell on the streets, and many group together in food courts.

Hawker-style food

Flavours of Asia

Asian cuisine is bursting with the warm tingling of spices, the sensuous perfume of lemongrass, ginger and herbs, the heat of chillies, the tang of lime and the soothing touch of coconut.

The essence of Asian cuisine is harmony; ingredients with strongly contrasting features are balanced in dishes in which no one flavour overpowers the other. Flavours and textures are balanced: sweet and sour, hot and mild, wet and dry, crisp and soft. Tasty dishes use ingredients such as chillies, ginger, lemongrass, curry powders, coriander, black beans, cardamom, sambal oelek, coconut, Kaffir limes, fish sauce, ketjap manis, soy sauce and palm sugar.

Thai flavour wheel

Ingredients and flavourings of Asia
Herbs

- Anise – often called star anise because of its star formation; has a liquorice flavour; is one of the main ingredients in five-spice powder
- Coriander – also known as Chinese parsley; used fresh in Thai curries or as a garnish; seeds are dry-fried and added to Indian curries
- Kaffir lime and leaves – a variety of lime with a dark skin and strong citrus aroma; used in curry dishes; glossy green leaves are available dried or fresh, and are added to curries
- Lemongrass – a long, tough-stemmed herb with a citrus, lemony flavour; the stem is bruised then chopped finely for use in Thai curries
- Mint – several varieties are available; used to flavour soups and as a component in rice-paper rolls.

Spices

- Cardamom – a spice used to give flavour to dishes; purchased in the pod, ground or as seeds
- Cinnamon – the fragrant bark of a tree; available ground or as sticks
- Cloves – dried flower buds; available ground or whole
- Cumin – seeds; roasted and ground and used in Indian curries.

Vegetables

- Bamboo shoots – tender shoots of the bamboo plant that add texture to a dish; available in cans
- Bean sprouts – from soy beans or mung beans; available fresh
- Chillies – available fresh, dried whole, powdered or as chilli sauce. Fresh chillies are available in many types and sizes – the smaller the chilli, the hotter the flavour. It is important to use rubber gloves when chopping to avoid burning the skin. Chilli sauce comes as a hot Chinese variety or a sweet, milder sauce
- Chinese broccoli – also known as Chinese kale
- Chinese cabbage – also known as wong bok
- Chinese chard – also known as bok choy or pak choy
- Galangal – variety of ginger with a distinct taste; available fresh or dried; used in Asian curry pastes, stir-fries and soups
- Ginger – fresh rhizome or root; the skin is removed by scraping, then the root is grated or chopped; used in a variety of Asian dishes
- Snake beans – also known as dau kok or dau que
- Water chestnut – also known as mai tai or may thay.

Asian flavouring ingredients

Flavourings

- Black bean sauce – made from fermented soy beans; used to flavour stir-fries
- Curry – available as paste or powder; prepared from a combination of ground seeds, particularly coriander, cumin, fennel and anise; other herbs and spices, such as galangal, turmeric, chillies and cinnamon, are also included
- Fish sauce – thin, brown sauce made from fermented fish; gives a salty, fishy flavour to Thai and Vietnamese dishes
- Palm sugar – coarse-grained brown sugar used in South-East Asian and Indian dishes
- Peanut – underground rhizome or nut that is crushed and used in satay sauce, or chopped, fried and sprinkled on salads and grilled foods

- Saffron threads – reddish to orange threads from the stigma of the crocus flower; when infused in liquid, they give off a vivid colour and impart a subtle flavour to rice and curries.

Tools and equipment

- Wok – the shape of a wok allows for only a small section of the base to contact the heat source and allow the tossing and moving of food while cooking; perfect for stir-frying, steaming, and deep-and shallow-frying

Wok   **Bamboo skewers**

- Bamboo skewers – used for threading and spearing food pieces; for example, satays and kebabs. Soak for several hours before use to prevent burning while cooking
- Bamboo steamer – bamboo steamers are available in a number of sizes and can be stacked on top of each other. The base of each has a grid of latticed bamboo and the top layer is lidded. The steamer is placed over a saucepan or wok of boiling water.

Bamboo steamer   **Mortar and pestle**

- Mortar and pestle – used to grind ingredients for dry and moist curry pastes
- Wok chan – long-handled, shovel-like utensil made from metal; used for lifting, tossing and stirring food to keep it in motion over the hottest part of the wok
- Wooden spatula – square-ended alternative to the chan
- Ginger grater – used to grate ginger and carrot.

Chinese cleaver   **Wok chan**

- Chinese cleaver – large, sharp, strong knife with a wooden handle; used for cutting and chopping
- Rice cooker – electrical appliance that automatically cooks rice. Washed rice is placed in the removable pan and cold water added to the prescribed level. The cooker automatically turns off when the rice is cooked, and keeps it warm for up to five hours.

Rice cooker

Asian cooking processes

- Chargrill – food is cooked by dry, radiant heat directed from above or below
- Deep-fry – food is submerged in hot oil. The oil must be hot (just below smoking point) to seal the food quickly and prevent it from absorbing too much oil. Extreme care must be taken when deep-frying to prevent water from coming in to contact with the hot oil, or the oil catching alight

Deep-frying Asian food

- Steam – food is cooked by heat in the form of steam, either atmospheric or under pressure; the food to be steamed is suspended over a boiling liquid
- Shallow-fry – oil covering the base of a wide-topped, heavy-based pan or wok is heated; the food is gently and slowly cooked and turned at least once during the cooking time
- Stir-fry – a small amount of oil is heated in a wok or wide-topped pan; the food is stirred and lifted during the short, hot cooking process.

Testing knowledge

11 Explain how Chinese food was introduced to Australia.

12 Describe how travel to Asian countries has influenced Australians' eating habits.

13 Why do Asian-style meals appeal to people in the community who are health-conscious?

14 What are hawker-style foods, and why are they so popular in Asia?

15 What is the significance of the Thai flavour wheel, and how can the information it provides be used when designing dishes and meals?

16 List six herbs and flavourings that provide the flavours characteristic of the Pacific Rim countries. Name one Asian recipe that you have made or eaten that includes at least one of these flavourings.

17 List five vegetables available in Australia that are of Asian origin. Place a tick against those that you have tried.

18 What is fish sauce, and why is it commonly used in Asian cuisine?

19 Explain why a wok and a bamboo steamer are both important pieces of equipment used in cooking many Asian-style foods.

20 Describe four cooking methods that are used in the cooking of Asian foods.

VIETNAMESE CUISINE

Vietnam shares its borders with three other countries: its northern border is shared with China, and its western border with Laos and Cambodia. While the influence of these three countries can be seen in the cuisine of Vietnam, its food still has a character and flavour of its own. Many forces – of climate, trade, religion, history and multiculturalism – have influenced Vietnamese cuisine.

Influences on Vietnamese cuisine

- Religion: the introduction of the Buddhist religion from China brought with it a rich and varied vegetarian diet with an emphasis on soy products for protein. Christianity – in particular, Catholicism – is another influence from the long French occupation.
- Celebrations: celebrations such as the Lunar New Year, or Tet festival, include sweet cakes made from sticky rice or sticky rice flour (Tet cakes, or moon cakes) and sweet delicacies made from bananas.

Vietnamese market

- Climate: in the northern areas, the climate is colder, and, consequently, a smaller variety of foods is grown. In southern areas, the climate is hotter and more humid and the rainfall higher. The area of the Mekong Delta is a very rich, fertile area for the growing of a greater variety of fruits and vegetables. Sugarcane is also grown in the south; as such, many of the foods there are sweeter.
- Chinese rule: the ingredients and cooking methods

of the food of Vietnam have been significantly influenced by early Chinese rule. The early Chinese culture brought noodles and chopsticks.
- French occupation: the Vietnamese learnt the art of bread-making from the French during their long occupation of the country. The French brought the baguette, pastries and filtered coffee.
- American occupation: ice-cream is now a popular food in South Vietnam. It was a favourite food imported for the American troops who were stationed in this area during the Vietnam War.

Baguettes for sale in a Vietnamese street

Characteristics of Vietnamese cuisine
- Fresh, well-flavoured ingredients are purchased daily from markets.
- Accompaniments such as chillies, fish sauce, lime wedges, fresh herbs and dipping sauces are used.
- Rice is the most important crop and is the country's staple food, forming the base for most dishes. Seventy-five per cent of the country's arable land is used to cultivate rice. A sticky, glutinous variety is used for sweets and cakes; spring roll wrappers and noodles are made from ground rice.
- Fish is the main protein source and a feature of the country's cuisine. Fishing is a major industry along the long coastline and in the rice paddies. Fish is available fresh daily in markets or is dried for later use or fermented to produce nuoc mon or fish sauce. Fish sauce is a very important flavouring for most Vietnamese food, and no meal is considered complete without it.

THAI CUISINE

Thailand is a small country in South-East Asia that shares a peninsula with Myanmar, Laos, Cambodia and Vietnam. Thailand forms a crescent around the Gulf of Thailand, and has a maze of rivers and canals running through it. Thai cuisine is renowned all around the globe for its distinctive characteristics, including its mix of spicy and sour flavours.

Influences on Thai cuisine
- Climate and location: geographically placed halfway between India and China, the hot, humid climate encourages the growth of a wide variety of plant foods.
- Religion: the philosophy of Buddhism influences the food customs of Thailand. Early in the morning, families offer monks food from tables outside their homes. This generous offering is an important means of gaining merit for their future incarnation. Buddhist religious days are celebrated by bringing specially cooked foods to the temples.
- Fuel: cooking fuel is expensive, so cooking usually employs quick methods, such as frying, grilling and stir-frying.

Cooking in Thailand

Characteristics of Thai cuisine

- The main meats eaten are pork, beef, chicken and water buffalo. The quantities of meat eaten are small, as it is considered an expensive food item.
- Fish, an important source of protein, is often used in Thai dishes.
- Dairy products are eaten in small amounts because the land space for raising cattle is limited.
- Food is purchased daily from markets.
- Fresh fruits are grown all year round in the hot, wet climate. Fruits such as jackfruit, pineapple, custard apple and mango are usually served at the end of a Thai meal.
- Desserts are generally based on coconut milk; for example, coconut custard or steamed coconut pudding.
- Fresh herbs, spices and seasonings such as basil, lemongrass and coriander are constantly used. Foods such as the Kaffir lime give Thai food a distinctive perfume and aroma. Other unique flavours are provided by galangal (a type of ginger), fish sauce, shrimp paste, coconut milk, tamarind, chillies and palm sugar. Curry pastes (red and green) are important ingredients.
- Garnishes are important in the presentation of Thai food. Garnishes such as roasted peanuts, shallots, toasted coconut and carved vegetables provide flavour and texture to dishes.
- The meal is built around rice. Southern Thai people eat long-grain rice, while northerners favour short-grain, or sticky rice. Noodles were introduced from China and also play a major role in Thai cuisine.
- Food is light and low in fat, with the major components being cooked vegetables with a small proportion of meat.
- The preparation of ingredients by pounding, chopping and blending is time-consuming compared with the final cooking process.
- The five flavours of salty, sweet, sour, spicy, and bitter are blended and balanced in each meal.

ACTIVITY 11.5

Investigating Thai curry pastes

Curry pastes

1. Purchase a red and a green Thai curry paste.
2. Investigate the ingredients used to produce red and green Thai curry pastes.
3. Record the uses for each type of curry paste.
4. Carefully read the label on each container and list the ingredients in their appropriate places on the flavour wheel.
5. Taste test the curry pastes in the following way.
   a. Stir-fry two teaspoons of red curry paste in a little oil in a wok.
   b. Wash the wok and repeat the process using green curry paste.
   c. Add half a cup of fresh noodles and toss through the paste.
   d. Evaluate each curry paste according to their aroma and flavour.

MEXICAN CUISINE

Mexican food, with its famous tacos, nachos and enchiladas, is one of the world's most popular cuisines. The food has experienced a wide variety of influences due to past colonisation and, later, from trade among people from various countries and colonies.

The staple foods of Mexico consist of native ingredients such as corn, beans, chilli peppers, tomatoes and avocados.

Influences on Mexican cuisine

The early culinary influences on Mexican cuisine were from the Mayan Indians, who were primarily hunter-gatherers who used native ingredients such as corn, bean, chilli peppers and herbs. Mexican cuisine has also been heavily influenced as a result of the Spanish conquest of the Aztec Empire in the 16th century. The Spanish conquistadors bought with them meats such as beef, pork, chicken, goat and sheep; dairy products, particularly cheese; and various herbs and spices. Today, locally-grown foods such as corn and beans form a large part of the foods.

Characteristics of Mexican cuisine

- The cuisine has sharp flavours with a wide variety of textures, combining fresh ingredients with quick, simple preparation methods.
- Corn is the most common starchy cereal and, while it is eaten fresh, it is also dried, treated with lime and ground to form masa, which is the dough used to make tortillas.
- The most common way to eat corn in Mexico is in the form of a tortilla, which accompanies a variety of dishes. Tortillas can be eaten plain or wrapped around a filling. They form the base for burritos, tacos, quesadillas and enchiladas.
- Chilli is a major flavouring ingredient in Mexican food. Chillis are used either fresh or dried, and vary greatly in heat and spiciness.
- Frijoles (pronounced fri-hole-lez) means 'beans'. Mexican foods use black, red or white beans or pintos.
- The main meal of the day usually consists of a meat served in a cooked sauce with salsa on the side, accompanied with beans and tortillas.
- Northern Mexico is famous for its beef production, while south-eastern Mexico is known for its spicy vegetable and chicken-based foods.
- Seafood is popular in the states that border the Pacific Ocean and the Gulf of Mexico.

THINKING SKILLS 11·1

1. Develop a timeline that shows the significant influences on eating in Australia.
2. Make predictions about what will be the next significant trend that will influence eating in Australia.

ACTIVITY 11·6
Salsa taste test

Salsa (which means 'sauce') is frequently used as a dip, spooned onto savoury dishes as a garnish, or served as an accompaniment to a main meal. Salsa can be either hot and spicy or cool and refreshing – it all depends on the combination of ingredients. The basic ingredients are dried or fresh peppers, onions, tomatoes, garlic and coriander. Oregano, vinegar and olive oil are sometimes added.

AIM
To carry out a taste test to compare the sensory properties of commercial salsas

METHOD
1. Purchase three varieties of the same brand of commercially prepared salsa with varying degrees of heat; that is, mild, medium and hot.
2. Use corn chips (unsalted) or teaspoons to taste the salsa.
3. Record your results in the table below.

RESULTS
Salsa taste test

Sensory properties	Mild salsa	Medium salsa	Hot salsa
Appearance			
Aroma			
Flavour			
Texture			

ANALYSIS

1. Which salsa had the most appealing colour? Why?
2. Did the texture of the salsas vary?
3. Was there a significant difference in the degree of hotness in each salsa?
4. Read the labels of the commercially prepared salsas and identify the ingredients that would make the salsas taste hot.
5. Some brands of salsa use a sketch of a chilli on the label. Do you think the use of the chilli image on the packaging is more useful for consumers than an ingredient list? Justify your answer.
6. Record the ingredients on the labels that are described as additives.
7. Why are additives included in salsas?
8. Why are unsalted corn chips more suitable for the taste test than salted ones?

CONCLUSION

Which salsa did you prefer? Why?

Testing knowledge

21. Discuss how the following factors have influenced modern Vietnamese food:
    - religion
    - celebrations
    - climate
    - French occupation.
22. Why is rice an important staple ingredient in Vietnam?
23. Explain why fish is the main source of protein in the Vietnamese diet.
24. List four herbs and spices that give Thai cuisine its unique flavour.
25. Draw a concept map to demonstrate the key characteristics of Thai cuisine.
26. Discuss how religion influences the food customs of Thailand.
27. Discuss how the Spanish have influenced the ingredients used in Mexican cuisine.
28. Explain how corn is treated before it is made into tortillas.
29. Why are tortillas an important ingredient in the Mexican diet?
30. Describe two main differences between Thai and Mexican cuisine.

Tacos

DESIGN ACTIVITY 11.1

Flavoured pasta and sauce

DESIGN BRIEF
A pasta manufacturer is planning to produce a range of flavoured pasta that reflects one of the influences on Australian cuisine such as Indigenous Australian, Italian or Asian. The flavoured pasta will be served with an accompanying sauce. The sauce should be quite simple, so that it complements but does not overpower the flavouring ingredient.

Worksheet

1. Write a design brief based on the 5 Ws.
   - Who – to whom will this new range of flavoured pasta be marketed?
   - What – what cultural influence will be the basis for the new pasta?
   - When – when will the new range of flavoured pasta be served?
   - Where – where will the pasta be served?
   - Why – why is the flavoured pasta being created?
2. Develop four evaluation criteria questions by which to judge the success of your end product.

INVESTIGATING
1. Complete Activity 11.4, 'Investigating pasta', on Page 250.
2. Undertake an internet search of pasta recipes to identify vegetables, herbs and bush flavourings suitable to include in a pasta and sauce.

GENERATING
1. Using your research into vegetables, herbs and bush flavourings suitable to include in a pasta and sauce, complete a recipe map like the one below.
2. Develop two possible design options for the pasta and sauce.
3. Select your preferred option. Justify your selection.

PLANNING AND MANAGING
1. Write up your new recipe. Use the Basic Pasta Dough recipe on Page 261 as the basis of your product.
2. Prepare a food order.
3. Write up a production plan.

PRODUCING
1. Prepare the product.

ANALYSING AND EVALUATING
1. Evaluate your product based on the previously established evaluation criteria questions.
2. What modifications would you make to the recipe if you were to produce it again? Why?
3. Describe the most challenging aspect of the production. Why was it challenging?

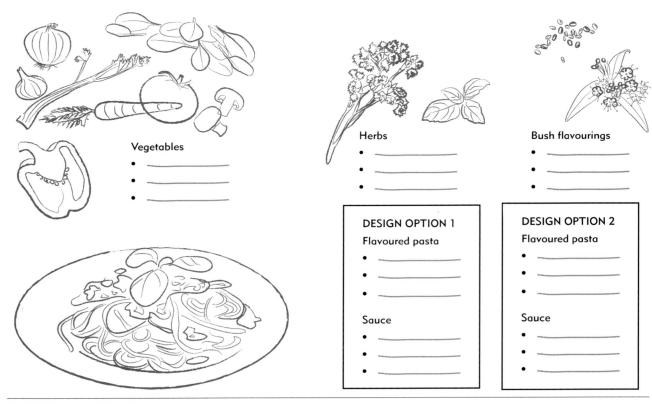

Recipe map for pasta and sauce

BASIC PASTA DOUGH

120 grams baker's flour or strong flour

¼–½ teaspoon salt

1–2 eggs

2 teaspoons olive oil

SERVES TWO

METHOD

1. Sift the flour and salt. Lightly beat the eggs.
2. Make a well in the centre of the flour and add some of the eggs.
3. Gradually mix the flour into the liquid, adding more liquid if necessary to form a firm dough.
4. Knead the dough for approximately 10 minutes or until very smooth and elastic, using extra flour as required. The dough should be able to stretch without cracking.
5. Cover the dough with plastic wrap. Allow to rest for 20 minutes.
6. Roll out by hand or use a pasta machine to cut into shapes.
7. When cut, hang over an oven rack or thin rod to dry.
8. Place in boiling water and boil for 2–3 minutes or until al dente, or tender to the bite. (You can test this by tasting a piece of the cooked pasta to check whether there is any uncooked starch remaining in the centre.)
9. Drain and serve as desired.

EVALUATION

1. Why is it important to gradually mix the dry ingredients into the liquid ingredients?
2. What is the purpose of kneading the dough for 10 minutes?
3. Why is it desirable to rest the dough for 20 minutes before using it?
4. Explain why it is essential to cook pasta in plenty of boiling water.
5. How could fresh pasta be kept for future use?

GNOCCHI WITH NAPOLI SAUCE

2 medium to large starchy potatoes
½ cup flour
½ teaspoon salt
1 teaspoon olive oil
½ egg, lightly beaten

SERVES TWO

METHOD

1. Boil the potatoes with their skins on until tender. Peel and put through a ricer (similar to a big Italian-style garlic press). Cool.
2. Mix together the potato, flour, salt, oil and egg into a soft dough.
3. Shape into a fat loaf and set on a floured board. Cut off pieces and roll very lightly into a cord 1 centimetre thick. Cut the cord into 2-centimetre lengths.
4. Lightly roll each segment in the centre under your forefinger to give the piece a bow shape.
5. Set the gnocchi aside. The pieces should not touch.
6. Bring a large saucepan of water to the boil and drop in the gnocchi one at a time. They are cooked when they rise to the top. Remove immediately to a heat-proof dish using a slotted spoon. Cook 8 to 10 at one time.
7. Place on serving plate. Top with hot Napoli Sauce (see recipe on Page 157). Garnish with parmesan cheese.

EVALUATION

1. Which potato variety is best suited for making gnocchi, and why?
2. Use the star diagram below to record the sensory properties of the gnocchi.
3. If you do not have a ricer, suggest another method you could use to process the potatoes.
4. Suggest an alternative method for shaping and rolling the gnocchi other than that used in the recipe.
5. Why is it important to drop the gnocchi into the boiling water one at a time and to remove each one when it rises to the top?
6. Write a paragraph that discusses the health benefits of the Gnocchi with Napoli Sauce.

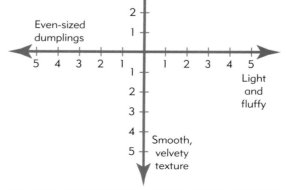

Star diagram

SPAGHETTI BOLOGNESE

Bolognese sauce

½ onion, finely diced

1 small clove garlic, crushed

1 tablespoon oil

125 grams minced beef

1 cup canned, diced tomatoes

2 tablespoons tomato paste

½ cup stock

¼ teaspoon dried basil

¼ teaspoon dried oregano

few shakes of pepper

1 tablespoon parmesan cheese, for garnish

Pasta

oil

200 grams dried or fresh pasta

salt

SERVES TWO

METHOD

Bolognese sauce

1. Heat the oil in a medium-sized saucepan. Add the onion and garlic and gently fry for 2–3 minutes until just beginning to brown.
2. Add the minced beef and stir continuously until well browned.
3. Add the tomatoes, tomato paste, stock, herbs and pepper.
4. Bring the sauce to the boil. Reduce heat to simmer, cover with a lid and cook very gently for approximately 30 minutes or until the sauce has thickened. Stir occasionally during cooking to prevent sauce from sticking to the bottom of the saucepan.

Pasta

1. Fill two-thirds of a large saucepan with water. Place the lid on the saucepan and bring to the boil.
2. Add a teaspoon of oil and a pinch of salt to the water – this will help to prevent the water from boiling over and stop the pasta from sticking together.
3. Gradually add the pasta to the boiling water. Stir once or twice only to separate the pasta.
4. Rapidly boil, uncovered, until the pasta is al dente. (You can check this by tasting a piece of the cooked pasta to check whether there is any uncooked starch remaining in the centre.) Cooking time will vary according to the type of pasta used. After the water has come to the boil, fresh pasta will take 2–4 minutes to cook; dried pasta will take 12–15 minutes.
5. Drain the pasta in a colander.
6. Serve immediately with the bolognese sauce. Garnish with parmesan cheese.

EVALUATION

1. Why is oil added to the cooking water when boiling pasta?
2. Explain why the sauce is simmered for 30 minutes.
3. How much dried pasta would you prepare for one serving?
4. Define 'al dente' in reference to pasta.
5. Discuss why spaghetti bolognese is a very popular family meal. Consider the preparation time, nutrition and sensory properties.

VIETNAMESE SPRING ROLLS WITH DIPPING SAUCE

Spring rolls

- 1 teaspoon peanut oil
- 1 teaspoon fresh ginger, grated
- 1 clove garlic, crushed
- 2 green shallots, chopped
- 125 grams minced pork
- 100 grams canned small prawns
- 1 tablespoon fresh mint, chopped
- 2 teaspoons fish sauce
- 1 tablespoon sweet chilli sauce
- 2 teaspoons lime juice
- pinch of sugar
- 15 spring roll wrappers
- 2 teaspoons cornflour
- 2 teaspoons water
- oil, for frying

Dipping sauce

- 2 tablespoons sugar
- 2 tablespoons hot water
- 1 tablespoon fish sauce
- 1 tablespoon lime or lemon juice
- 1 tablespoon sweet chilli sauce (or 1 small red chilli, sliced)
- 2 teaspoons white vinegar
- 2 teaspoons chopped fresh coriander leaves

MAKES 15 SPRING ROLLS

METHOD

Spring rolls

1. Heat the oil in a wok and add ginger, garlic and shallots. Cook until shallots are soft.
2. Add the pork and cook until the meat is tender and has changed colour. Add the prawns and toss through the meat mixture.
3. Stir in the mint, sauces, lime juice and sugar. Stir and cook for a few minutes until the liquid has been reduced. Spread on a plate and allow to cool.
4. Place 2 teaspoons of mixture in the middle of each spring roll wrapper. Fold each side inwards and roll up. Secure the end of the wrapper with a paste made from the cornflour and water.
5. Deep-fry the rolls in hot oil until golden brown. Drain well on absorbent paper. Alternatively, place the spring rolls on a greased baking tray, lightly spray with oil and bake at 200°C for approximately 10 minutes or until golden.
6. Serve with dipping sauce.

Dipping sauce

1. Dissolve the sugar in the hot water and boil, uncovered and without stirring, for 5 minutes or until the mixture slightly thickens.
2. Stir in the remaining ingredients and allow to cool.
3. Serve in a small bowl as a dipping sauce for the spring rolls.

EVALUATION

1. Why is it important to fry the garlic, ginger and shallots in Step 1 of the spring rolls recipe?
2. Explain why it is recommended that you allow the filling to cool before assembling the spring rolls.
3. Explain how the cornflour seals the spring rolls.
4. What are the advantages of baking the rolls in the oven, rather than deep-frying them?
5. Why is it important to boil the sugar and water mixture until it thickens when making the dipping sauce?

CHICKEN CURRY WITH COCONUT PANCAKES

Chicken curry

2 teaspoons oil

1 clove garlic, crushed

1 small piece fresh ginger, grated

1 chilli, seeds removed and finely chopped

½ teaspoon lemongrass, finely chopped

2 skinless chicken fillets, sliced

1 small onion, quartered and separated

1 teaspoon coriander, roughly chopped

1 tablespoon light soy sauce

1 cup coconut cream

½ cup chicken stock

1–2 teaspoons green curry paste (according to taste)

1 extra chilli, to prepare as chilli flower garnish

Coconut pancakes

1 egg, lightly beaten

½ cup coconut milk

½ cup flour

2 tablespoons shredded coconut

small quantity of butter, for frying

SERVES TWO

METHOD

Chicken curry

1. Heat the oil in a wok over a medium flame and sauté the garlic, ginger, chilli and lemongrass for 1–2 minutes.
2. Add chicken and allow to brown on both sides. Cook for 5–8 minutes. Do not turn the chicken pieces until each side has seared and the juices are sealed in.
3. Add the onion and quickly fry.
4. Stir in the remaining ingredients and simmer for 15 minutes or until the chicken is tender.
5. Serve with coconut pancakes and garnish with chilli flowers and chopped coriander.

Coconut pancakes

1. Whisk together egg and coconut milk.
2. Sift flour and add to the egg mixture with shredded coconut. Whisk until smooth.
3. Allow to stand for 30 minutes.
4. Melt a small amount of butter in a crepe pan or frying pan (6 centimetres in diameter) over medium heat.
5. Spoon in small quantities of batter.
6. Allow to cook until the surface appears dry and the underneath is golden brown.
7. Loosen edges and turn or flip.
8. Cook the underside until it too is golden brown.
9. Remove and keep warm by wrapping in foil wrap.
10. Serve as an accompaniment to the chicken curry.

EVALUATION

1. Explain why it is important to take great care when chopping chillies.
2. Why are the garlic, ginger, chilli and lemongrass fried before being combined with the chicken?
3. Explain what would happen if the chicken were turned before it had been completely seared on one surface.
4. List four other vegetables that could be added to the chicken curry recipe.
5. What effect does standing have on the pancake batter?

THAI FISHCAKES

185 grams canned tuna (in brine)

1 teaspoon Thai green curry paste

¼ cup coconut cream

2 tablespoons cornflour

1 egg

2 tablespoons coriander, chopped

1 cup fresh breadcrumbs

2 tablespoons peanut oil

sweet chilli dipping sauce

SERVES TWO

METHOD

1. Drain the tuna well.
2. Place the tuna in a food processor with the Thai curry paste, coconut cream, cornflour, egg and coriander.
3. Process until just combined – do not over-process or allow the mixture to become paste-like.
4. Add the breadcrumbs and combine in the food processor; if too wet, add more breadcrumbs.
5. Use a tablespoon of mixture and shape into small patties.
6. Rest the mixture in the refrigerator for 10 minutes to firm up.
7. Heat oil in frying pan and cook fishcakes for 2–3 minutes or until lightly browned all over. Turn once only.
8. Drain on absorbent paper before serving.
9. Serve in lettuce cups with a sweet chilli dipping sauce.

EVALUATION

1. Why is it important to drain the tuna well before adding it to the other ingredients?
2. What is the purpose of adding fresh breadcrumbs to the mixture?
3. List the important safety issues to consider when frying the fishcakes.
4. Explain why it is important to drain the fishcakes on absorbent paper before serving.
5. Suggest some accompaniments for the fishcakes other than the chilli dipping sauce.

MINI LAMB KEBABS WITH MINTED GARLIC DIP

Mini lamb kebabs

125 grams minced lamb

¼ onion, grated

¼ teaspoon ground cumin

¼ teaspoon ground coriander

2 teaspoons parsley, chopped

1 small egg, lightly beaten

2 teaspoons vegetable oil

Minted garlic dip

¼ cup plain yoghurt

2 teaspoons mint, chopped

1 clove garlic, crushed

½ Lebanese cucumber, finely diced

MAKES 10–12 MINI KEBABS

METHOD

1. Mix together the minced lamb, onion, cumin, coriander, parsley and egg until smooth and well combined.
2. Take 2 level teaspoons of the mixture and, with wet hands, shape into balls; repeat until all mixture has been used.
3. Chill the kebabs in the a refrigerator for 15 minutes.
4. Prepare the yoghurt dip by combining all ingredients and mixing well.
5. Chill the dip for 15 minutes to allow flavours to develop.
6. Heat oil in a frying pan, add half of the kebabs and cook until browned and cooked through.
7. Drain on absorbent paper and spear each kebab with a toothpick. Cook remaining kebabs.
8. Serve with dip.

EVALUATION

1. Why is the onion grated and not diced before being added to the kebab mixture?
2. Why is it important to use wet hands when shaping the uncooked kebab mixture?
3. What role does the egg have in the kebab mixture?
4. What rules are important to remember when heating and cooking with oil?
5. Why is it important to drain the cooked kebabs on absorbent paper before serving?

CHAPTER 11 GLOBAL GOODIES

TANDOORI CHICKEN BITES

½ large chicken fillet
¼ cup yoghurt
2 teaspoons tandoori paste
2 spring onions

MAKES 6–8 CHICKEN BITES

METHOD

1 Cut chicken into bite-sized pieces.
2 Combine yoghurt and tandoori paste in a ceramic or glass bowl. Add the chicken pieces and coat with the paste.
3 Cover the bowl with plastic wrap and refrigerate for 30 minutes.
4 Prepare the spring onion garnishes by cutting off 2-centimetre lengths of the green stem and making fine cuts halfway down each length.
5 Place onion lengths in iced water to curl. Drain and pat-dry.
6 Cover the grill tray with foil and grill chicken pieces until lightly browned and tender.
7 Assemble on toothpicks with spring onion frills at the top.

EVALUATION

1 Why is it important to use a glass or ceramic bowl when allowing the coated chicken to stand?
2 Why is the coated chicken allowed to stand for 30 minutes?
3 What might happen to other foods in the refrigerator if the coated chicken was not covered?
4 Why is it important to cover the grill tray with foil when grilling the chicken pieces?
5 Suggest another suitable garnish for the Tandoori Chicken Bites.

CHICKEN BURRITOS

To poach chicken

1 cup chicken stock

¼ onion

8 centimetres celery, chopped

½ bay leaf

1 chicken breast, skinned

Filling

1 chicken breast, poached, or 1 cup cooked chicken

2 teaspoons oil

½ onion, diced

2 teaspoons chilli sauce

3 teaspoons cornflour

150 millilitres light evaporated milk

Assembling the burritos

4 tortillas

lettuce, shredded (as needed)

tomato, diced (as needed)

avocado, chopped (as needed)

salsa

SERVES TWO

METHOD

1. Add the chicken stock, onion, celery and bay leaf to a pan with a tight-fitting lid. Bring to boil.
2. Add the chicken breast and reduce heat to simmer. Cover with lid and cook for 8 minutes.
3. Turn off heat and allow chicken to cool in the liquid.

METHOD

1. Preheat oven to 180°C.
2. Dice the cooked chicken into 1-cm cubes.
3. Heat oil in pan, add onion and cook for 1 minute. Remove from heat.
4. Add chilli sauce, cornflour and light evaporated milk. Stir to combine.
5. Return to heat and bring to boil, stirring constantly. Simmer for 1 minute – the sauce should have thickened. Remove saucepan from heat.
6. Fold chicken through the sauce.
7. Wrap the tortillas together in foil and warm for 10 minutes in preheated oven.
8. Fill each burrito with chicken mixture, lettuce, tomato, avocado, salsa and extra chilli sauce (if desired), then roll up and serve.

EVALUATION

1. What are the advantages and disadvantages of poaching your own chicken for this recipe?
2. When making the sauce for the filling, why is it important to bring the mixture to the boil in Step 5 of the recipe?
3. Why are the tortillas wrapped and warmed in the oven before being filled?
4. Which part of this production did you find the most challenging? Why?
5. Plot the ingredients of the Chicken Burritos on a diagram of the Healthy Eating Pyramid or the Australian Guide to Healthy Eating. Comment on the the nutritional rating you would give this dish.

This recipe uses a light white sauce to bind the filling ingredients. This has health benefits because the sauce is much lower in fat than sour cream.

TACOS

Spicy meat filling

2 teaspoons oil

¼ onion

½ clove garlic, crushed

100 grams minced meat

1 tablespoon tomato paste

⅛–¼ teaspoon chilli powder

pinch salt, pepper and sugar

¼ teaspoon ground cumin

¼ teaspoon ground coriander

½ teaspoon Worcestershire sauce

¼ cup water

1 tablespoon canned kidney beans, rinsed and roughly chopped

Assembling the tacos

2 king-sized taco shells

½ tomato, diced

lettuce leaf, shredded

30 grams cheese, grated

chilli sauce

SERVES ONE

METHOD

1. In a small saucepan, heat the oil and lightly brown the onion and garlic.
2. Add minced meat and stir until brown. Mash with a fork or potato masher to break the meat into small pieces.
3. Add the tomato paste, spices, flavourings, water and kidney beans.
4. Simmer for 5–10 minutes until the water has evaporated and the mixture has thickened and is almost dry.

To assemble

1. Preheat oven to 180°C.
2. Place the taco shells on a baking tray with the openings facing down so they remain open during cooking. Heat shells in oven for 4 minutes.
3. Fill with the spicy meat filling, tomato, lettuce and cheese. Top with chilli sauce.

EVALUATION

1. Why are the onion and garlic lightly browned before the meat is added to the spicy meat filling?
2. Why are the taco shells heated upside down?
3. Describe two safe work practices you should consider when removing food from the oven.
4. What changes would you make to the recipe if you were to make it again?
5. Plot the ingredients of the Tacos on to a diagram of the Healthy Eating Pyramid or the Australian Guide to Healthy Eating. Comment on their nutritional rating.

NACHOS

2 teaspoons oil
¼ onion, finely diced
½ clove garlic, crushed
⅛ teaspoon chilli powder
¼ teaspoon paprika
¼ teaspoon cumin
1 tomato, skinned and diced
2 tablespoons canned kidney beans, rinsed and roughly chopped
1 tablespoon tomato paste
¼ teaspoon brown sugar
¼ cup water
30 grams tasty cheese, grated
extra pinch of paprika, for garnish
corn chips

SERVES ONE

METHOD

1. Preheat oven to 180°C. Omit this step if the final cooking is done in the microwave.
2. Heat the oil in a small saucepan and sauté the onion, garlic and spices. Cook for 2–3 minutes but do not brown.
3. Add the tomato, beans, tomato paste, sugar and water. Cover and gently simmer for about 5 minutes or until the mixture is thick. Stir occasionally.
4. Arrange the corn chips on an ovenproof plate and spoon mixture into the centre.
5. Sprinkle with cheese and a pinch of paprika.
6. Cook in moderate oven until cheese has melted or microwave on medium-high for 2–3 minutes.

How to skin the tomato

1. Remove the stem end from the tomato using the point of a vegetable knife.
2. Place the tomato in a large bowl. Boil enough water to cover the tomato.
3. Pour the water over the prepared tomato and let it stand for 30 seconds or until the skin begins to split. Drain off the hot water.
4. Cover in cold water. Stand for 30 seconds. Drain off cold water.
5. Allow the tomato to cool, then peel.

EVALUATION

1. Identify the ingredients in the recipe that are classified as spices.
2. Why are the kidney beans an important part of the sauce?
3. What processed product could you use instead of having to peel and dice a fresh tomato?
4. Why is the sauce simmered with the lid on the saucepan in Step 3?
5. What are some of the health concerns about serving corn chips with dips at a party?

12 BEST BAKING

KEY KNOWLEDGE

- The art of baking
- The ingredients in baked products
  - Flour
  - Sugar
  - Eggs
  - Butter
  - Other ingredients
- How baking works
  - Aeration in baked products
  - Processes used to make baked products
  - Top tips for cake-making
  - Top tips for making biscuits
- Pastry
  - Functional ingredients in pastry
  - Types of pastry
  - Top tips for making pastry
- Chocolate
  - Fair trade chocolate
  - Top tips for cooking with chocolate
- Presenting and decorating with chocolate
- Decorating with sugar

KEY TERMS

aeration, or leavening the trapping of air in a mixture that then expands as the product cooks and causes the product to rise and be light in texture

dextrinisation a process that occurs during cooking in which dry heat in an oven breaks down starch molecules, forming a brown crust on the outside of baked products

Maillard reaction a browning reaction that occurs when sugar or starch and a protein are present during baking

AUSTRALIAN CURRICULUM LINKS

DESIGN AND TECHNOLOGIES
- Knowledge and Understanding
- Processes and Production Skills

GENERAL CAPABILITIES
- Critical and creative thinking

THE ART OF BAKING

Baking is the process used to produce a range of cakes, biscuits, slices, sweet treats and pastry. The main ingredients used in these foods are flour, sugar, eggs and butter.

Baked products are cooked in an oven and exposed to dry heat, which is circulated around the food by convection currents. These foods are usually sweet in nature – although some have a savoury flavour profile – and are often high in fat and/or sugar, so they should only be eaten as special treats or indulgences, and for celebrations.

THE INGREDIENTS IN BAKED PRODUCTS

Flour

Flour is an important ingredient in baked products because it provides volume and structure to the end product. Flour can absorb milk, eggs, water or butter during the mixing process to form batters and doughs. When flour is exposed to heat in an oven, the protein or gluten in it sets, preventing the final product from collapsing when it is cooled. Two other processes that involve flour in baked products are dextrinisation and the Maillard reaction. **Dextrinisation** occurs when the dry heat in an oven causes the starch in flour to dextrinise, contributing to the browned surface and delicious aroma that develops during baking. The **Maillard reaction** also assists in colour development of baked products; it occurs when starch from the flour, and/or the sugar and the protein from the eggs, are exposed to heat in the oven, and creates a golden-brown colour.

Flours that are made from soft wheats – in other words, low in gluten – which are sometimes known as cake flour, are most suitable for baking products, because they produce a soft texture. White, wholemeal or self-raising flour can be used to make cakes, biscuits, slices and pastry.

Sugar

In Australia, sugar is made by crushing and refining sugarcane. During the refining process, the juice is squeezed out of the cane and then boiled. Molasses – a dark, thick, sticky liquid – is separated from the sugarcane juice, and the remaining juice is crystallised to create a range of different types of sugar. Sugar is the common name for sucrose, which is one of the many natural sweeteners found in plant food.

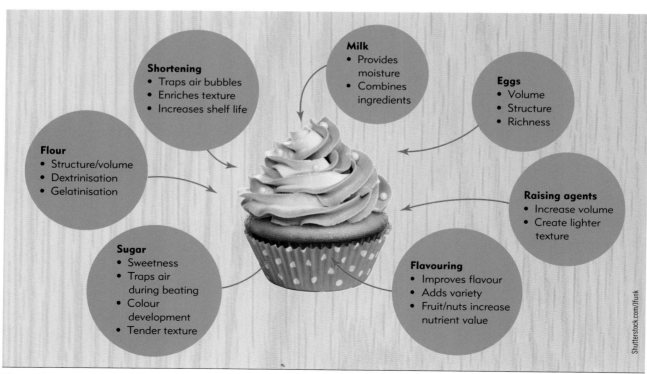

Functional properties of ingredients in baked products

Macarons

Types of sugar

In food preparation, different types of sugars have been designed for specific purposes and to create different sensory properties in food products. Sugar can:
- add colour
- add flavour
- improve texture
- aerate or increase volume when creamed with butter
- act as a preservative
- help yeast ferment in bread-making.

Types of sugar

A1, white, granulated sugar is the most common form of sugar, and is sometimes called table sugar. It has medium-sized crystals. Caster sugar has finer, smaller crystals than A1 sugar, and is often used in cakes, biscuits and desserts because it dissolves more effectively into the mixture and gives a more even appearance and texture. Icing sugar looks and feels like a white powder, and is usually used to ice cakes and biscuits. Soft icing sugar has an anti-caking agent added to it so that it does not go lumpy. Brown sugar is a soft, moist sugar with very fine crystals. Its distinctive flavour and colour are usually made by adding molasses to white sugar. Raw sugar is the sugar crystals before the refining process takes place. Demerara sugar, used for baking, has a pale-golden colour and slight toffee flavour. Golden syrup is a golden-coloured, thick, sweet syrup made by processing molasses.

Eggs

Eggs fulfil many complex functions in baked products. They assist flour in building the structure of a product when they coagulate, or become firm, during baking. During the mixing process, eggs trap bubbles of air, which then expand in the heat of the oven and aerate the mixture to produce a lighter texture. Separating, then beating or whisking egg whites enables a large volume of air to be incorporated into the mixture, creating the very light texture that is the feature of products such as sponge cakes and meringues. The yolk of the egg gives the baked products a rich, yellow colour. Most recipes featuring eggs are based on eggs weighing 55–60 grams.

Butter

Butter is made from the fat component of milk being separated and churned. Salted butter is used in most recipes because the salt helps to balance the overall flavour of the product. Unsalted, or cultured butter has a slightly softer flavour. In cakes and muffins, butter creates a moist, tender texture. In biscuits and pastry, it is responsible for the characteristic short, crumbly texture and delicate flavour. In baked products made by the creaming method, butter helps to aerate the mixture and lighten the texture. During the creaming process, beating causes fat from the butter to trap small bubbles of air around the individual crystals of sugar, and so increase the volume of the mixture. Butter is a perishable food and should be stored in the refrigerator to prevent the fats becoming rancid or going 'off'. Margarine or blends of butter and margarine can be used as an alternative to butter in most recipes, but may slightly alter the flavour and texture.

Other ingredients

Milk

Milk is a liquid, or wet ingredient that is used in cake batters to help to combine dry ingredients. The small fat component of milk contributes to the tender texture of cakes. In cakes made by the melt-and-mix method, milk keeps the batter soft and moist, allowing it to be beaten to incorporate air.

Raising agents

Raising agents are the ingredients that are responsible for the *aeration, or leavening* in many doughs, cakes and biscuits. During baking, they give lift to a dough or cake batter to produce a light, airy texture in the final product. Baking powder is a white powder that, when it comes in contact with moisture and heat, releases carbon dioxide. Bicarbonate of soda looks similar to baking powder. Because it is alkaline, it must be combined with an acidic ingredient as well as moisture and heat before it produces the gas carbon dioxide. As the bubbles of carbon dioxide are produced in a mixture, it begins to rise and its volume increases.

Spices

Spices are flavouring ingredients made from the buds, bark, roots, berries or aromatic seeds of certain plants. In baking, they are usually used in their dry form and ground into a powder. Vanilla is a spice with a warm, floral aroma that is often used to flavour baked products. It is available in liquid form as vanilla essence, or as a paste. Some other examples of spices often used in baked products are cinnamon, nutmeg and ginger.

Honey

Honey is a sweet, thick, sticky liquid made by bees from the nectar of flowers. It can be used to bind and sweeten baked products. The flavour and colour of honey depends on the flowers from which the nectar was collected. In Australia, some flowers used as a source of honey include clover, which is a grass, and flowering trees such as orange blossom, yellow box and leatherwood. Honey is judged by its aroma, flavour and density (how long it takes an air bubble to travel from the bottom to the top of an upturned jar).

ACTIVITY 12.1

Getting to know honey

1. Select four varieties of honey; for example, blue gum, leatherwood, yellow box and stringybark.
2. Draw up a table to record the following properties of honey.
   a. Before opening the jar, test for density. Record the time (in seconds) that it takes for an air bubble to travel from the bottom of the upturned jar to the top.
   b. To check the colour, hold up the glass jar to the light so that it is easier to record the colour differences.
   c. Spread a teaspoon of honey on a small, plain white piece of bread, then taste test. Record the flavour in a table similar to the one below.

Variety	Density	Colour	Flavour
Blue gum			
Leatherwood			
Yellow box			
Stringybark			

Honey varieties

3   Rank the honey varieties according to density, from most dense to least dense. (The longer the air bubble takes to rise to the top, the denser the honey.)
4   Rank the honey varieties according to colour, from darkest to lightest.
5   Rank the honey varieties according to flavour, from strongest to most delicate flavour.
6   Record the variety of honey you preferred. Justify your answer.

HOW BAKING WORKS

During baking, exposure to dry heat in the oven causes many chemical changes to the ingredients of cakes, biscuits and pastry. The firm structure of baked products is formed when the protein in the egg becomes solid as it begins to coagulate, and the gluten – the protein in flour – sets to form a framework around the air bubbles. The starch in the flour of baked products undergoes the process of dextrinisation. The dry heat in the oven breaks down the starch molecules and creates a brown crust on the outside of the product. This chemical change also causes the development of the delicious aromas of cake- and bread-baking. Sugar also combines with the protein and starch in the baked product to assist with the development of the crust and the golden-brown colour of a cooked product. This process is known as the Maillard reaction.

Aeration in baked products

Sifting, creaming together butter and sugar, whisking and beating are all processes that mechanically trap bubbles of air within a cake batter in order to create a light, airy texture in the finished product. Including chemicals that are food-safe, such as bicarbonate of soda and baking powder, or using self-raising flour, is another way to aerate a batter to achieve a good-quality cake. Sometimes, both chemical and mechanical raising agents are combined in the production so that tiny air bubbles can be created and then trapped and held by specific ingredients within the structure of a cake.

When using a chemical raising agent, it is important to work quickly and bake the product as soon as possible, because the raising agent is activated as soon as the mixture becomes moist. When you work efficiently, you maximise the agent's aeration capacity, and the mixture is pushed up and out to increase the product's volume. When using bicarbonate of soda, it is very important to accurately measure the raising agent, since too much will cause the baked product to taste bitter. To overcome the bitter after-flavour, bicarbonate of soda is used in combination with sweet ingredients, such as brown sugar or golden syrup, which have a strong, distinctive flavour that masks the bitterness. In contrast, baking powder, which has a mild flavour, is used in scones and muffins, as it does not affect their delicate flavour.

Chocolate cake

Electric hand beaters

Processes used to make baked products

The key ingredients in baked products such as cakes, biscuits, scones and puddings are flour, sugar, eggs and, usually, a fat such as butter. The processes used to combine the key ingredients directly effects the size of the air bubbles, the tenderness of the cake crumb and the final sensory properties of the baked product. Common methods used to make baked products are:

- Creaming – the butter and sugar are beaten together, then the egg is added and, finally, the dry and wet ingredients are folded in alternately
- Quick mix – the butter is melted and added with other wet ingredients; together, they are stirred through the dry ingredients
- Beating – whole or separated eggs are beaten until they are light and fluffy; the dry ingredients are then incorporated into the mixture
- Rubbing-in – the butter is rubbed into the flour until it resembles fresh breadcrumbs; the wet ingredients are then stirred through.

The chemical and mechanical processes used to combine ingredients in baking are summarised in the following table.

	METHOD	INGREDIENTS	PROCESS	PRODUCTS
Chemical	Bicarbonate of soda	Alkaline white powder with a bitter flavour	• Becomes effervescent when mixed with a liquid • Often combined with golden syrup or brown sugar to balance flavours	• Anzac biscuits • Gingerbread • Steamed puddings
	Baking powder	Combination of: • bicarbonate of soda • an acid (either tartaric or cream of tartar) • starch filler	When mixed with wet ingredients, produces carbon dioxide to create a leavening action	• Scones • Muffins
Mechanical	Sifting	Dry ingredients	Air is incorporated by passing dry ingredients through a sieve	• All cakes and biscuits
	Creaming	Butter or margarine combined with sugar	• Air bubbles are trapped by the fat and sugar crystals • Bubbles expand during cooking, causing the cake to rise	• Butter cakes • Fruit cakes • Patty cakes • Puddings
	Whisking	• Eggs and sugar • Egg whites • Egg whites and sugar	Protein in egg white stretches and traps tiny air bubbles	• Sponge cake • Meringue • Pavlova
	Beating	Dry and wet ingredients in a cake	Air bubbles are introduced into the mixture using a circular motion	• Quick mix cakes

Testing knowledge

1. Identify the two functional roles of flour in baked products.
2. Describe the two processes that occur during baking that cause browning in baked products.
3. Outline the different roles sugar may have in a product that is baked in the oven.
4. Discuss the similarities and differences in the properties of white A1 sugar, caster sugar and icing sugar.
5. Explain how the treatment of eggs can contribute to the aeration of a cake when mixing.
6. What sensory properties does the inclusion of butter influence in biscuits and pastry?
7. Describe the functional role of milk in a cake batter.
8. Explain how chemical raising agents aerate cakes.
9. Why are spices included in recipes for baked products? Give two examples of spices that are often used in cake and biscuit recipes.
10. Sifting, creaming, whisking and beating are processes that incorporate air into a cake. Describe how each process achieves aeration.

Top tips for cake-making

1. Accurately measure ingredients, as the ratio between butter, sugar, flour and eggs has been carefully calculated for success.
2. Butter and line cake tins before starting to mix ingredients to minimise loss of volume from the mixture.

3. Sift flour and dry ingredients such as cocoa and spices before adding them to the mixture. This removes lumps that may not break up during mixing, incorporates air into the dry ingredients to help create a light texture and combines all the ingredients.
4. Do not over-mix a cake batter; this causes the gluten to form long strands that will make the cake dense and heavy. Just mix until all the ingredients are combined and the batter is smooth.
5. In quick mix recipes, melt the butter over low heat to prevent it from burning.
6. In recipes where the butter and sugar are creamed together, remove the butter from the refrigerator a few hours before creaming so that it will be at room temperature; alternatively, soften it in a microwave for 10 to 15 seconds. When the butter is soft, it is easier to beat with sugar and aerate the mixture.
7. When you cook a sponge, it is important to be well organised and to have all ingredients measured and tins prepared before you start to mix the sponge. The mixture can lose volume if the procedure takes too long before baking.
8. Check cakes at least five minutes before the end of the specified cooking time to prevent overcooking.

Biscuits

Top tips for making biscuits

1. Allow space for the biscuits to spread when you place them on the baking trays.
2. Bake biscuits at a low temperature so that they can cook through without burning on the bottom.
3. If baking more than one tray of biscuits, swap the shelves halfway through baking and turn the trays around, front to back. This helps to achieve even cooking if the oven being used has 'hot spots'.
4. After removing biscuits from the oven, loosen them with a metal spatula and allow them to cool on the tray.
5. Store biscuits in an airtight container so they retain their crisp texture.

Making biscuits

PASTRY

Pastry is a simple mixture of flour, fat, salt and a small amount of water. Flexible sheets of raw pastry can be used to wrap or contain single ingredients or fillings with a mixture of ingredients. The golden crust of pastry has a crisp or flaky texture, and usually provides contrast to the soft filling it protects. Pastry products are very filling and satisfy hunger because they are high in fat. Pastry contains carbohydrates, from the flour, and a high proportion of fat, which places it in the 'Eat only sometimes and in small amounts' group of foods in the Australian Guide to Healthy Eating. For this reason, it should be only eaten very occasionally.

Functional ingredients in pastry

Pastry is a mixture of flour, fat and liquid. Plain flour with a low gluten content is generally used for making pastry. Minimising the amount of gluten helps to prevent the pastry from becoming tough. Butter is considered to be the best form of fat to use in pastry, because it has good flavour and keeping qualities.

Pastry – Linzer torte

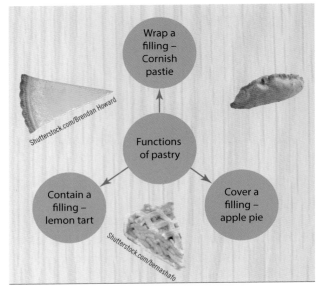

Functions of pastry

It is usually rubbed into the flour so that it coats the granules of starch, resulting in a crisp, textured pastry. A small amount of water is used in most pastry types; this should always be added cold to avoid causing the fat to melt. An egg yolk or a squeeze of lemon juice is added to some pastries to make them more tender. Savoury pastry products are glazed with egg and milk; sweet pastries are glazed with sugar syrup to provide a shiny, brown finish when cooked.

Types of pastry

The ratio, or amount, of butter or fat to flour varies depending on the type of pastry being made. The method used to combine the ingredients in pastry also influences the sensory properties of the final product.

Characteristics of different types of pastry

SHORTCRUST PASTRY	PUFF PASTRY
• Contains shortening, usually butter and flour • Butter is added by rubbing into the flour • Has a short, crumbly texture • Used for making pasties, meat and fruit pies	• Contains shortening, usually butter and flour • Butter is added by rubbing in and layering • Has a light, flaky texture and rich, buttery flavour • Used for sausage rolls, pies and vanilla slices
FILO PASTRY	CHOUX PASTRY
• Dough is kneaded, then rolled and stretched to be paper thin • Contains very thin layers of flour and water dough • Each layer is brushed with melted butter or oil • Has a dry, flaky texture • Used for spinach triangles and apple strudel	• Contains flour, water, butter and eggs • Dough is cooked, and then the eggs are beaten in • Has a crisp, light texture • Puffs up and increases in volume during baking • Used for eclair, profiteroles and cream puffs

Top tips for making pastry

1. Prepare pastry in a cool environment to prevent the butter from melting.
2. Knead pastry dough only lightly and as little as possible to minimise the development of gluten.
3. Add liquid only gradually, because the amount required will vary depending on the flour's ability to absorb moisture.
4. Rest pastry in the refrigerator to allow it to become more evenly hydrated and for the gluten to relax. This process makes the pastry easier to roll out and prevents it from shrinking during baking.
5. If the pastry becomes too soft to work with, simply return it to the refrigerator until it becomes firm again.
6. Pastry is cooked at a high temperature (200°C) to ensure that the fat is quickly absorbed into the flour.

ACTIVITY 12.2
Effect of cooking on puff pastry

AIM

To observe the effect of heat on puff pastry

Worksheet

METHOD

1. Make one quantity of Curry Curlies from the recipe on Page 301.

Curry Curlies

2. During baking, watch the Curry Curlies cook through the oven's glass door. Draw up the following table in your workbook and record your observations.

RESULTS

	Sketch of one Curry Curly	Observations of changes in: • colour development • shape • size • surface texture.
Before baking		
After four minutes of baking		
After seven minutes of baking		

CHAPTER 12 BEST BAKING

ANALYSIS

1. Describe the changes in the size and shape of the pastry over the duration of the cooking time.
2. Why have the size and shape of the pastry changed during the duration of the cooking time?
3. Identify the ingredient that leaked out of the pastry during cooking.
4. Which ingredients in the pastry form its flaky layers?
5. Why do the Curry Curlies need to be cooked in a hot oven (210°C)?

CONCLUSION

Describe the changes that occurred to the colour, shape and size of puff pastry when it was baked in a hot oven.

CHOCOLATE

The word 'chocolate' comes from the Aztec language, and means 'bitter water'. Chocolate comes from cocoa beans, which are picked then fermented, dried, roasted and cracked to remove the cocoa butter from the shell. This is then refined into chocolate liquor. Other ingredients such as sugar and milk powder are added to the liquor to make the product we know as chocolate. There are many grades of chocolate and, usually, the higher the percentage of cocoa butter, the better the texture, aroma and flavour. In compound chocolate, vegetable oils are substituted for cocoa butter. This product has different sensory properties from chocolate that contains cocoa butter, but is sometimes easier to use when preparing decorations.

Fair trade chocolate

Cocoa is widely grown throughout the world, especially in tropical climates in West Africa, South America and Asia. Seventy-three per cent of the world's cocoa crop is grown in African countries. However, many small cocoa farmers find it almost impossible to compete with large multinational producers and often struggle to sell sufficient cocoa beans to make a sustainable living. They face a wide range of problems, none greater than that the low prices they receive for their crops may not even cover their costs of production. Farmers need to purchase tools, fertilisers and pesticides, as well as provide food and clothing for their families. Many farmers lack the education or financial stability to be able to work in other, more lucrative occupations. An increase in child and slave labour in cocoa-producing West African countries is another major problem facing many communities.

The introduction of the fair trade movement has meant that many small farmers can now make a real living and plan for their future. There are now over 140 000 cocoa farmers around the world involved in the fair trade movement. The fair trade movement enables farmers and workers in the developing world to achieve better prices and decent working conditions, to ensure the sustainability of their environment and to receive fair terms of trade. Buying fair trade products also assists many families by giving them access to education, a safer environment, improved healthcare and improved nutrition.

Chocolate

Fair trade chocolate

ACTIVITY 12.3
Sensory and chemical comparison of commercial chocolate

Melting chocolate

AIM
To analyse the sensory and chemical properties of various types of dark chocolate to determine the best-quality product

Worksheet

INGREDIENTS
- Dark compound chocolate
- Dark cooking chocolate
- Dark eating chocolate

METHOD
1 Draw the table below and record the ingredients listed on the label of each product.

	Compound chocolate	Couverture (or cooking) chocolate	Dark eating chocolate
Ingredients			
Gloss			
Snap			
Mouth feel			
Flavour			
Overall rating: 5 = excellent; 3 = OK; 1 = unsatisfactory			

2 Record a description of the gloss for each sample.
3 Break a sample of each chocolate and record the sound of the snap. Was the sound sharp or dull?
4 Taste a sample of each chocolate. Remember to let the sample melt on your tongue rather than chew it to experience all of its sensory properties.
5 Give each chocolate an overall rating.

ANALYSIS
1 Discuss the similarities and differences between the ingredients in each type of chocolate.
2 What information does the order in which the ingredients appear on the food label provide? Identify the ingredients that appear in some products but not in others. What reasons can you suggest for this?
3 How much fat is in each type of chocolate? What type of fat is it?
4 Which chocolate scored highest in terms of its sensory properties? Why?
5 Compare the sensory properties of a high cocoa butter chocolate and a compound (vegetable fat) variety.

CONCLUSION
With other members of the class, develop a list of characteristics that denote good quality chocolate and explain which properties influenced your decision.

ACTIVITY 12.4
Methods of melting and resetting chocolate

AIM
To evaluate various methods of melting chocolate and resetting chocolate

INGREDIENTS
- 50 grams compound chocolate
- 50 grams couverture (or, cooking) chocolate
- 50 grams dark chocolate melts
- 50 grams dark eating chocolate

METHOD
Work in small groups to melt and reset the different types of chocolate. Record your observations in the results table below.

MICROWAVE
1. Grate the chocolate into a china or plastic bowl, then microwave on defrost for one minute.
2. Remove from the microwave and stir.
3. Return to the microwave and heat on defrost for a further 30 seconds. Remove and stir as before.
4. Continue to heat the chocolate and stir, as in Step 3, until the chocolate is melted.
5. Record your observations and results in your table.
6. Spread half of the chocolate onto a flat surface covered with foil and allow to set. Pour the remaining chocolate into a small container or mould and allow to set.
7. Record your observations and results in your table.

OVER HOT WATER
1. Grate the chocolate and place it in to a small basin. Bring a small saucepan of water to the boil.
2. Reduce the heat to very low.
3. Place the bowl containing the chocolate over the simmering water. Ensure that the bowl does not touch the water.
4. Using a spatula, gently stir until chocolate is melted.
5. Record your observations and results in your table.
6. Spread half of the chocolate onto a flat surface covered with foil and allow to set. Pour the remaining chocolate into a small container or mould and allow to set.
7. Record your observations and results in your table.

DIRECT HEAT
1. Grate the chocolate and place it in a small saucepan.
2. Place the saucepan over a very low heat.
3. Using a spatula, gently stir until melted.
4. Record your observations and results on your table.
5. Spread half of the chocolate onto a flat surface covered with foil and allow to set. Pour the remaining chocolate into a small container or mould and allow to set.

RESULTS
Record your observations and results in a table like the one below.

	Compound chocolate		Couverture (or, cooking) chocolate		Dark chocolate melts		Dark eating chocolate	
	Melting	Resetting	Melting	Resetting	Melting	Resetting	Melting	Resetting
Microwave								
Over hot water								
Direct heat								

ANALYSIS
1. According to your results, which method was the best for melting each type of chocolate?
2. Did each type of chocolate reset to its original consistency, texture and flavour?

CONCLUSION
Which method was the most effective for melting and resetting chocolate?

Top tips for cooking with chocolate

1. When melting chocolate, grate or chop the chocolate into small pieces first.
2. Make sure there is no water in the bowl before melting chocolate; water or steam will affect the chocolate's consistency and may cause it to 'seize'.
3. Keep the heat as low as possible when melting chocolate to prevent it from becoming granular in texture.
4. Do not overheat or heat the chocolate for longer than necessary, as it may burn.
5. Melt only small quantities of chocolate at a time.
6. Do not wash chocolate moulds in hot water – rinse them in warm water and polish them with a clean, dry tea towel.
7. Set and store at room temperature. Do not refrigerate; this will cause the chocolate to sweat and lose its gloss.
8. Store chocolate away from direct heat, which this causes a white discolouration on its surface (blooming).

PRESENTING AND DECORATING WITH CHOCOLATE

Chocolate is a very useful decoration for cakes, biscuits and desserts because it can be prepared in many creative ways. It can be grated, curled, moulded, cut into shapes, used for dipping, piped into shapes and used for coating.

Chocolate cup with coffee mousse

ACTIVITY 12·5

Making chocolate decorations

AIM

To pipe, make cut-out shapes or leaves out of chocolate.

EQUIPMENT

- 100 grams of chocolate, to melt
- a double boiler or a microwave

PIPED SHAPES

1. Make a piping bag from baking paper.
2. Half-fill the piping bag with melted chocolate and snip off the point the bag to allow the chocolate to flow through it.
3. Pipe shapes onto baking paper or foil and leave to set.
4. Peel off the baking paper or foil and use to decorate.

Piping chocolate shapes

CHOCOLATE BASKET

1. Lightly spray a small bowl with oil.
2. Pipe melted chocolate over the bowl, covering its base, and allow to set.
3. When firm, carefully separate the chocolate basket from the mould.

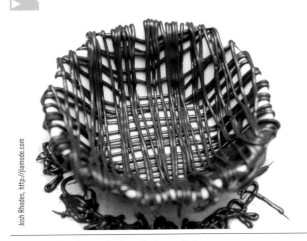

Dessert basket of piped chocolate

CUT-OUT SHAPES

1. Spread a thin layer of melted chocolate onto foil and allow to set at room temperature.
2. Using a sharp cutter or knife, cut out the shapes.

CHOCOLATE LEAVES

1. Select a range of suitable leaves (however, note that some leaves are poisonous). Wash and dry them before using.
2. Paint the surface of each leaf with melted chocolate. Leave to set either at room temperature or in the refrigerator.
3. Gently peel leaf away from chocolate.

Making chocolate leaves

DECORATING WITH SUGAR

Sugar is an inexpensive ingredient that has a long shelf life, provided it is stored in a dry environment in a sealed container. Sugar is the main ingredient in toffee, and it can be worked to create interesting shapes such as toffee spears and lace, and products such as praline that can be used to decorate cakes and desserts. Icing sugar can be used to dust cakes and to form a pattern. A1 sugar is rolled over the surface of a Swiss roll or on to the surface of biscuits such as ginger nuts to give a crunchy, crystalline texture.

Using icing sugar to decorate

ACTIVITY 12·6
Creating decorations with toffee

AIM

To practise making decorations with toffee

TOFFEE INGREDIENTS
- ½ cup sugar
- ¼ cup water

METHOD

1. Combine the sugar and water in a saucepan over a medium heat, stirring continuously with a wooden spoon or until the sugar dissolves. Brush down the sides of the saucepan with cold water to remove the sugar crystals during this process.

2. After the sugar syrup has dissolved, remove the spoon and bring to the boil. Do not stir again. Boil rapidly until it turns pale-golden colour. Remove from heat and let the toffee stand until the bubbles disappear.

3. Use a metal spoon to spoon out the toffee.

   Instructions for creating toffee spears, toffee lace and praline are shown in the table that follows.

   Hint: the saucepan will retain its heat, meaning that the syrup will continue to cook once it is removed from the heat. Allow for this to prevent the syrup from burning.

Toffee spears	Toffee lace	Praline
• Run toffee off a metal spoon onto a foil-lined tray, forming 10-centimetre long spear shapes. • Allow to harden before removing from foil.	• Drizzle hot toffee from a metal spoon onto a foil-lined tray, forming small lattice patterns. • Allow to harden before removing from foil.	• Spread three tablespoons of nuts in a single layer on a foil-lined tray. Pour hot toffee over them and set at room temperature. • When firm, chop or process until fine. • Sprinkle onto the surface or fold into ice-cream. • Store in an airtight container.

Testing knowledge

11. Explain why it is important to accurately measure ingredients when making cakes.
12. Outline why dry ingredients are sifted before being added to a cake or biscuit mixture.
13. Explain why it is important for butter to be at room temperature before creaming.
14. Why should biscuits be stored in an airtight container?
15. List three uses of pastry, and give an example of each use.
16. Flour, butter and liquid are the key ingredients in pastry. Draw a diagram that shows each of their functional roles.
17. Explain why pastry is rested in the refrigerator before rolling out and/or baking.
18. Why is pastry baked at a high temperature?
19. Which ingredient in chocolate is a key indicator of quality?
20. How does setting melted chocolate in the refrigerator affect its physical properties?

THINKING SKILLS 12·1

1. Develop a graphic organiser to identify the functional properties of ingredients used in making cakes and biscuits. Include an illustration that demonstrates each property.

2. Develop a graphic organiser to summarise the different processes that can be used to make baked products.

DESIGN ACTIVITY 12.1

Cupcakes

DESIGN BRIEF

Design six cupcakes that could be given as a thank-you gift to a friend or family member. The cupcakes should be suitable to serve with tea or coffee. They should be visually appealing and showcase your creativity.

1 Develop a design brief based on the five Ws.
- Who – who will be receiving the cupcakes? Explain their relationship to you.
- What – explain why cupcakes are appropriate to give as a gift.
- When – outline the time of day at which you will be giving the gift.
- Where – where are the cup cakes going to be served?
- Why – why are you preparing a thank-you gift of cupcakes?

2 Develop five evaluation criteria questions to judge the success of your end product.

INVESTIGATING

1 Use the Cupcakes recipe on Page 289 to prepare your gift.

2 Research information about the following types of icing that could be used as part of the decoration for your cupcakes. Record your findings in a table like the one below.

	Main ingredients in the icing and how they are combined	Describe how you would apply the icing to the cupcakes; e.g., spread, pour, pipe, cut out
Butter icing		
Glacé icing		
Fondant icing		
Ganache		
Cream cheese		

GENERATING

1 Sketch and annotate three ideas for decorating your cupcakes to suit the person who is receiving the gift. Commercial decorations such as confectionary should not be used.

PLANNING AND MANAGING

1 Write out a food order for your cupcakes and decoration ingredients.

2 Make a list of the aspects of the production task that will rely on you and your bench partner sharing and working collaboratively.

PRODUCING

1 Prepare the product using the recipe for Cupcakes on Page 289.

2 Record any changes or modifications you made during production.

EVALUATING

1 Answer the five evaluation criteria questions you developed earlier.

2 List all the processes you used to make the cupcakes and decorations.

3 Describe the sensory properties – appearance, aroma, flavour and texture – of your decorated cupcakes.

4 Describe how you were able to make your cupcakes visually appealing and demonstrate your creativity.

5 If you were to make the product again, what changes would you make to the cupcakes or the decoration?

6 Explain how your final product met the needs of the design brief.

CUPCAKES

2 eggs
½ cup caster sugar
½ cup thickened cream
½ teaspoon vanilla essence
1 cup self-raising flour, sifted

MAKES 10–12 CUPCAKES

METHOD

1. Preheat the oven to 200°C.
2. Lay out 12 cupcake papers in a patty cake tray – ⅓ cup size.
3. Place the eggs and cream in a medium bowl and beat with electric hand beaters until combined.
4. Add caster sugar and vanilla essence and beat on high speed for 4 minutes or until the mixture is light and fluffy.
5. Sift in the flour and gently fold in using a metal spoon until there are no lumps.
6. Spoon equal quantities of cake batter into the cupcake papers.
7. Bake for 15–20 minutes or until pale-golden and just firm to the touch.
8. Remove from oven and cool on a cake rack.
9. Decorate with glacé or butter icing.

EVALUATION

1. Describe the sensory properties – appearance, aroma, flavour and texture – of the Cupcakes.
2. Explain how beating the eggs, cream, caster sugar and vanilla for four minutes contributes to the aeration of the cupcakes.
3. Why is the flour folded into the recipe in Step 5 instead of being beaten in?
4. Identify the processes that are responsible for the cupcakes becoming golden brown during baking.
5. Describe two safety practices you followed when using the electric hand beaters.

UPSIDE-DOWN CAKE

Caramel topping

⅓ cup sugar

1 ½ tablespoons water

1 ½ tablespoons boiling water

Fruit layer

½ Granny Smith apple, peeled, cored and thinly sliced

Cake batter

60 grams butter

⅓ cup brown sugar

1 egg

¼ cup light sour cream

⅔ cup self-raising flour

¼ teaspoon baking powder

¼ teaspoon ground ginger

¼ teaspoon ground cinnamon

SERVES 4

The fruit in this recipe is baked on the bottom but when it is turned out of the tin, the fruit is on the top.

METHOD

1. Preheat oven to 190°C. Grease a 21 × 9-centimetre-long cake tin. Line base with baking paper.
2. Prepare caramel topping. Combine the sugar and 1 ½ tablespoons water in a small saucepan and stir over medium heat until sugar dissolves. Bring to boil, without stirring, until the syrup turns golden brown. Brush down the sides of the saucepan with a pastry brush dipped in water to prevent the syrup from crystallising.
3. Immediately remove from heat and carefully add 1 ½ tablespoons boiling water. Take care, as the syrup will spit. Stir until smooth and pour into the base of the greased cake tin. Allow to cool.
4. Arrange the fruit. Consider the overall design and arrange the fruit so that it is easy to cut the cake into serving portions.
5. Using electric hand beaters, cream the butter and brown sugar until light and fluffy. Beat in egg, then add sour cream and mix well.
6. Fold in the sifted flour, baking powder and spices. Mix thoroughly.
7. Place batter on top of fruit in the cake tin. Bake for 25–30 minutes.
8. Allow to stand in the tin for 3–4 minutes before inverting on serving dish. Cut into portions and serve with ice-cream.

EVALUATION

1. Describe the sensory properties appearance, aroma, texture and flavour – of the end product.
2. Why was the cake left to stand on the bench for a short time before turning it out of the tin?
3. List four significant processes you used in this production, and the equipment required to carry them out successfully.
4. Suggest some modifications you could make, to either the ingredients or the processes, to improve the Upside-Down Cake the next time you make it.
5. Describe the safe work practices you followed when preparing the caramel topping and the fruit for your cake.

CHOCOLATE SPONGE

- spray oil and extra flour, for preparing the tins
- ½ cup wheaten cornflour
- 1 teaspoon plain flour
- ½ teaspoon bicarbonate soda
- ½ teaspoon cream of tartar
- 1 tablespoon cocoa
- 1 tablespoon golden syrup
- 3 eggs (60–70 grams), at room temperature
- ½ cup caster sugar

Sponge cakes are light, delicate cakes that are often cooked for a special afternoon tea or as a celebration cake such as a birthday cake. A 'sponge sandwich' refers to two layers of sponge cake that have been filled, or 'sandwiched', with whipped cream, jam or lemon filling.

METHOD

1. Cut two circles of greased paper and line two 20-centimetre round cake tins. Lightly grease with spray oil and dust with plain flour, tapping to remove extra flour.
2. Preheat oven to 180°C. Check the racks to make sure both tins can sit side by side on shelves near the centre of the oven.
3. Sift the flours, bicarbonate of soda, cream of tartar and cocoa twice.
4. Slightly warm the golden syrup in a microwave before measuring. (This makes it easier to fold through the mixture.)
5. Separate the egg whites from the yolks, placing the whites in a large, clean, dry bowl and retaining the yolks.
6. Beat the egg whites until stiff, then beat in the egg yolks one at a time.
7. Gradually add the sugar and then the golden syrup, and beat until combined.
8. Add the sifted dry ingredients and lightly fold through with a metal spoon.
9. Divide the mixture between the two greased and lined tins; tap lightly to remove any air bubbles.
10. Bake for 15–20 minutes. When cooked, the cakes will begin to leave the sides of the tins and will spring back when lightly touched with the fingers.
11. When cooked, tip out each cake onto a sheet of greased paper and remove the paper lining. Quickly turn the cakes over and allow to cool.
12. The cakes can now be filled with whipped cream. The tops can be spread with cream and topped with strawberries or iced with chocolate icing.

EVALUATION

1. Why is it important to grease and line the cake tins?
2. Identify all the processes you followed when preparing and beating the egg whites to maximise the volume of the sponge cakes.
3. Why were the dry ingredients sifted twice before being folded into the egg mixture?
4. Explain the role of the bicarbonate of soda, the cream of tartar and the golden syrup in the mixture.
5. Describe the tests you used to determine when the sponge cakes were cooked.

ANZAC BISCUITS

1 cup flour
¾ cup caster sugar
1 cup rolled oats
1 cup coconut
125 grams butter
2 tablespoons golden syrup
1 teaspoon bicarbonate of soda
2 tablespoons boiling water

MAKES 12–16 BISCUITS

METHOD

1. Preheat oven to 180°C.
2. Sift flour into a basin; add the sugar, rolled oats and coconut.
3. Melt the butter and golden syrup in a small saucepan.
4. Dissolve the bicarbonate of soda in boiling water and add to melted butter and syrup.
5. Add the melted mixture to the dry ingredients and mix well.
6. Roll into 3-centimetre balls and place on an oven tray. Flatten with a fork or spatula.
7. Cook for 15–20 minutes in the preheated oven.
8. Allow to cool for 5 minutes on the tray and then lift onto a cooling rack.

EVALUATION

1. Describe the appearance, aroma, flavour and texture of your Anzac Biscuits.
2. Why is it important to heat the golden syrup with the butter in this recipe?
3. What is the purpose of the rolled oats in these biscuits?
4. What would be the effect of cooking the biscuits at 220°C rather than 180°C?
5. What modifications would you make to the Anzac Biscuits if you were to make them again?

RICOTTA AND ORANGE BISCUITS

65 grams butter, softened

⅓ cup caster sugar

rind of one orange, finely grated

½ teaspoon vanilla

1 egg

100 grams ricotta

1 cup plain flour

1 teaspoon baking powder

pinch of salt

½ cup icing sugar, for coating

MAKES 18–20 BISCUITS

METHOD

1. In a large bowl, cream together the softened butter, caster sugar, orange rind and vanilla. Beat until pale and fluffy.
2. Add the egg and beat until well combined.
3. Fold in the ricotta.
4. Sift together the flour, baking powder and salt and stir into the mixture until a soft dough is formed. Turn onto a plate, spread the mixture out evenly, cover with plastic wrap then refrigerate for 15 minutes.
5. Preheat oven to 180°C and grease a biscuit baking tray.
6. Sift the icing sugar onto a plate and remove biscuit dough from the refrigerator.
7. Roll tablespoons of mixture into balls and roll through the icing sugar to coat. Place on the greased baking tray about 3 centimetres apart.
8. Bake for approximately 25 minutes or until golden and crackled on the surface.

EVALUATION

1. Describe the sensory properties – appearance, aroma, flavour and texture – of your Ricotta and Orange Biscuits
2. Identify the ingredients used to add flavour to the biscuits.
3. Why are the butter and sugar creamed together in this recipe?
4. Why is the mixture refrigerated in Step 4 of the recipe?
5. How does rolling the biscuits in icing sugar before baking add to the sensory properties of the cooked biscuits?

VANILLA BISCUITS

½ cup plain flour

½ cup self-raising flour

60 grams butter, at room temperature

¼ cup caster sugar

1 egg, lightly beaten

1 teaspoon vanilla essence

MAKES 12–18 BISCUITS, DEPENDING ON THE SIZE

EVALUATION

1. What tools and small electrical equipment can be used to cream butter and sugar?
2. Why is the biscuit dough kneaded in Step 6?
3. What are the benefits of using a cutter to shape the biscuits?
4. Suggest some methods you could use to decorate your biscuits to improve their appearance.
5. How can you test the Vanilla Biscuits to check if they are cooked?

METHOD

1. Preheat oven to 160°C. Grease an oven tray.
2. Sift together flours.
3. Cream butter and sugar until light and creamy in colour.
4. Add the egg and vanilla and beat well.
5. Add sifted flour and mix to a firm dough.
6. Lightly flour the bench and knead the dough until it is smooth.
7. Roll out the dough to 3–5 millimetres in thickness and cut into shapes.
8. Bake shapes on a lightly greased baking tray at 160°C for 10–15 minutes, or until a pale-golden colour.
9. Allow to cool on tray for 5 minutes, then place on a cooling rack until cold.

Variations

1. To make chocolate biscuits, replace the 1 ½ tablespoons of flour with 1 ½ tablespoons of cocoa. Sift the cocoa with the flours.
2. To make lemon and coconut biscuits, add 1 teaspoon of grated lemon rind and 1 tablespoon of desiccated coconut to the creamed mixture before adding the flours.
3. To decorate with colours, lightly whisk ½ an egg white and remove the biscuits from the oven after 10 minutes. Brush the egg white over the biscuits in a thin layer and sprinkle with ¼ cup of hundreds and thousands. Return biscuits to the oven and continue baking until the base is pale gold.

GINGERBREAD

90 grams butter

½ cup soft brown sugar

1 egg, lightly beaten

1 tablespoon golden syrup

¼ teaspoon bicarbonate of soda

1 teaspoon milk

1 ¾ cups plain flour

½ teaspoon cinnamon

½ teaspoon ground ginger

glacé cherries, nuts, currants, chocolate buds, lollies, for decoration

MAKES 10–12 PEOPLE SHAPES

This ginger-spiced biscuit mixture can be cut into a range of shapes and decorated with piped icing and/or sweets. The best-known shapes are gingerbread people.

METHOD

1. Preheat oven to 180°C.
2. Cream the butter and sugar.
3. Add the beaten egg and golden syrup.
4. Dissolve the bicarbonate of soda in milk and add to the mixture.
5. Add sifted flour and spices.
6. Flour the bench, then lightly knead the mixture. Roll out to a 3–5 millimetre thickness, then cut out shapes.
7. Re-roll the leftover dough and cut more shapes until all the dough is used. Add decorations such as currants or chocolate buds.
8. Bake for 10–15 minutes or until just beginning to brown on the edges. Remove from oven, then cool on a tray for 5 minutes.
9. Lift onto a rack to cool completely before decorating.

EVALUATION

1. Why is bicarbonate of soda used in this recipe?
2. Why is the egg lightly beaten before it is added to the mixture?
3. Why are the flour and spices sifted before they are added to the mixture?
4. How do you prevent the dough sticking to the bench when you roll it out into a thin layer?
5. Describe two safety rules to follow when baking biscuits in an oven.

CHOCOLATE-COATED BISCUITS

- 30 grams butter
- ⅓ cup golden syrup
- ¾ cup plain flour
- ½ teaspoon bicarbonate of soda
- ¼ teaspoon ground ginger
- ¼ teaspoon cardamom
- ¼ teaspoon cinnamon
- pinch of ground cloves
- ¼ teaspoon cocoa
- 2 teaspoons mixed peel, finely chopped
- 2 teaspoons milk
- ¼ cup flour (extra as needed, for bench)
- 60 grams dark cooking chocolate
- 2 tablespoons raspberry jam

MAKES 12–16 BISCUITS

These biscuits highlight the use of ground spices in baked products. Chocolate can be used creatively to decorate the biscuits.

METHOD

1. Preheat oven to 180°C and lightly grease a biscuit tray.
2. In a small saucepan, heat the butter and golden syrup and bring to boil. Remove from heat and allow to cool.
3. When cool, stir in the dry ingredients and the mixed peel and milk. Spread onto a plate and refrigerate to cool for 5 to 10 minutes. (The mixture will become thicker as it rests.)
4. Flour the bench with the extra flour and knead until the dough is no longer sticky.
5. Roll out the dough to 4-millimetre thickness and cut into shapes using a biscuit cutter. Place on greased tray.
6. Bake for 8–10 minutes or until golden brown.
7. Allow biscuits to cool on the tray. Melt chocolate over double boiler or in a microwave.
8. Match together pairs of biscuits and sandwich with a thin layer of jam. Decorate with piped chocolate or spread the base of single biscuits with the melted chocolate. Refrigerate until the chocolate is firm.

EVALUATION

1. Why is bicarbonate of soda used in this recipe?
2. Why is the biscuit dough cooled and rested before it is rolled and cut?
3. Describe the physical changes that occur to the biscuits as they cool on the tray.
4. What did you find the most challenging aspect of this production? Why?
5. How should the Chocolate-Coated Biscuits be stored to maintain their crisp texture?

GREEK BISCUITS

¼ cup caster sugar

125 grams butter, at room temperature, chopped

1 egg yolk

½ teaspoon lemon rind, grated

60 grams roasted almonds or macadamia nuts, finely chopped or processed

1 cup plain flour

icing sugar, for dusting

MAKES APPROXIMATELY 16 BISCUITS

These biscuits are nutty-flavoured with a very light texture.

METHOD

1. Preheat oven to 180°C and lightly grease the oven tray.
2. Place sugar and chopped butter into a medium-sized bowl and beat with a wooden spoon until creamed.
3. Add the egg yolk and lemon rind and beat well.
4. Add the nuts and mix well.
5. Add the sifted flour and mix to a soft dough.
6. For each biscuit, shape 2 level teaspoons of mixture into a crescent shape. Place the crescents on the prepared trays.
7. Bake for 12–15 minutes or until a pale, golden-brown colour.
8. Dust with icing sugar while still warm.
9. Allow crescents to stand for 5 minutes before moving to a wire rack to cool completely.

EVALUATION

1. Why is the butter removed from the refrigerator and brought to room temperature before it is used in this recipe?
2. How did you know when the butter and sugar had been sufficiently creamed?
3. What is the role of lemon rind in the Greek Biscuits?
4. Why are the biscuits left to cool on the baking tray for a few minutes?
5. Why is it advisable to eat these biscuits only as a treat?

FRUITY BALLS

2 tablespoons skimmed milk powder

¼ cup desiccated coconut

2 tablespoons crunchy peanut butter

2 tablespoons honey

1 cup dried fruit medley

3 plain sweet biscuits, crumbed

3 teaspoons fresh orange juice

¼ cup extra desiccated coconut, for outside coating

MAKES APPROXIMATELY 12 FRUITY BALLS

METHOD

1. In a food processor, combine the milk powder, coconut, peanut butter and honey. Process until combined.
2. Add the dried fruit and biscuit crumbs and process for 30 seconds. Add orange juice and process for another 10 seconds or until the mixture is combined. Remember to scrape down the sides of the bowl in between processing.
3. If mixture is too dry, add a little more orange juice.
4. Transfer mixture to a bowl, then roll teaspoonfuls of the mixture into balls. Toss each fruity ball through the extra coconut. Place on a tray.
5. Cover with plastic wrap and refrigerate for 2–3 hours.

CHOCOLATE DELIGHTS

125 grams dark cooking chocolate chips (not melts)

⅓ cup condensed milk

1 cup desiccated coconut

¼ teaspoon vanilla

⅓ cup chocolate sprinkles

MAKES APPROXIMATELY 12 CHOCOLATE DELIGHTS

METHOD

1. Heat the chocolate chips and condensed milk in a double boiler until the mixture is smooth.
2. Remove from heat and add coconut and vanilla. Stir until combined.
3. As soon as the mixture is cool enough to handle, roll teaspoonfuls of the mixture into balls.
4. Roll each ball through the chocolate sprinkles, then place on a tray.
5. Chill until firm.

SCOTCH MINTS

8 chocolate ripple biscuits
8 hard peppermints
2–4 tablespoons condensed milk
⅓ cup desiccated coconut

MAKES APPROXIMATELY 10 SCOTCH MINTS

METHOD

1. Crush the biscuits and mints in a food processor or electric food mill, or place in a clean plastic bag and crush with a rolling pin.
2. Combine biscuits and mints with condensed milk and mix well.
3. Roll teaspoonfuls of the mixture into balls.
4. Roll each ball through the coconut, then place on a tray.
5. Chill until firm. These are best made the day before they are required.

EVALUATION

1. Identify the dry ingredients in the three recipes.
2. Suggest two methods that could be used to make biscuit crumbs from whole biscuits.
3. Why are honey and sweetened condensed milk valuable ingredients in no-bake sweet treat recipes?
4. Why is it important to melt the chocolate over a double boiler?
5. Explain how you could store these Scotch Mints if you wanted to serve them at a later date.

GLUTEN-FREE PATTY CAKES

30 grams butter
½ cup selected flour
¼ teaspoon baking powder
¼ cup caster sugar
1 egg
1 tablespoon milk
½ teaspoon vanilla essence

MAKES 6 SMALL PATTY CAKES

METHOD

1. Preheat oven to 200°C. Grease a patty cake tray with butter, or line with paper cups.
2. Melt the butter, either in the microwave for 30 seconds on high or over a gentle heat without browning.
3. Sift all the dry ingredients into a bowl and add the egg, milk and vanilla essence.
4. Stir in the melted butter.
5. Mix well and pour into prepared patty tray.
6. Bake in the preheated oven for 12 minutes. Do not overbake, because this can cause the cakes to become dry.

CURRY CURLIES

1 sheet of ready-rolled frozen puff pastry
1 tablespoon butter, melted
½ teaspoon curry powder

MAKES 6 CURRY CURLIES

METHOD

1. Preheat oven to 210°C.
2. Cut the sheet of pastry in half.
3. Melt the butter in a small saucepan or a microwave, then stir in curry powder.
4. Brush each piece of pastry with the curry butter and place the second piece on top of the first, buttered sides together.
5. Cut the pastry into 2-cm wide strips.
6. Twist each strip into a spiral shape and place on an oven tray.
7. Bake in the preheated oven for 8 minutes or until golden brown.

FREEFORM FRUIT TART

Pastry

½ cup self-raising flour

2 tablespoons plain flour

30 grams butter, chilled and cubed

2 teaspoons caster sugar

¼ teaspoon cinnamon

1–2 tablespoons iced water

Filling

½ egg white, lightly beaten

2 teaspoons fine polenta or semolina

200 grams fresh fruit – apricots, nectarines or peaches

or

1 Granny Smith apple

2 teaspoons sugar

extra sugar

SERVES 2

Light, buttery pastry filled with a luscious fresh fruit makes a beautiful dessert.

METHOD

Pastry

1. Process the flour, butter, sugar and cinnamon until the mixture resembles fine breadcrumbs.
2. Add iced water and process until the mixture starts to come together.
3. Turn the dough onto a lightly floured bench and bring together until smooth. Do not over-knead.
4. Flatten into a disc, cover with plastic wrap and rest in the refrigerator for 20–30 minutes.
5. Roll the disc into a 3-millimetre-thick round. Turn the pastry as you roll so that it does not stick. Transfer to a lightly oiled tray.

Filling

1. Preheat oven to 200°C.
2. Brush the pastry with egg white and sprinkle with polenta.
3. Prepare fruit. If using apples, peel, core and thinkly slice.
4. In a bowl, toss fruit in the sugar.
5. Place fruit in centre of pastry, leaving a 4–5 centimetre border.
6. Bring back pastry over the fruit, pleating to fit.
7. Brush pastry with egg white and sprinkle with extra sugar.
8. Bake at 200°C for 15–20 minutes.
9. Serve warm with cream or ice-cream.

EVALUATION

1. Describe the sensory properties – appearance, aroma, flavour and texture – of your Freeform Fruit Tart.
2. What are the advantages and disadvantages of making a free-form tart compared with a tart in a traditional flan or pie tin?
3. What are the advantages and disadvantages of using a food processor to make the pastry?
4. Why is the pastry rested in the refrigerator?
5. Plot the ingredients of your Freeform Fruit Tart on a diagram of the Healthy Eating Pyramid. Rate your product as not very healthy, healthy or very healthy.

ICINGS

Butter icing

2 tablespoons milk
50 grams butter, softened
1 teaspoon vanilla essence
2 cups icing sugar
food colouring

Lemon cream cheese Icing

125 grams creamed cheese, at room temperature
60 grams butter, at room temperature
½ lemon, zest and juice
½ teaspoon vanilla
125 grams soft icing sugar

🍴 MAKES ENOUGH TO ICE 24 MUFFINS

Chocolate icing

1 cup icing sugar
1 tablespoon cocoa
10 grams (roughly 2 teaspoons) butter
1–2 tablespoons boiling water

🍴 MAKES ENOUGH TO DECORATE 1 20-CENTIMETRE CAKE OR 9-12 PATTY CAKES

Orange icing

10 grams (roughly 2 teaspoons) butter
¾ cup soft icing sugar
1 tablespoon orange juice

🍴 MAKES ENOUGH TO ICE 6-9 PATTY CAKES

Soft (or pure) icing sugar can be used for these recipes. Soft icing sugar is more convenient because it is lump-free and so does not require sifting before use.

Butter icing is often used as a filling for biscuits or for piping onto cakes.

METHOD

Butter icing

1. Warm the milk in a microwave for 20 seconds.
2. Cream the butter for 1–2 minutes or until soft and smooth.
3. Add half the icing sugar and the warm milk and vanilla, and beat for approximately 3 minutes or until the mixture is pale and fluffy.
4. Add the remaining icing sugar and beat for a further 2–3 minutes until the mixture is fluffy and of a smooth spreading consistency.
5. Add a little food colouring a few drops at a time until you reach the desired colour.

Note: If you do not wish to use the icing immediately, place plastic wrap directly on the surface of the icing and press down lightly to form a seal.

Lemon cream cheese icing

1. Whip together the creamed cheese and butter in a food processor until light and fluffy.
2. Add the lemon zest, lemon juice and vanilla and mix well.
3. Keep the processor running and slowly add the icing sugar.
4. Use as a topping for muffins.

Chocolate icing

1. Sift the icing sugar and cocoa into a medium-sized bowl.
2. Add the melted butter to a well in the centre, and then a very small quantity of boiling water until a spreading consistency is reached. Be very careful to add the boiling water in very small amounts; only a small quantity is used.
3. Using a small spatula or palette knife, spread the icing over the top of a cake and smooth. The knife can be dipped into hot water if the icing is difficult to smooth.
4. The top can be decorated with chocolate sprinkles or nuts.

Orange icing

1. Melt the butter in the microwave for 10–20 seconds.
2. Combine all the ingredients and mix until combined.
3. Spread onto cooled cakes.

Lemon cream cheese icing is a smooth, creamy topping with a tangy flavour. A spoonful served on top of a muffin makes the muffin a little bit more special.

GLOSSARY

aeration, or leavening the trapping of air in a mixture that then expands as the product cooks and causes the product to rise and be light in texture

anthocyanins substances that produce red to purple to blue colours in fruits and vegetables

breakfast the first meal you eat soon after waking up from your night's sleep

bush tucker the huge variety of edible native Australian herbs, spices, mushrooms, fruits, flowers, vegetables, animals, birds, reptiles and insects

cardiovascular disease a general term used to describe a range of diseases, including heart disease, stroke and blood vessel disease

carotenoid a substance that produces the orange colour in fruit and vegetables

cereals edible seeds of certain grasses, including millet, oats, barley, wheat, rye, rice and corn

chlorophyll a substance that produces the green colour in fruit and vegetables

coagulation the permanent change of the physical and chemical structure of protein

coeliac disease a disease of the small intestine associated with permanent intolerance or hypersensitivity to gluten

connective tissue the tissue in meat that links and holds together muscles

considerations factors in the design brief, such as the season of the year or the skills of the chef, which are more flexible than constraints but may also influence the design and development of the product

constraints factors in the design brief with which the product must comply

cross-contamination the transfer of harmful bacteria from uncooked food to food that has been cooked or prepared

danger zone the temperature – between 5°C and 60°C – at which bacteria can multiply very quickly

design brief specific information about the type of product to be developed and the audience at which the new product is aimed

design process the process of investigating and designing, producing, and analysing and evaluating

dextrinisation a process that occurs during cooking in which dry heat in an oven breaks down starch molecules, forming a brown crust on the outside of baked products; the process that occurs when the starch in flour is exposed to dry heat and is broken down into dextrin, resulting in a change in colour to golden brown

diabetes a disease where the pancreas is unable to produce sufficient insulin to enable the glucose produced during digestion to be absorbed into the bloodstream

dietary fibre a nutrient found in the cell walls of all plant foods that improves the health of the digestive system by adding bulk to faeces

electric oven uses radiant and convection heat produced by electricity to cook food

enzymatic browning a process that occurs when the enzymes in cut or peeled fruits cause browning when exposed to oxygen in the air

fast a period of time during which we eat nothing

fermentation the process of yeast growing and reproducing by budding, then converting carbohydrates into carbon dioxide, alcohol and water

fire blanket an insulated blanket used to extinguish small fires in the kitchen

food allergy an abnormal immunological reaction to food

food any substance that we eat or drink that provides the body with chemical substances called nutrients

food hypersensitivity a reaction to food that is of a similar type to food allergies, but generally less severe

food poisoning an illness caused by eating food that has been contaminated with harmful bacteria

functional ingredients recipe ingredients that flavour, create texture, colour, help other ingredients to combine or increase overall nutrient value

gas oven uses radiant and convection heat produced by gas to cook food

gelatinisation the process that occurs when starch granules in the endosperm of cereals absorb liquid in the presence of heat and thicken the liquid

gluten the main protein in wheat flour

glycaemic index (GI) a ranking of carbohydrate foods based on the immediate effect they have on blood sugar levels

ground oven an oven made by lining a pit in the ground with hot stones, placing wrapped food on the stones and covering the hole over with grasses, leaves and sand

hawker-style foods small portions of food traditionally prepared, cooked and served in front of the customer on the street

in season the time of year when a fruit or vegetable has its best sensory properties

kneading a process in which air bubbles are evenly distributed and the gluten strengthened in a yeast dough

Maillard reaction a browning reaction that occurs when sugar or starch and a protein are present during baking

marbling even distribution of deposits of fat cells in red muscle tissue

marinate to tenderise and/or enhance the flavour of meat or other food

metric measuring tools spoons, cups, jugs and scales that have been calibrated to accurately measure ingredients by weight and volume using the metric system

microwave oven an appliance used for thawing frozen foods and reheating pre-cooked foods

modified atmosphere packaging (MAP) a method of packaging that causes change in the levels of gases inside a package in order to extend a product's shelf life of a product

modified atmosphere packaging (MAP) a system that changes or modifies the atmosphere or gas inside a package, in order to extend the shelf life of the food

monounsaturated fats fats found in olives, olive oil, avocados and nuts that have been shown to reduce blood cholesterol levels

muscle fibres cells that are bound into thin sheets of connective tissue; these bundles then form groups to create muscles

no-till farming leaving the stubble from last year's crop to enrich and stabilise the soil; the new crop is planted by direct drilling in between the rows of the previous crop without tilling the soil

nutrients chemical substances in food that are broken down during digestion, including protein, carbohydrates, fat, vitamins and minerals

osteoporosis occurs when calcium is lost from the bones, making them very fragile and easily broken

Pacific Rim cuisine foods originating from countries around the edges of the Pacific Ocean

pomme fruit fruit with crisp, juicy flesh surrounding a core that contains seeds; for example, apples and pears

preferred option the design option that best meets the requirements set out in the design brief

processed breakfast cereals grains such as corn, wheat and rice that have been softened by precooking and then dried; most become fortified or have had vitamins and minerals added during processing

proving a process in which a yeast dough is rested to allow time for fermentation to take place – the gas bubbles are trapped in the structure, so the volume of the dough increases

qualitative or sensory analysis the evaluation of the sensory properties of food, such as appearance, aroma, flavour and texture

quantitative measures ways to measure the physical, chemical or nutritional properties of food

recipe a list of ingredients and instructions for preparing food

risk factors habits that can increase the chance of developing a disease or health condition

rolled oats wholegrains of oats that have been cooked in water to soften them and make them easier to digest, and then rolled

saturated fats fats found mainly in foods of animal origin such as meat, cheese and butter that are linked to raised cholesterol levels; coconut oil and palm oil are also high in saturated fats

sensory properties the appearance, aroma, flavour and texture of food

small appliances pieces of equipment such as toasters, food processors, handheld beaters or blenders

solanine a toxin found in green potatoes

specifications the considerations and constraints within the design brief

sustainable farming farming practices that maintain the land's productivity so that it will be available for future generations

Thai flavour wheel a diagram that shows the five flavours – salty, sweet, sour, spicy and bitter – that are blended in Thai foods

trans fats bad fats that can lead to serious health concerns and should be avoided; found mainly in hydrogenated vegetable oil used by food manufacturers in processed and fast foods

tuber a vegetable that grows under the ground and has a high starch content; for example, potato

yeast a single-celled, microscopic fungus

INDEX

A

abbreviations (recipes) 44
Aboriginal and Torres Strait Islander people
    diabetes and 203
    foods 229–43, 246, 247
absorption method (rice) 145
active dried yeast 137
adolescents
    bone mass 209–10
    food choices and 192–3, 207
    food needs 64–9
adults, food needs 65
advertising 169, 192, 193–4
aeration 273, 275, 276, 277, 279, 289
Africa 142, 247, 282
air bubbles 277, 278
akudjura 236, 238
alcohol 136, 138, 195
allergies 213, 214, 247
almonds 151, 238, 297
American foods 246
amino acids 136, 165, 197, 216
animal welfare 165
animal-sourced foods 56, 60–2
    Aboriginal and Torres Strait Islander people and 231, 232
aniseed myrtle 236, 238
anthocyanins 99, 110
antibodies 214
ants 230, 231
apple and blueberry muffin 7
apple juices 31
apple pie 280
apples 40, 90, 101–4, 151, 160, 212, 221, 290, 302
arborio rice 144, 218, 222
Asia 170, 282
Asian foods 208, 251–7
asthma 172, 214
attitudinal descriptors 4
Australia 101, 116, 142, 143, 202, 232
    no-till farming 135
    red meat 169
    wheat 132
Australian Bureau of Statistics (ABS) 66
Australian Capital Territory (ACT) 205–6

Australian Dietary Guidelines 195, 196, 206, 220
Australian food, evolution of 245–71
Australian Guide to Healthy Eating 14, 16, 59, 70, 72, 73, 74, 75, 88, 90, 100, 107, 120, 121, 125, 135, 143, 147, 151, 158, 165, 166, 170, 179, 180, 187, 189, 195, 218, 219, 220, 223, 225, 227, 234, 269, 270, 280
Australian Health Survey 66
Australian Institute of Health and Welfare (AIHW) 202
Australian Nutrition Foundation 59
Australian Standards 4
avocado 92, 152, 198, 257, 269

B

bacon 70, 74, 96, 97, 166, 223, 226
bacteria 22, 24, 145, 172
'baked for' date 141
baked potatoes 70, 112
baking 47, 118, 138–9, 237, 273–303
baking powder 136, 243, 276, 277, 278, 290, 293, 300
balanced diet 59, 65, 195
balanced flavours 252
bamboo 35, 37, 71, 176, 183, 253, 254
bananas 36, 98, 104–5, 255
barbecuing 166, 167, 168, 173, 175, 178
barley 132, 136, 215
barn-laid eggs 85–6
barramundi 172, 231
basal metabolic rate (BMR) 197
basil 157, 158, 182, 221, 227, 257, 263
batter 136, 274, 276
bean shoots 73, 253
beans 77, 212, 216, 257, 258, 258
beating 47, 277, 278, 289
beef 164, 165, 170, 235, 236, 257, 258, 263
    basic cuts 167
    marketing 169
    production of 177
beef curry 168
beef roasts 172
beef stock 181
bees 230, 231, 276
berry fruit 100, 230, 232

best before date 141
Better Health Channel 81, 200
bicarbonate of soda 136, 276, 277, 278, 291, 292, 295, 296
binding 47
birds 231, 232
biscuits 207, 215, 237, 238, 239, 274, 275, 276, 278, 279–80, 299
black beans 252, 253
black pepper 78, 222
blade cut (beef) 167
blanching 47, 114
blending 47
blood sugar 199, 212
body fat 196, 204
    see also fats
body tissues 164
boiled eggs 91
boiled potato 112, 118
boilers 170–1
bolognese sauce 12, 43, 263
bone, health of 209–10
boneless beef 168
bowel health 212, 249
boys, breakfast eating habits 81
braising 167, 175
bran 133, 135, 143, 199, 212
bread 37, 60, 67, 81, 93, 97, 98, 126, 135, 140–2, 148–9, 153, 182, 197, 215, 224, 237, 256
bread flour 133, 136, 154
bread improvers 140, 153
breadcrumbs 123, 161, 184, 188, 241, 266
breakfast 61, 66, 79–98
breast (chicken) 171
breastfeeding 195
brisket cut (beef) 167
broccoli 76, 183
brown sugar 36, 243, 271, 275, 277, 295
Buddhism 255, 256
burritos 208, 258
bush tucker 229–43, 260
    see also Aboriginal and Torres Strait Islander peoples, foods
butter 17, 36, 38, 39, 54, 61, 62, 75, 94, 95, 96, 97, 98, 123, 124, 126, 127, 130, 161, 185, 186, 221, 222, 240, 243, 265,

INDEX

274, 275–6, 277, 278, 279, 280, 281, 290, 292, 293, 294, 295, 296, 297, 300, 301, 302, 303
butter cakes 278

caffeine 210
cakes 133, 197, 207, 215, 237, 274, 275, 276, 278, 279, 290
calcium 60, 61, 64, 65, 83, 85, 87, 104, 111, 172, 209, 223
   food sources 210–11
Caliban potatoes 116, 117
Cambodia 247, 255, 256
cancers 202, 204, 206, 234
canning 119, 246
canola 135, 154
carbohydrates 56, 60, 65, 67, 80, 82, 83, 87, 100, 103, 113, 116, 133, 135, 142, 143, 166, 197, 199–200, 210, 214, 230, 233, 236, 249, 250, 252, 280
carbon dioxide 115, 136, 137, 138, 139, 169, 276
cardiovascular disease (CVD) 191, 233, 234, 249
cardomom 252, 253, 296
carotenoids 99, 110
carrots 15, 16, 74, 75, 76, 124, 127, 181, 183, 185, 188, 221, 224
case-ready packaging 169
casseroles 167, 175
caster sugar 17, 39, 40, 98, 242, 243, 275, 289, 291, 292, 293, 294, 297, 300, 302
cayenne pepper 182, 225
celery 74, 76, 160, 188, 221, 269
cereals 60, 80, 81, 82–3, 119, 131–62, 197, 210, 212, 215, 216, 232, 249
chapatti 140, 141
cheese 3, 15, 43, 44, 54, 60, 61, 70, 77, 78, 84, 95, 96, 97, 127, 159, 195, 211, 216, 220, 223, 226, 248, 258, 265, 269, 270, 271
chicken 72, 77, 164, 170, 186–7, 236, 252, 257, 258, 268
chicken cuts 171
chicken drumettes 186–7
chicken kebabs 48, 63, 71
chicken marinade 171, 185
chicken noodle soup 126
chicken nuggets 198
chicken skin 170
chicken stock 75, 222
chickpeas 215, 216, 225

children
   food needs 65, 81
   obesity and 194, 202
chillies 237, 252, 253, 256, 257, 258, 265, 266
chilli powder 173, 270, 271
chilli sauce 73, 77, 183, 264, 269, 270
China 142, 144
   Thai food and 257
   Vietnamese food and 255–6
Chinese cabbage 253
Chinese cleaver 254
Chinese food 246, 247, 251
chives 75, 126
chlorophyll 99, 110, 116
chocolate 197, 207, 209, 214, 243, 282–6, 291, 296, 298, 303
chocolate cake 277, 291
chocolate liquor 282
chocolate ripple 299
Choice 82
cholesterol 61, 87, 166, 198, 203, 211, 212, 235
chopping board 24, 25, 26, 47, 173
chopsticks 256
choux pastry 281
chuck cut (beef) 167
chump cut (lamb) 168
cinnamon 36, 37, 40, 98, 123, 130, 161, 253, 276, 290, 295, 296, 302
citrus fruits 100, 105–7
climate
   Thai food and 256
   Vietnamese 255
cloves 122, 253, 296
coagulation 79, 87, 175, 277
cocoa 279, 282, 291, 296, 303
cocoa butter 282, 283
coconut 35, 52, 123, 252, 265, 292, 298, 299
coconut cream 265, 266
coconut milk 76, 257
coconut oil 198
coeliac disease 191, 214–15, 217, 227
coffee 81, 210, 256
coffee mousse 285
coffee substitute 237
cola drinks 205, 206, 210, 214, 246
colander 42
collagen 167, 175
colours
   baked products 274
   meat, poultry and fish 175

   vegetables 110–11
commercial products
   breads 155
   cordial 106
   fast foods 213
   salsas 258, 259
community factors, food choices and 193
compound chocolate 282, 283, 284
compressed fresh yeast 137
condensed milk 298, 299
connective tissue 163, 166–7, 172, 175
considerations and constraints 1, 5
consumers, vegetables and 115
convenience foods 247
cooking apples 104
cooking chocolate 283, 284, 296, 298
cooking fuel, Thai 256
cooking methods
   Aboriginal and Torres Strait Islander people's traditional 232–3
   Asian 254–5
   vegetables 112, 118
cooking times 43, 279
coriander 73, 160, 225, 252, 253, 257, 258, 264, 265, 266, 267, 270
corn 82, 132, 135, 215, 257, 258
   see also maize
corn chips 258, 259, 271
cornflour 241, 264, 266, 269, 291
cornstarch 133, 156
corrugated cardboard cartons 108, 109
couscous 142, 160
couverture chocolate 283, 284
crabs 231
cream 61, 62, 241, 242, 289
cream cheese 288, 303
creaming 47, 275–6, 278, 279
cream of tartar 291
creamy sauces 249
cross-contamination 19, 24, 173
crumbing 174
crushing, fruit juice 110
crust 277
cucumbers 15, 152, 214
cultural background, foods and 246–8
cumin 160, 225, 253, 267, 270, 271
cupcakes 63, 207, 288, 289
cups 42
curries 181, 237, 253
curry paste 257, 265, 266
curry powder 76, 162, 181, 252, 301
custard 162

INDEX

custard apple 257
cutting 26, 114
cycling 201

D

dairy products 198, 216, 219, 246, 248, 257, 258
danger zone 19, 24, 25
Daube, Mike 66–7
decision table 9, 180
decorations 44
    chocolate 285
    cupcakes 288
    gingerbread 295
    sugar 286–7
deep-frying 254
defreezing 173
design brief 1, 5–9, 14
Desiree potatoes 116, 117
desserts 119–20, 236, 257, 275
detergents 23
Devonshire tea 33
dextrinisation 131, 139, 273, 274, 277
diabetes type 1 203
diabetes type 2 67, 191, 195, 200, 202, 203, 206, 212, 233, 234, 249
diarrhoea 214, 215
dicing 26, 47, 160
diet
    cattle 164
    obesity and 202–3
dietary fibre 55, 56, 60, 82, 83, 100, 103, 104, 106, 111, 116, 133, 135, 148, 149, 211–13, 226, 249
digestive system 58
dipping sauces 73, 256, 264
dishwasher 23
dough 136–9, 140, 141, 274, 276
doughnuts 207
dried beans 216
dried fruit 53, 160, 216
dried gluten 133
drip irrigation system 104
drumstick (chicken) 171
dry cooking 45, 175
duck 164, 170, 231
Dutch Cream potatoes 116, 117

E

eating chocolate 283, 284
eating tips, adolescents 66, 67
eggplant 159, 248

eggs 15, 16, 17, 42, 49, 51, 54, 60, 61, 72, 77, 78, 85–98, 124, 136, 152, 174, 182, 184, 185, 186, 188, 189, 214, 216, 226, 227, 230, 241, 242, 243, 250, 251, 261, 262, 265, 266, 267, 274, 275, 277, 278, 279, 280, 289, 290, 291, 293, 294, 295, 297, 300, 302
egg tomato 158
electric hand beaters 277, 289
electric juicers 31
electric ovens 19, 26, 27, 28, 246
enchiladas 257, 258
endosperm 133, 136, 142, 144, 250
energy 82, 87, 200–1, 216, 238
energy-dense foods 196–7, 201, 204, 207, 249
energy drinks 206
English muffins 89
environmental sustainability, meat industry and 164–5
enzymatic browning 99, 104, 105
enzymes 58, 84, 104, 214
European settlement, food changes 233, 246
evaluation 6, 7, 8, 12–14
evaporated milk 223, 269
exercise 209, 247

F

faeces 58, 215
fair trade movement (chocolate) 282
family influence, food choices 192, 193
family meal, two-course 220
farmers' markets 247
fast foods 194, 207, 247
fast freezing 115
fasting 79, 80
fats 56, 61, 62, 82, 83, 84, 85, 87, 104, 133, 143, 166, 167, 172, 175, 176, 197–8, 204, 207, 208, 209, 210, 212, 230, 232, 235, 236, 238, 247, 249, 257, 278, 280, 281
fermentation 131, 136, 137, 138, 139, 140
fertilisers 104, 165
fibre *see* dietary fibre
filo pastry 78, 120–1, 130, 281
fire safety 19–21, 177
fish 60, 61, 172–7, 195, 198, 214, 220, 232, 233, 247, 256, 257
    cooked test 176
fish patties 174, 188–9
fish sauce 73, 252, 253, 256, 257, 264
five food groups 195

flavourings 49, 51, 69
    Asian 252–4
    baking 274
    bread 149
    bush food 236–9
    hotness 258, 259
    meat, poultry and fish 175
    pasta sauce 260
    pizzas 147
    risotto 218
    stir-frying 179
flour 39, 49, 51, 119, 124, 130, 140, 149, 154, 174, 181, 184, 185, 186, 221, 224, 246, 251, 262, 265, 274, 278, 279, 280, 292, 296, 300
floury potatoes 117, 118
flow chart 43
flowers 231
fluoride 57
folding 47, 289
food 55, 56
    dating 141
    labelling 214
    manufacturing 197, 207
    preservation 115
    properties 4
    role 2
    storage 25
    temperature 24, 25
    variety of 66
food allergies 191, 213, 214
Food and Nutrition Policy (1992) 233
food balance, Thai 257
food choices 191–227
food colouring 128, 240, 303
food courts 252
food evolution, Australia 245–71
food gathering, Aboriginal and Torres Strait Islander peoples 230–2
food groups 195
food hypersensitivities 191, 214–15
food models 59
food poisoning 19, 21, 24–5, 32, 172
food preparation terms 47–8
food processing 29, 39, 40, 42, 246, 302
food safety 19–40, 117–21, 172–3, 183, 189, 229
Food Standards Australia New Zealand (FSANZ) 141, 174, 208, 214
Food Standards Code 4–5, 174
food supply disruption, European settlement and 233

INDEX

food truck 8
foreleg shin cut (beef) 167
forequarter cut (lamb) 168
free-form fruit tart 302
free-range chicken 170
free-range eggs 86
freezing 114, 246
French dressing 188
fresh fruit 85, 89
freshwater bream 231, 232
fried foods 93, 198, 208
fried rice 72, 68–9, 145
frozen beans 114, 115
frozen bread 142
frozen meat 172
frozen peas 124, 225
fruit 35, 60, 64, 66, 80, 89, 100, 107, 110, 114, 119, 128, 195, 197, 199, 207, 212, 214, 220, 230, 231, 232, 234, 236, 246, 248, 255, 257, 290, 298, 302
fruit bread 37
fruit cakes 278
fruit juices 31, 66, 85
frying 47, 93, 96, 98, 167, 177, 256
Fuji apples 101, 102

G

galangal 253, 257
Gallagher, Katy 206
garlic 16, 42, 71, 72, 76, 77, 126, 157, 158, 159, 173, 183, 184, 214, 221, 225, 226, 248, 258, 263, 264, 265, 267, 270, 271
garnishes 44, 47, 128, 149, 184, 185, 257, 258
gas ovens 19, 27, 28, 246
gelatin 131, 144, 167, 175
gender, body fat and 197
generation 7, 10, 14
generational food survey 248
germ layer 133, 135, 143
ginger 71, 77, 183, 185, 252, 253, 254, 264, 265, 276, 290, 295, 296
gingerbread 278, 295
ginger nuts 286
glacé 128, 288
glaze 38, 47, 124, 153, 154, 158, 181, 224, 280
glucose 80, 136, 197, 203, 211
gluten 131, 133, 136, 138, 139, 140, 155, 274, 277, 279, 280, 281
gluten-free foods 143, 247, 300
gluten intolerance 214–15

glycaemic index (GI) 191, 199–200, 249
glycogen 80, 197, 199
gnocchi 248, 251, 262
Golden Delicious apples 101, 102
golden syrup 52, 275, 277, 291, 292, 295, 296
GPS technology 135
grain (meat) 166
grains 82, 197, 230
    see also cereals
Granny Smith apples 101, 102, 122, 151, 221, 290, 302
graphic organiser 287
grating 42, 47, 186, 284
Great Depression 246
Greek foods 247, 297
green capsicum 16, 71
green curry paste 257, 265, 266
greengrocer 99–130, 246
greenhouse gases 165
green vegetables 69, 113, 114–15, 210, 216
Grenfell, Rob 202, 203
griddle oatcakes 141
grilling 27, 47, 166, 167, 175, 176, 256
ground oven 229, 232
growth spurt 64, 65
grubs 230, 231

H

ham 15, 152, 248
hamburger 178, 184, 197, 198, 208, 246
handheld electric beater 29, 42
hand washing 22, 29, 30, 173
hard-boiled eggs 87, 91
harvesting 109, 114, 134, 135
hash brown potato 112
hawker-style foods 245, 252
health risk factors 233
Healthy Eating Pyramid 14, 16, 59–69, 70, 72, 73, 74, 75, 76, 88, 90, 107, 111, 120, 121, 125, 135, 147, 151, 153, 158, 179, 180, 187, 189, 269, 270, 302
healthy foods 55–78, 195, 197–8, 203, 208, 247
healthy weight 204
heart disease 172, 195, 198, 199, 202, 203
Heart Foundation 67, 87, 202–3, 204
hedonic scale 16
height increase 64

height loss 209
herbs 121, 126, 127, 218, 237, 252, 253, 256, 257, 258, 260
Hiddens, Les 235
high blood pressure 202, 203
high-density lipoprotein (HDL) 198
high-fat foods 166, 194
high-intensity activities 201
hind shank cut (lamb) 168
hind shin cut (beef) 167
homemade products 106, 155, 213
homogenisation 83–4
honey 37, 52, 90, 276–7, 298
hypersensitivities 213, 214–15, 247

I

ice cakes 275
ice-cream 36, 236, 237, 238, 256
icing sugar 40, 275, 286, 293, 297, 303
immigration, post-war 246, 247
immune system 203
incidental exercise 206
independence, food choices and 192, 193
India 140, 142, 144
Indigenous food movement 229–43, 247
    see also Aboriginal and Torres Strait Islander peoples
Indonesian foods 251
infants, food needs 65
infusion 221, 237
Inge, Kieran 202
in-season fruits 99, 100
insoluble dietary fibre 212
instrument cleanliness 173
insulin 203
intensive cage production (eggs) 85
international food companies 246
intestinal flora 212
investigation 7, 8–10, 14
involuntary life processes 200
iron 60, 61, 64, 87, 100, 104, 166, 170, 216, 217, 235
Italian foods 218, 246, 248–51

J

Japan 101, 142, 144, 164
Japanese foods 251
joints (chicken) 171
Jolly, Kellie-Ann 67
jugs 42
junk food 67, 194

INDEX

K

Kaffir limes 252, 253, 257
kangaroo meat 230, 231, 232, 235–6, 238, 241
Kellogg's cereals 82, 83
Kelly, Bridget 194
Kennebec potatoes 116, 117
key terms 47–8
kidney beans 270, 271
kidney disease 203, 233
kilojoules 64, 201, 202, 207, 208
Kipfler potatoes 116, 117
kitchen tools 26, 42–3
kneading 47, 131, 138, 139, 140, 240, 281

L

lacto-ovo-vegetarians 216
lactose 61, 83
lactose intolerance 214, 247
laksas 252
lamb 164, 165, 168, 169, 170, 172, 235, 236, 267
land clearing 233
Laos 255, 256
large intestine 58, 215
lasagne 12, 219, 221
late adulthood, food needs 65
leafy vegetables 179
lean meats 61, 165, 170, 195, 220, 235, 236
leavening 139, 140, 273, 276
leg of lamb 168
legumes 60, 165, 212, 214, 216
lemongrass 73, 252, 253, 257, 265
lemon juice 35, 39, 77, 90, 124, 128, 129, 173, 188, 280
lemon myrtle 237, 238
lemons 92, 105, 128, 185, 297, 303
lemon tart 280
lemon zest 151
lentils 215, 216, 217
lettuce 15, 16, 74, 269, 270
lifestyle, food choices and 192, 193
lifestyle diseases 202–3
limes 73, 105, 252, 256, 258, 264
Linzer torte 280
liquids 45, 49, 51
Livelight campaign 204, 205
loin cut (beef) 167
long-grain rice 143–4, 150, 257
low-density lipoprotein (LDL) 198
low-fat diets 252
low-fat yoghurt 209
low-intensity activities 201

M

macadamia nuts 231, 238, 243, 297
macarons 275
macronutrients 64
magnesium 11, 104, 116
Maillard reaction 273, 274, 277
maize 132, 215
    see also corn
Malaysian food 251
malnutrition, Aboriginal and Torres Strait Islander people 233
manufacturers, Asian foods and 252
marbling 163, 164, 167
margarine 61, 62, 276
marinating 47, 163, 173–4, 176, 178, 183, 185
marine pollution 172
marketing
    food choices and 192, 193–4
    meat 169–70
    pasta 260
Martin, Jane 67, 206
mashed potato 112
measurements 44–6, 279
Measure up! campaign 204
meat 60, 163–90, 198, 216, 218, 246, 257, 258
meat mallet 42
meat sauce 248, 251
melt-and-mix method 276
melting chocolate 283–4, 285
meringues 275, 278
mesh bags 108, 109
metric measurements 41, 44
Mexican foods 208, 257–9
micronutrients 64
microorganisms 173, 175
microwaving 19, 28, 36, 112, 118, 145, 162, 173, 284
middle loin cut (lamb) 168
milk 17, 36, 38, 40, 53, 54, 60, 61, 66, 70, 75, 81, 82, 83–4, 94, 98, 124, 127, 130, 151, 162, 182, 188, 189, 195, 212, 214, 216, 220, 221, 224, 226, 240, 274, 275, 276, 280, 295, 296, 300, 303
milk products 52, 214, 282
milling 134
minced meats 124, 127, 169, 172, 180, 181, 182, 184, 263, 270
minerals 57, 60, 100, 103, 111, 133, 135, 230, 232
mint 73, 253, 264, 267
mixing 47
moderate-intensity activities 201
modified atmosphere packaging (MAP) 99, 115–16, 163, 169
moist cooking 175
molasses 274, 275
monosodium glutamate (MSG) 174
monounsaturated fats 191, 198, 235
mortar and pestle 254
mould inhibitors 142
mozzarella cheese 158, 221, 227, 248
muesli 52, 90, 199
muesli bars 209
muffins 49–50, 51, 53, 89, 197, 207, 275, 277, 278
Murray–Murrumbidgee area 105, 143
muscle fibres 64, 163, 166, 172, 173
mushrooms 71, 159, 221, 222

N

naan 140, 141, 155
nachos 257, 271
Napoli sauce 157, 159, 227, 262
native foods, rediscovery of 235–9, 321
    see also Aboriginal and Torres Strait Islander peoples, food
neck cut (lamb) 168
nectar 230, 276
New South Wales (NSW) 101, 105, 143, 238
niacin 116, 166
nomadic lifestyle, Aboriginal and Torres Strait Islander people and 230, 233
noodles 60, 209, 252, 256
Northern Territory 143, 234
no-till farming 131, 135, 146
nutmeg 36, 78, 123, 221, 276
nutrients 55, 56, 204, 210
    Aboriginal and Torres Strait Islanders' diet 232, 234
    breakfast drinks 85
    eggs 87
    milk 83
    snack foods and 207–9
nutrition
    age and 64–9
    breakfast and 80–1
    cereals and 135
    couscous and 142
    fish and 172

INDEX

fruit and 100, 118
green beans and 113
kangaroo meat and 235–6
meat and 165–6
pasta and 250
poultry and 170
rice and 143
vegetables and 111, 113, 118
nutrition information 195, 246, 247
marketing and 193
meat, poultry and fish 175
sweet potatoes 113
nutrition pack 103, 104, 106, 116
nuts 60, 119, 195, 198, 216, 220, 231

O

oats 52, 82, 132, 136, 199, 215
obesity 194, 195, 197, 198, 202–3, 233, 234, 249
Obesity Policy Coalition (OPC) 194, 205–6
oils 61, 62, 71, 72, 76, 77, 78, 127, 130, 153, 160, 173, 174, 176, 177, 181, 183, 184, 185, 186, 187, 188, 225, 226, 230, 237, 238, 241, 250, 254, 263, 264, 265, 269, 270, 271, 291
olive oil 16, 156, 157, 158, 159, 186, 198, 221, 248, 258, 261, 262
Olympic Games (1956) 246
omega-3 fatty acids 61, 172, 198
omelettes 87, 96
onions 72, 73, 75, 76, 77, 97, 112, 124, 126, 127, 157, 158, 160, 181, 182, 183, 184, 185, 186, 221, 222, 223, 224, 225, 226, 241, 258, 263, 265, 267, 269, 270, 271
'Only sometimes and in small amounts' foods 166, 280
orange juice 85, 89, 216, 298, 303
oranges 98, 105–7, 128, 214, 293
oregano 221, 258, 263
organic foods 86, 247
osteoporosis 191, 209–11
Outback Pride Project 235
outdoor cooking 247
ovens 26–8, 138, 139, 175, 277
overcooking 235
overgrazing 165
oxygen 115, 169

P

Pacific Rim cuisine 245, 247
packaging
beans 114
eggs 86
fruit 107–9
salad mixes 115, 117
paella 144
palm sugar 252, 253, 257
pancakes 13–14, 17, 29
pancreas 203, 204
pan-frying 175
paperbark 232, 233
pappadams 28, 141, 225
paprika 173, 271
parmesan cheese 182, 221, 222, 223, 227, 263
parsley 16, 70, 75, 94, 96, 124, 126, 127, 182, 188, 223, 267
passata sauce 157, 182
pasta 60, 199, 215, 216, 219, 223, 248–51, 260, 261, 263
pasteurisation 83, 84, 110
pastry 54, 198, 208, 256, 274, 275, 280–2, 302
patty cakes 215, 278, 300
peanut butter 76, 298
peanut oil 264, 266
peanuts 52, 160, 214, 253
peas 181, 212, 222, 225
peeling 112
peer group influence, food choices and 192, 193
pepper 70, 75, 92, 94, 96, 126, 127, 157, 159, 182, 186, 188, 221, 223, 226, 227, 237, 241, 263, 270
peppermints 299
peppers 258
personal feelings, food choices and 192, 193
personal hygiene 21–2
phosphorus 61, 100, 104, 209, 210
physical activity 67, 202, 203, 206
pies 215
pilafs 142, 144
pineapple juice 128
pineapples 71, 214, 257
Pink Lady apples 101, 102
piping 285–6
pitta bread 15, 140, 141, 155, 238
pizza 136, 146–7, 156, 159, 208, 215, 227, 248
plain flour 158, 188, 189, 243, 280, 291, 293, 294, 295, 296, 297
plant foods 56, 60, 135, 150, 165, 212, 230, 231, 232
plastic bags 108, 109
poaching 47, 92, 161
pocket bread 141
polenta 156, 248, 302
polysaccharides 199
polyunsaturated fat 170, 235
pomme fruit 99, 100, 101
Pontiac potatoes 116, 117
pop-up food trucks 247
pork 164, 236, 257, 258, 264
pork buns 252
portion control 171, 204, 207
potassium 61, 87, 104, 106, 116
potato crisps 112, 209
potatoes 16, 42, 70, 111–12, 116–17, 118, 126, 127, 130, 159, 181, 185, 186, 188, 189, 225, 251, 262
pot-roast 175
poultry 60, 61, 165, 170–1, 172, 173, 175, 195, 218, 220
pralines 286, 287
prawns 73, 264
preferred option 1, 8
pregnancy, food needs 65
prepackaged salad mixes 115–16
preserved fruit 120
preventative health measures 206
primary processing, fruit 109, 114–15
processes 7, 11, 193, 204
baked products 278, 279
biscuits 279–80
breakfast cereals 82–3
chocolate cooking 285
fruit 109–10
meat, poultry and fish 172–4
milk 83–4
rice 143
vegetables 114–15
vegetarian meals 216–17
wheat products 134–5
proteins 56, 60, 61, 69, 83, 85, 87, 133, 135, 136, 138, 139, 165, 166, 170, 172, 173, 174, 175, 179, 197, 209, 210, 214, 216, 217, 230, 232, 247, 250, 255, 256, 257, 274, 277
proving 131, 138, 139, 140
puddings 238, 278
puff pastry 181, 224, 281, 301
pulses 216
purée 48, 151

Q

qualitative measures 1, 4, 165, 170, 213
quantitative measures 1, 4–5, 213

INDEX

Queensland 164, 238
quesadillas 258
quiche 10, 54, 226
quick mix cakes 278, 279

R

rabbits 236, 246
rack of lamb 168
raising agents 49–50, 51, 274, 276
raspberry and vanilla muffin 7
raspberry jam 296
ready-to-eat meals 247
recipe maps 10, 33, 44
    muffins 51
    risotto 217–18
    spaghetti 260
recipes *see* Recipe Index
red capsicum 73, 77, 152, 159, 160, 183, 188, 221, 222
Red Delicious apples 101, 102
red meat 61, 164, 166, 169, 173
reduced-fat milk 84
refrigeration 145, 172, 173, 246
religion
    Thai food and 256
    Vietnamese food and 255
research 8–9
rib cut 167, 168
riboflavin 166, 170
rice 60, 68–9, 71, 72, 76, 82, 132, 135, 142–6, 150, 151, 152, 215, 216, 217–18, 225, 248, 252, 256, 257
rice cooker 42, 145, 254
rice paper rolls 73
ricotta 78, 223, 243
rigid plastic packs 108, 109
risotto 144, 217–18, 222
roasted potato 112
roasting 159, 166, 167, 171
rolled oats 52, 79, 82, 90, 123, 292
Roma tomatoes 158, 248
roots 230, 231
Royal Gala apples 101, 102
rubbing in 48, 278
rump cut (beef) 167, 183
rye 132, 136, 140, 215

S

salad dressings 238
salads 11, 15, 16, 115, 248
salami 158, 248
salmon 172, 188

salmonella 172, 173
salsa 258–9, 269
salt 3, 61, 62, 70, 75, 82, 84, 92, 94, 124, 127, 140, 143, 153, 154, 155, 156, 157, 158, 159, 182, 185, 186, 188, 195, 207, 208, 209, 210, 212, 221, 223, 227, 233, 246, 251, 261, 262, 270, 280, 293
salted butter 275
Sanitarium cereals 82, 83, 217
satays 76, 183, 252
saturated fats 61, 62, 67, 166, 170, 191, 195, 198, 233, 235, 246
sausage rolls 213, 224
sausages 166, 172
sauté 48, 160, 222
savoury dishes 97, 209, 236, 280
scales 42, 45
school cafetaria 182, 205–6
scones 32–3, 38, 207, 277, 278
seafood 231, 258
searing 48
seasonality, fruit 100, 103, 105, 106
Sebago potatoes 116, 117
Second World War 246, 248
secondary processing
    beans 114
    fruit 109
    meat 165
seeds 135, 230, 231, 232
self-raising flour 17, 38, 39, 53, 123, 124, 133, 155, 226, 240, 274, 277, 289, 290, 294, 302
self-service stores 246
semolina 142, 156, 250, 251, 302
sensory properties 1, 2, 4, 12
    Anzac biscuits 292
    apples 103
    Australian biscuits 239
    baked fruit dessert 120
    baking 278
    breads 153
    breakfast drinks 85
    bush chips 238
    cheese 211
    cheesy pasta bake 223
    chocolate 282, 283
    crustless quiche 226
    cupcakes 289
    damper 240
    family meal 220
    flatbreads 155
    fruit tarts 302
    gnocchi 262

    hamburgers 178
    margherita pizza with cauliflower crust 227
    meat 167
    pancakes 17
    pastry 280
    potatoes 118
    ricotta and orange biscuits 293
    risotto 222
    salsa 258
    sausage rolls 213, 224
    twirl dessert 162
    upside-down cake 290
    vegetables 113
sesame seeds 52, 130, 154
shallow-frying 185, 255
shanks 168
shaped pastas 249
Share a Taste of Australia 239
shark (flake) 172
shelf life 109, 169
shortcrust pastry 124, 181, 281
shortening 49, 51, 53, 274
short-grain rice 144, 151, 152, 257
shredding 48
shrink-wrapping 108, 109
sieve 42
sifting 48, 277, 278, 279
simmering 48
skim milk 84, 298
slices 207, 274
slicing 48
slotted spoon 42
small appliances 19, 29–34
small game 230
small intestine 58, 215
small-portion restaurants 247
smoking 203
snack foods 61, 62, 63, 64, 65, 103, 107, 199, 204, 206–9, 216
snake beans 113, 253
snipping 114
snow peas 16, 183, 185
social media 193
soft-boiled eggs 87, 91
soft drinks 62, 81, 206, 207
soil, health of 165
solanine 99, 116
soluble dietary fibre 212
soups 75, 252
sour cream 70, 290
sourdough breads 142

INDEX

outh America 116, 142, 247, 282
outh-East Asia 142, 144
y products 215, 216, 255
y sauce 71, 72, 74, 77, 152, 183, 184, 186, 187, 252, 265
aghetti 182, 248, 260
aghetti bolognese 12, 43, 263
anish Mexican food 258
ecifications 1, 5
ices 119, 121, 218, 233, 253, 257, 258, 276, 279
ider map 192–3
inach 78, 249
onge cakes 275, 278, 279
onge sandwich 291
oons 42
orts drinks 206
orts sponsorship 193, 194
raying 134
ring onions 54, 70, 78, 188, 268
aple foods 132, 142
ar diagram 262
arch 111, 133, 136, 138, 139, 144, 199, 274, 277
teak 127, 172
teaming 48, 118, 187, 255, 278
tewing 48, 122, 167, 175
tick mixer 42
tick-on labels 107
ticky rice 252, 255
tir-frying 48, 77, 166, 168, 175, 176, 178–9, 255, 256
tirring 284
tone fruit 100
torage
   beans 114
   bread 142
   fruit 103, 105, 106, 109
   grains 134
   meat, poultry and fish 172–3
   potatoes 116
   sweet potatoes 113
toves 26
tovetop desserts 162
tructure
   eggs 86
   meat 166–8
ugar 53, 61, 62, 67, 75, 82, 85, 104, 119, 122, 123, 128, 143, 151, 153, 154, 155, 156, 157, 158, 161, 162, 182, 138, 195, 197, 199, 204, 207, 208, 209, 212, 233, 242, 246, 247, 264, 270, 274–5, 277, 278, 279, 282, 286–7, 290, 302

sugarcane 255, 274
sugar syrup 280
sugary drinks 194, 205–6
sultanas 40, 52, 90, 160
Sundowner apples 101, 102
Sunrice 144, 217–18
sun-ripened vegetables 248
supermarkets 247, 250
super-sized meals 207
sushi 144, 145, 152
sustainable production 99, 104, 135, 143, 164–5
sweet biscuits 197, 298
sweeteners 3, 49, 51
sweet potatoes 113, 130, 159, 225
sweets 161, 230, 239, 274, 280
SWOT (strengths, weaknesses, opportunities, threats) analysis 9, 146
syrups 237

T

tacos 257, 258, 259, 270
takeaway foods 61, 62, 246, 247
tap water 57
Target 100 164–5
tarts 238
taste testing 3, 83
tea 81, 210, 233, 246
television 193, 201, 246, 247
temperature (cooking) 177
tender cuts (meat) 167
tenderising 173, 174
tenderloin cut (beef) 167
10 000 Steps Campaign 206
texture
   baking 275, 276, 277
   meat, poultry and fish 175
   pastry 280
Thai food 245, 251, 252–3, 256–7, 266
thiamine 116, 166, 170
thickeners 215
thigh (chicken) 171
toast 27, 30, 92
toffee 287
tofu 216
tomato paste 127, 181, 249, 263, 270, 271
tomato sauce 127, 241, 251
tomatoes 15, 16, 71, 75, 157, 214, 221, 225, 236, 248, 257, 258, 263, 269, 270, 271
tongs 42
Toolangi Delight potatoes 116, 117

toppings, pizza 147, 227
topside cut (beef) 167
tortillas 141, 258, 269
tough cuts (meat) 166–7
traditional diet, Aboriginal and Torres Strait Islander people 232–3
trans fats 191, 198
transportation, fruit 107–8, 114
tropical fruit 100
tubers 99, 113, 116
tuna 16, 76, 188, 198, 266
turkey 15, 164, 170
turmeric 225, 240
turtles 231
twirl desserts 161–2

U

unbleached flour 155
unpolished rice 144
unsalted butter 275
unsaturated fat 62
'upsizing' 207
use-by date 141, 172
utensils 42–3, 50

V

vanilla 151, 161, 242, 243, 276, 289, 293, 294, 298, 300, 303
vegans 216
vegetable fat 283
vegetable oils 52, 77, 93, 198, 267, 282
vegetable omelette 89
vegetable parcel 120–1
vegetable pizza 216
vegetable protein 216
vegetable stock 76, 160, 225
vegetables 60, 64, 66, 69, 72, 73, 76, 77, 80, 89, 110–21, 178, 179, 181, 195, 197, 199, 212, 218, 220, 221, 226, 231, 246, 247, 248, 252, 255, 257, 258, 260
   Asian 253
   pizzas and 147
   stir-frying 176
vegetarian diets 216–19
Victorian Government 200, 204
Vietnam 247, 256
Vietnamese foods 251, 255–6, 264
villi 58, 215
vine fruit 100
vinegar 16, 92, 152, 173, 183, 242, 258, 264
vitamin A 61, 100, 111, 113, 210
vitamin B group 60, 135, 142, 143, 166, 170, 175, 249

INDEX

vitamin C 85, 87, 100, 106, 112, 113, 116, 210, 216, 236
vitamin D 61, 209, 210
vitamins 57, 60, 83, 84, 100, 103, 104, 111, 133, 175, 197, 230, 232
voluntary life activities 200

W

waist circumference 204
walking 201, 206–7
washing-up techniques 22–3, 25
water 39, 57, 61, 66, 72, 76, 81, 82, 87, 96, 103, 111, 122, 124, 126, 128, 136, 139, 140, 143, 144, 145, 150, 151, 153, 154, 155, 156, 158, 161, 165, 166, 169, 182, 183, 195, 210, 242, 250, 251, 261, 264, 270, 271, 280, 284, 285, 290, 292, 302, 303
water chestnuts 183, 253

water-soluble nutrients 112
wattleseed 237, 240, 242
waxy potatoes 117, 118
weight increase 64, 65, 207
weight loss 196, 199
weight measurement 45
Western Australia (WA) 101, 105, 143, 204
wheat 82, 132, 133–5, 140, 142, 214, 215, 216, 250, 274
wheat flour 133, 136
whisking 42, 48, 277, 278
white flour 133, 233
white meat 164
white vegetables 116–17
wholegrain bread 54, 80, 199, 226
wholegrain cereals 89, 135, 195, 199, 220
wholemeal flour 133, 249, 274
wholemeal pizza dough 156, 159

wild rice 144, 231
wine 173, 246
winglets 171
wok 27, 176, 252, 254
Worcestershire sauce 127, 270
World Health Organization (WHO) 194, 206

Y

yams 230, 231, 232
yeast 131, 136–9, 140, 153, 154, 155, 156, 158, 275
yoghurt 35, 36, 60, 61, 90, 155, 195, 209, 216, 220, 267, 268

Z

zinc 61, 64, 87, 111, 166, 170
zucchini 159, 178, 183, 184, 221, 224

RECIPE INDEX

A

almonds 151
    Creamed rice with almonds and apple puree 151
    Fruit crumble 123
    Greek biscuits 297
Anzac biscuits 292
Apple and cinnamon turnovers 40
apple juices 31
apples
    Apple and cinnamon turnovers 40
    apple juices 31
    Creamed rice with almonds and apple puree 151
    Fragrant, fruity couscous salad 160
    Quick apple muesli 90
    Stewed apples 122
    Upside-down cake 290
apricots
    Fragrant, fruity couscous salad 160
    Freeform fruit tart 302
    Toasted muesli 52
Asian
    Basic fried rice 72
    Beef noodle stir-fry 174, 183
    Chicken and vegetable stir-fry 77
    Chicken curry with coconut pancakes 265
    Plain rice 145, 150
    Prawn and vegetable rice paper rolls 73
    Ricey lettuce parcels 74
    Satay vegetables and tuna 76
    Satay vegetables and tuna 76
    Sushi roll 152
    Thai curry pastes 257
    Thai fishcakes 266
    Vietnamese spring rolls with dipping sauce 264
Australian
    Anzac biscuits 292
    Bonus: Fish in paperbark ONLINE
    bush chips 238
    Damper with eucalyptus or gumleaf oil butter 240
    Kangaroo meatballs 241
    tiger toast 27
    Wattleseed pavlovas with macadamia cream and sugar bark 242
avocado
    Chicken burritos 269
    Poached egg 92
    Sushi roll 152

B

bacon 96
    Cheese omelette with bacon 96
    Cheesy pasta bake 223
    Crustless quiche 10, 226
    Dressed baked potatoes 70
    Mini quiches 54
    Ricey lettuce parcels 74
    Savoury egg roll 97
Baked meatballs and spaghetti 182
bananas
    Banana toast 37
    Fabulous French toast 98
    Frozen banana whiz 36
    Microwaved banana 36
Banana toast 37
Basic bread 153
Basic fried rice 72
Basic muffins 53
Basic pasta dough 261
Basic risotto 217–8, 222
Basic scones 38
beef
    Baked meatballs and spaghetti 182
    Beef noodle stir-fry 174, 183
    Cornish pasties 124–5, 280
    Curried meat and vegetable pie 181
    Shepherd's pie 127
    Spaghetti bolognese 263
    Zucchini barbecue burgers 178, 184
Beef noodle stir-fry 174, 183
beverages
    apple juices 31
    Citrus cordial 106, 128–9
    Raspberry punch 128–9
    Sparkling pineapple punch 128–9
biscuits
    Anzac biscuits 278, 292
    Chocolate and macadamia biscuits 243
    Chocolate delights 298
    Chocolate-coated biscuits 296
    Fruity balls 298
    Gingerbread 295
    Greek biscuits 297
    Ricotta and orange biscuits 293
    Scotch mints 299
    Vanilla biscuits 294
Boiled eggs 91
Bonus recipes
    Carrot cake ONLINE
    Fish in paperbark ONLINE
    Grilled chops with rosemary ONLINE
    Orange and poppyseed patty cakes ONLINE
    Spanish omelette ONLINE
bread
    Basic bread 153
    bush chips 238
    Calzone 158
    Damper with eucalyptus or gumleaf oil butter 240
    Fabulous French toast 98
    Flatbreads 140–1, 155
    Multigrain cob loaf 154
    Pizza bases 156
    Pumpkin noodle soup with herb and garlic bread 126
breakfast
    Banana toast 37
    Boiled eggs 91
    Bonus: Spanish omelette ONLINE
    Cheese omelette with bacon 96
    Fabulous French toast 98
    Fried egg 93
    Poached egg 92
    Quick apple muesli 90
    Scrambled eggs 94
    Shirred egg 87, 95
    tiger toast 27
    Toasted muesli 52
broccoli
    Beef noodle stir-fry 174, 183
    Satay vegetables and tuna 76
    Spicy 'shake and bake' chicken with potato latkes and steamed broccoli or beans 186–7
bush flavours

RECIPE INDEX

Bonus: Fish in paperbark ONLINE
bush chips 238
Chocolate and macadamia biscuits 243
Damper with eucalyptus or gumleaf oil butter 240
Kangaroo meatballs 241
Wattleseed pavlovas with macadamia cream and sugar bark 242

burger
Zucchini barbecue burgers 178, 184

butter
icing 288, 303
Damper with eucalyptus or gumleaf oil butter 240

C

cakes
Basic muffins 53
Bonus: carrot cake ONLINE
Bonus: orange and poppyseed patty cakes ONLINE
Chocolate sponge 291
Cupcakes 288, 289
Gluten-free patty cakes 300
Upside-down cake 290

Calzone 158

caramel
Upside-down cake 290

carrots
Beef noodle stir-fry 174, 183
Bonus: carrot cake ONLINE
Cornish pasties 124–5
Creamy carrot and tomato soup 75
Fish patties with coleslaw 188–9
Lemon chicken served with chat potatoes, carrot straws and snow peas 185
Minestrone soup 43
Potato salad with egg and tuna 16
Prawn and vegetable rice paper rolls 73
Ricey lettuce parcels 74
Salad roll-up 15
Satay vegetables and tuna 76

cauliflower
Margherita pizza with a cauliflower crust 227

cheese
Baked meatballs and spaghetti 182
Basic risotto 217–8, 222
Bonus: Spanish omelette ONLINE
Calzone 158
Cheese and spinach pastries 78
Cheese omelette 11

Cheese omelette with bacon 96
Cheesy pasta bake 223
Crustless quiche 10, 226
Dressed baked potatoes 70
Gnocchi with Napoli sauce 262
Lemon cream cheese icing 303
Margherita pizza with a cauliflower crust 227
Mini quiches 54
Nachos 257, 271
Ricotta and orange biscuits 293
Roasted vegetable pizza 159
Salad roll-up 15
Savoury egg roll 97
Shepherd's pie 127
Shirred egg 95
Spaghetti bolognese 263
Tacos 270
tiger toast 27
Vegetable lasagne 219, 221

Cheese and spinach pastries 78
Cheese omelette 11
Cheese omelette with bacon 96
Cheesy pasta bake 223

chickpeas
Spicy potatoes with chickpeas 225

chicken 186–7, 268
Chicken and noodle stir-fry 3
Chicken and sweet corn soup 43
Chicken and vegetable stir-fry 77
Chicken burritos 269
Chicken curry with coconut pancakes 265
Chicken kebabs 48, 63, 71
Lemon chicken served with chat potatoes, carrot straws and snow peas 185
Oriental chicken kebabs 71
Spicy 'shake and bake' chicken with potato latkes and steamed broccoli or beans 186–7
Tandoori chicken bites 268

Chicken and noodle stir-fry 3
Chicken and sweet corn soup 43
Chicken and vegetable stir-fry 77
Chicken burritos 269
Chicken curry with coconut pancakes 265
Chicken kebabs 48, 63, 71

chilli
Chicken and vegetable stir-fry 77
Chicken burritos 269
Chicken curry with coconut pancakes 265

Nachos 271
Prawn and vegetable rice paper rolls 73
Tacos 270
Vietnamese spring rolls with dipping sauce 264

chocolate
Chocolate and macadamia biscuits 24
chocolate basket 285–6
Chocolate delights 298
Chocolate icing 303
chocolate leaves 286
Chocolate sponge 291
Chocolate-coated biscuits 296
cut-out shapes 286
melting and resetting chocolate 284
piped shapes 285
Scotch mints 299

Chocolate and macadamia biscuits 243
chocolate basket 285–6
Chocolate delights 298
Chocolate icing 303
chocolate leaves 286
Chocolate sponge 291
Chocolate-coated biscuits 296

cinnamon
Apple and cinnamon turnovers 40
Banana toast 37
Chocolate-coated biscuits 296
Fabulous French toast 98
Fruit crumble 123
Gingerbread 295
Microwaved banana 36
Swirling twirl dessert in a glass 161–2
Upside-down cake 290

Citrus cordial 128

coconut
Anzac biscuits 292
Chicken curry with coconut pancakes 265
Chocolate delights 298
Fruit crumble 123
Fruity balls 298
Scotch mints 299
Summer or winter fruit kebabs with yoghurt and coconut dip 35
Thai fishcakes 266
Toasted muesli 52

coleslaw
Fish patties with coleslaw 188–9

Cornish pasties 124–5, 280

corn chips
Nachos 271

RECIPE INDEX

couscous
   Fragrant, fruity couscous salad 160
Creamed rice with almonds and apple puree 151
Creamy carrot and tomato soup 75
crumbing 174
Crustless quiche 10, 226
Cupcakes 288, 289
   Gluten-free patty cakes 300
   Bonus: orange and poppyseed patty cakes ONLINE
Curried meat and vegetable pie 181
curry
   Chicken curry with coconut pancakes 265
   Curried meat and vegetable pie 181
   Curry curlies 281–2, 301
   Satay vegetables and tuna 76
   Thai curry pastes 257
   Thai fishcakes 266
Curry curlies 281–2, 301
Custard powder sauce 162

D

Damper with eucalyptus or gumleaf oil butter 240
decorations
   chocolate basket 285–6
   chocolate leaves 286
   cut-out shapes 286
   piped shapes 285
dessert
   Anzac biscuits 278, 292
   Apple and cinnamon turnovers 40
   Basic muffins 53
   Basic scones 38
   Bonus: orange and poppyseed patty cakes ONLINE
   Chocolate and macadamia biscuits 243
   Chocolate delights 298
   Chocolate sponge 291
   Chocolate-coated biscuits 296
   Creamed rice with almonds and apple puree 151
   Cupcakes 288, 289
   Freeform fruit tart 302
   Frozen banana whiz 36
   Fruit crumble 123
   Fruity balls 298
   Gingerbread 295
   Gluten-free patty cakes 300
   Greek biscuits 297
   Microwaved banana 36
   Pancakes 17
   Ricotta and orange biscuits 293
   Scotch mints 299
   Stewed apples 122
   Summer or winter fruit kebabs with yoghurt and coconut dip 35
   Swirling twirl dessert in a glass 161–2
   Swirly scones 32–4
   Upside-down cake 290
   Vanilla biscuits 294
   Wattleseed pavlovas with macadamia cream and sugar bark 242
Dressed baked potatoes 70

E

eggs
   Boiled eggs 91
   Bonus: Spanish omelette ONLINE
   Cheese omelette 11
   Cheese omelette with bacon 96
   Crustless quiche 10, 226
   Fabulous French toast 98
   Fried egg 93
   Mini quiches 54
   Poached egg 92
   Potato salad with egg and tuna 16
   Savoury egg roll 97
   Scrambled eggs 94
   Shirred eggs 95

F

Fabulous French toast 98
fish
   Bonus: Fish in paperbark ONLINE
   Fish patties with coleslaw 188–9
   Potato salad with egg and tuna 16
   Prawn and vegetable rice paper rolls 73
   Thai fishcakes 266
Fish patties with coleslaw 188–9
Flatbreads 155
Food-processor sweet shortcrust pastry 39
Fragrant, fruity couscous salad 160
Freeform fruit tart 302
Fried egg 93
Frozen banana whiz 36
fruit
   Apple and cinnamon turnovers 40
   apple juices 31
   Apple purée 151
   Banana toast 37
   Basic muffins 53
   Citrus cordial 106, 128–9
   Creamed rice with almonds and apple puree 151
   Fragrant, fruity couscous salad 160
   Freeform fruit tart 302
   Frozen banana whiz 36
   Fruit crumble 123
   Fruity balls 298
   Microwaved banana 36
   Pineapple punch 128, 129
   Quick apple muesli 90
   Raspberry punch 128, 129
   Ricotta and orange biscuits 293
   Stewed apples 122
   Summer or winter fruit kebabs with yoghurt and coconut dip 35
   Swirling twirl dessert in a glass 161–2
   Upside-down cake 290
Fruit crumble 123
Fruity balls 298

G

ginger
   Chicken and vegetable stir-fry 77
   Chicken curry with coconut pancakes 265
   Chocolate-coated biscuits 296
   Gingerbread 295
   Oriental chicken kebabs 71
   Vietnamese spring rolls with dipping sauce 264
Gingerbread 295
Gluten-free patty cakes 300
Gnocchi with Napoli sauce 251, 262
Greek
   Greek biscuits 297
   Mini lamb kebabs with minted garlic dip 267
   Pitta bread 155
Greek biscuits 297
grilling 176

H

honey 276–7
   Banana toast 37
   Damper with eucalyptus or gumleaf oil butter 240
   Fruity balls 298
   Quick apple muesli 90
   Toasted muesli 52

I

icing 303
   butter 303

RECIPE INDEX

caramel topping 290
chocolate 303
lemon cream 303
orange 303
toffee 287
Indian
    Curry curlies 301
    microwaved pappadams 28
    Naan 155
    Tandoori chicken bites 268
Italian
    Baked meatballs and spaghetti 182
    Basic pasta dough 261
    Calzone 158
    Flavoured pasta and sauce 260
    gnocchi 251
    Gnocchi with Napoli sauce 262
    Margherita pizza with a cauliflower crust 227
    Napoli sauce 157
    Pizza bases 156
    Roasted vegetable pizza 159
    Spaghetti bolognese 263
    Vegetable lasagne 219, 221

K

Kangaroo meatballs 241
kebabs
    Mini lamb kebabs with minted garlic dip 267
    Oriental chicken kebabs 71
    Summer or winter fruit kebabs with yoghurt and coconut dip 35

L

lamb
    Bonus: Grilled chops with rosemary ONLINE
    Mini lamb kebabs with minted garlic dip 267
lasagne
    Vegetable lasagne 219, 221
legumes
    Beef noodle stir-fry 174, 183
    Chicken and vegetable stir-fry 77
    green beans 115
    Lemon chicken served with chat potatoes, carrot straws and snow peas 185
    Minestrone soup 43
    Nachos 271
    Spicy 'shake and bake' chicken with potato latkes and steamed broccoli or beans 186–7
    Spicy potatoes with chickpeas 225
    Tacos 270
Lemon chicken served with chat potatoes, carrot straws and snow peas 185

M

macadamia nuts
    Chocolate and macadamia biscuits 243
    Greek biscuits 297
    Wattleseed pavlovas with macadamia cream and sugar bark 242
Margherita pizza with a cauliflower crust 227
Mexican
    Chicken burritos 269
    Nachos 271
    Tacos 270
Microwaved banana 36
mince
    Baked meatballs and spaghetti 182
    Cornish pasties 124–5
    Shepherd's pie 127
    Spaghetti bolognese 263
    Traditional sausage rolls 224
    Zucchini barbecue burgers 178, 184
Minestrone soup 43
Mini lamb kebabs with minted garlic dip 267
Mini quiches 54
mint
    Mini lamb kebabs with minted garlic dip 267
    Prawn and vegetable rice paper rolls 73
    Scotch mints 299
mushrooms
    Basic risotto 217–8, 222
    Oriental chicken kebabs 71
    Roasted vegetable pizza 159
    Vegetable lasagne 219, 221
Multigrain cob loaf 154

N

Nachos 271
Napoli sauce 157
noodles
    Beef noodle stir-fry 174, 183
    Chicken and vegetable stir-fry 77
    Prawn and vegetable rice paper rolls 73
    Satay vegetables and tuna 76
nuts
    Chocolate and macadamia biscuits 243
    Prawn and vegetable rice paper rolls 73
    Wattleseed pavlovas with macadamia cream and sugar bark 242

O

oats
    Anzac biscuits 292
    Fruit crumble 123
    Quick apple muesli 90
    Toasted muesli 52
omelette
    Bonus: Spanish omelette ONLINE
    Cheese omelette 11
    Cheese omelette with bacon 96
Oriental chicken kebabs 71

P

Pancakes 17
    Chicken curry with coconut pancakes 265
pappadams
    microwaved pappadams 28
pasta
    Baked meatballs and spaghetti 182
    Basic pasta dough 261
    Cheesy pasta bake 223
    Flavoured pasta and sauce 260
    Gnocchi with Napoli sauce 262
    Minestrone soup 43
    Spaghetti bolognese 263
    Vegetable lasagne 219, 221
pastry
    Apple and cinnamon turnovers 40
    Cheese and spinach pastries 78
    Cornish pasties 124–5
    Curried meat and vegetable pie 181
    Curry curlies 301
    Food-processor sweet shortcrust pastry 39
    Freeform fruit tart 302
    Sweet potato parcel 130
    Traditional sausage rolls 224
patty cakes
    Bonus: orange and poppyseed patty cakes ONLINE
    Gluten-free patty cakes 300
pavlova
    Wattleseed pavlovas with macadamia cream and sugar bark 242
peanut butter
    Fruity balls 298
    Satay vegetables and tuna 76
penne
    Cheesy pasta bake 223

RECIPE INDEX

pears
- Swirling twirl dessert in a glass 161–2

pizza
- Calzone 158
- Margherita pizza with a cauliflower crust 227
- Pizza bases 156
- Roasted vegetable pizza 159

Pizza bases 156
Plain rice 150
Poached egg 92

pork
- Vietnamese spring rolls with dipping sauce 264

Potato salad with egg and tuna 16

potatoes 118
- Bonus: Spanish omelette ONLINE
- Cornish pasties 124–5
- Curried meat and vegetable pie 181
- Dressed baked potatoes 70
- Fish patties with coleslaw 188–9
- gnocchi 251
- Gnocchi with Napoli sauce 262
- Lemon chicken served with chat potatoes, carrot straws and snow peas 185
- Potato salad with egg and tuna 16
- Pumpkin noodle soup with herb and garlic bread 126
- Roasted vegetable pizza 159
- Shepherd's pie 127
- Spicy 'shake and bake' chicken with potato latkes and steamed broccoli or beans 186–7
- Spicy potatoes with chickpeas 225

Prawn and vegetable rice paper rolls 73
Pumpkin noodle soup with herb and garlic bread 126

Q

quiche
- Crustless quiche 10, 226
- Mini quiches 54

Quick apple muesli 90

R

Raspberry punch 128
Roasted vegetable pizza 159

rice
- Basic fried rice 72
- Basic risotto 217–8, 222
- Creamed rice with almonds and apple puree 151
- Oriental chicken kebabs 71
- Plain rice 145, 150
- Ricey lettuce parcels 74
- Satay vegetables and tuna 76
- Sushi roll 152

Ricey lettuce parcels 74

ricotta
- Cheese and spinach pastries 78
- Cheesy pasta bake 223
- Ricotta and orange biscuits 293

Ricotta and orange biscuits 293

S

salad
- Fragrant, fruity couscous salad 160
- Potato salad with egg and tuna 16
- Salad roll-up 15

Salad roll-up 15
Satay vegetables and tuna 76

sauce
- Napoli sauce 157

Savoury egg roll 97
Scotch mints 299
Scrambled eggs 94
Shepherd's pie 127
Shirred egg 95

snow peas
- Beef noodle stir-fry 174, 183
- Lemon chicken served with chat potatoes, carrot straws and snow peas 185
- Potato salad with egg and tuna 16

soup
- Chicken and sweet corn soup 43
- Creamy carrot and tomato soup 75
- Minestrone soup 43
- Pumpkin noodle soup with herb and garlic bread 126

spaghetti
- Baked meatballs and spaghetti 182
- Spaghetti bolognese 263

Spaghetti bolognese 263
Sparkling pineapple punch 128
Spicy 'shake and bake' chicken with potato latkes and steamed broccoli or beans 186–7
Spicy potatoes with chickpeas 225

sponge
- Chocolate sponge 291

Stewed apples 122
stir-frying 176

strawberries
- Chocolate sponge 291

Summer or winter fruit kebabs with yoghurt and coconut dip 35
Sushi roll 152

sweet potato
- Roasted vegetable pizza 159
- Sweet potato parcel 130

Sweet potato parcel 130
Swirling twirl dessert in a glass 161–2

T

Tacos 270
Tandoori chicken bites 268

tart
- Freeform fruit tart 302

Thai
- Chicken curry with coconut pancakes 265
- Thai curry pastes 257
- Thai fishcakes 266

Thai fishcakes 266
tiger toast 27

toast
- Banana toast 37
- Fabulous French toast 98
- Fried egg 93
- Poached egg 92

Toasted muesli 52
toffee 287

tomato
- Baked meatballs and spaghetti 182
- Calzone 158
- Chicken burritos 269
- Creamy carrot and tomato soup 75
- Gnocchi with Napoli sauce 262
- Margherita pizza with a cauliflower crust 227
- Minestrone soup 43
- Nachos 271
- Napoli sauce 157
- Oriental chicken kebabs 71
- Potato salad with egg and tuna 16
- Roasted vegetable pizza 159
- Salad roll-up 15
- Shepherd's pie 127
- Spaghetti bolognese 263
- Spicy potatoes with chickpeas 225
- Tacos 270
- Vegetable lasagne 219, 221

Traditional sausage rolls 224

tuna
- Fish patties with coleslaw 188–9
- Potato salad with egg and tuna 16

RECIPE INDEX

Satay vegetables and tuna 76
Thai fishcakes 266

U

Upside-down cake 290

V

Vanilla biscuits 294
vegan
- Citrus cordial 128
- Fragrant, fruity couscous salad 160
- Fruit punches 128
- Pitta bread 155
- Plain rice 145, 150
- Spicy potatoes with chickpeas 225
- Stewed apples 122

Vegemite
- tiger toast 27

vegetables
- Basic fried rice 72
- Basic risotto 217–8, 222
- Beef noodle stir-fry 174, 183
- Cheese and spinach pastries 78
- Chicken and sweet corn soup 43
- Chicken and vegetable stir-fry 77
- Cornish pasties 124–5
- Creamy carrot and tomato soup 75
- Crustless quiche 10, 226
- Curried meat and vegetable pie 181
- Fish patties with coleslaw 188–9
- Fragrant, fruity couscous salad 160
- Lemon chicken served with chat potatoes, carrot straws and snow peas 185
- Margherita pizza with a cauliflower crust 227
- Minestrone soup 43
- Oriental chicken kebabs 71
- Potato salad with egg and tuna 16
- Prawn and vegetable rice paper rolls 73
- Pumpkin noodle soup with herb and garlic bread 126
- Ricey lettuce parcels 74
- Roasted vegetable pizza 159
- Salad roll-up 15
- Satay vegetables and tuna 76
- Shepherd's pie 127
- Spicy 'shake and bake' chicken with potato latkes and steamed broccoli or beans 186–7
- Spicy potatoes with chickpeas 225
- Sushi roll 152
- Sweet potato parcel 130
- Traditional sausage rolls 224
- Vegetable lasagne 219, 221
- Zucchini barbecue burgers 178, 184
- Vegetable lasagne 219, 221
- Vietnamese spring rolls with dipping sauce 264

W

Wattleseed pavlovas with macadamia cream and sugar bark 242

Y

yoghurt
- Frozen banana whiz 36
- Mini lamb kebabs with minted garlic dip 267
- Quick apple muesli 90
- Summer or winter fruit kebabs with yoghurt and coconut dip 35
- Tandoori chicken bites 268

Z

zucchini
- Beef noodle stir-fry 174, 183
- Crustless quiche 226
- Minestrone soup 43
- Roasted vegetable pizza 159
- Traditional sausage rolls 224
- Vegetable lasagne 221
- Zucchini barbecue burgers 178, 184
Zucchini barbecue burgers 178, 184